Mathematics

AND THE PHYSICAL WORLD

Mathematics

AND THE PHYSICAL WORLD

BY MORRIS KLINE

THOMAS Y. CROWELL COMPANY

NEW YORK

The illustration on page 118 is reproduced from *A History of Astronomy from Thales to Kepler,* by J. L. E. Dryer, second edition, Cambridge University Press, after Apelt, *Johann Keplers astronomische Weltansicht,* p. 93.

 Manufactured in the United States of America by the Vail-Ballou Press, Inc., Binghamton, New York. Library of Congress Catalog Card No. 59-5252. First Printing.

TO *Douglas Mann Kline*

PREFACE

In every department of physical science there is only so much science, properly so-called, as there is mathematics.

—IMMANUEL KANT

MATHEMATICS is a model of exact reasoning, an absorbing challenge to the mind, an esthetic experience for creators and some students, a nightmarish experience to other students, and an outlet for the egotistic display of mental power. But historically, intellectually, and practically, mathematics is primarily man's finest creation for the investigation of nature. The major concepts, broad methods, and even specific theorems have been derived from the study of nature; and mathematics is valuable largely because of its contributions to the understanding and mastery of the physical world. These contributions are numerous.

Insofar as it is a study of space and quantity, mathematics directly supplies information about these aspects of the physical world. But, going far beyond this, mathematics enables the various sciences to draw the implications of their observational and experimental findings. It organizes broad classes of natural phenomena in coherent, deductive patterns. And today mathematics is the heart of our best scientific theories, Newtonian mechanics, the electromagnetic theory of Maxwell, the Einsteinian theory of relativity, and the quantum theory of Planck and his successors. Indeed, physical science has reached the curious state in which the firm bold essence of its best theories is entirely mathematical whereas the physical meanings are vague, incomplete, and in some instances even self-contradictory. Science has become a collection of mathematical theories adorned with a few physical facts. Further, if one can speak of *the* goal of modern scientific theory it is to subsume all its results under one mathematical principle whose implications would describe the multifarious operations of nature.

It is true that great mathematicians have often transcended the immediate problems of science. But because they were great and therefore understood fully the meaning of mathematics, they could discern directions of investigation which must ultimately prove significant in the scientific enterprise or shed light on the mathematical concepts that were already clearly instrumental in the investigation of nature. With this understanding and assurance they felt warranted in pursuing their own ideas and occasionally left the applications to others.

Unfortunately the relationship of mathematics to the study of nature is not presented in our dry and technique-soaked textbooks. Moreover, the fact that mathematics is valuable primarily because of its contributions to the understanding and mastery of nature has been lost sight of by some present-day mathematicians who wish to isolate their subject and offer only an eclectic study. An undue emphasis on abstraction, generality, rigor, and logically perfect deductive structures has caused a number of mathematicians to overlook the real importance of the subject. In the last fifty years a schism has developed between those who would hew to the ancient and honorable motivations for mathematical activity—the motivations which have thus far supplied the substance and most fruitful themes—and those who, sailing with the wind, would investigate what strikes their fancy. But history favors only one view. To supply the historical evidence within the limits of a preface it must suffice to mention that the greatest mathematicians—Eudoxus, Apollonius, Archimedes, Fermat, Descartes, Newton, Leibniz, Euler, Lagrange, Laplace, Hamilton, Cauchy, Gauss, Riemann, and Poincaré, to mention just a few—were also giants of science. All of these men would have earned enduring fame for their physical researches alone.

To display the role of mathematics in the study of nature is the purpose of this book. Subordinate, but by no means incidental, objectives may also be fulfilled. We may see mathematics in the process of being born—that is, see how physical problems, when idealized and formulated in the language of number and geometry, become mathematical problems and see how, guided still by intuition and physical thinking, the mathematician creates a new method or a new branch. The precise manner in which mathematics produces answers to physical problems—mathematics at work, so to speak—may also be evident. As we follow the gradual development of mathematical power and the increasing absorption of mathematics in the scientific enterprise, we may learn how and why mathematics has become the essence of scientific theories. Though mathematics has been intimately involved with science—on a Platonic basis of course—mathematics is not physics, astronomy, or chemistry. Hence the distinctions between mathematics and science will also be presented.

Finally, a study of mathematics and its contributions to the sciences exposes a deep question. Mathematics is man-made. The concepts, the broad ideas, the logical standards and methods of reasoning, and the ideals which have been steadfastly pursued for over two thousand years were fashioned by human beings. Yet with this product of his fallible mind man has surveyed spaces too vast for his imagination to encompass; he has predicted and shown how to control radio waves which none of our senses can perceive; and he has discovered particles too small to be seen with the most powerful microscope. Cold symbols and formulas completely at the disposition of man have enabled him to secure a portentous grip on the universe. Some explication of this marvelous power is called for. A presentation of these features of mathematics may also make clear the true meaning of the subject and the reason that it is regarded as of supreme importance in human thought.

The presentation will be confined to elementary mathematics, and in fact we shall not presume much prior knowledge. These limitations are not significant because the major characteristics of mathematics and the ways in which mathematics interacts with science can still be fully displayed. Perhaps all that we forego by restricting ourselves to elementary mathematics is a fuller indication of the extent to which mathematics enters into modern science.

It is a pleasure to turn from affirmations and contentions to acknowledge some positive facts. That two times two remains four throughout this book is due to the conscientious checking and proofreading of Leon Wilson and Kay Anderson. To Carl Bass's work are due the excellent figures. I am happy to express my obligation to these people and to the staff of the publishers.

MORRIS KLINE

CONTENTS

1: THE WHY AND WHEREFORE

Mathematics is the gate and key of the sciences. . . . Neglect of mathematics works injury to all knowledge, since he who is ignorant of it cannot know the other sciences or the things of this world. And what is worse, men who are thus ignorant are unable to perceive their own ignorance and so do not seek a remedy.

—ROGER BACON

PERHAPS the most unfortunate fact about mathematics is that it requires us to reason, whereas most human beings are not convinced that reasoning is worth while. Indeed, it is not at all obvious that reasoning in general and mathematical reasoning in particular are valuable. People do not use reasoning to learn how to eat or to discover what foods maintain life. One can engage in very enjoyable sports without calling upon reasoning at all. Learning to get along with the opposite sex is an art rather than a science, and this art is not mastered by reasoning. Insofar as making a living is concerned, many a shrewd businessman has amassed a fortune without knowing more than the rudiments of arithmetic. One's social life is not more successful because one possesses a knowledge of trigonometry. In fact, entire civilizations in which reasoning and mathematics played no role have endured and even flourished. If one were willing to reason, he could readily conclude from such observations that reasoning is not a natural or even a necessary activity for the human animal.

But a little investigation of the ways of our society does show that some reasoning can be useful. The doctor who seeks to diagnose an illness reasons from symptoms to the cause. The lawyer who wishes to convince a jury often resorts to reason. The engineer reasons continually in order to design or produce a new device. The scientist observes or experiments and draws conclusions from the evidence he obtains. All of these people utilize and even depend upon reasoning. The purpose and value of all such reasoning is the derivation of knowledge that is not otherwise obtainable or is obtainable by other means but only with considerably greater expense and effort.

Whereas in the professions of medicine, law, engineering, and science reasoning is employed somewhat, in mathematics reasoning is the sum and substance. Since the case for mathematics rests entirely upon what can be achieved by reasoning, perhaps we should convince ourselves at the very outset that mathematical reasoning can yield desirable knowledge. The examples we shall consider at the moment are not the most impressive but may furnish grounds for the further exploration of the subject.

Let us be practical and consider first a matter of dollars and cents. Suppose that a young man has a choice between two jobs. Each offers a starting salary of $1800 per year, but the first one would lead to an annual raise of $200 whereas the second would lead to a semiannual raise of $50. Which job is preferable? One would think that the answer is obvious. A raise of $200 per year seems better than one that apparently would amount to only $100 per year. But let us do a little arithmetic and put down what each job offers during successive *six-month* periods.

The first job will pay

900, 900, 1000, 1000, 1100, 1100, 1200, 1200 . . .

The second job, which bears a semiannual increase of $50, will pay

900, 950, 1000, 1050, 1100, 1150, 1200, 1250 . . .

It is clear from a comparison of these two sets of salaries that the second job brings a better return during the second half of each year and does as well as the first job during the first half. The second job is the better one. With the arithmetic before us it is possible to see more readily why the second job is better. The semiannual increase of \$50 means that the salary will be higher at the rate of \$50 for six months or at the rate of \$100 for the year because the recipient will get \$50 more for each of the six-month periods. Hence two such increases per year amount to an increase at the rate of \$200 per year. Thus far the two jobs seem to be equally good. But on the second job the increases start after the first six months, whereas on the first job they do not start until one year has elapsed. Hence the second job will pay more during the latter six months of each year.

Let us consider another simple problem. Suppose that a merchant sells his apples at two for 5 cents and his oranges at three for 5 cents. Being somewhat annoyed with having to do considerable arithmetic on each sale, the merchant decides to commingle apples and oranges and to sell any five pieces of fruit for 10 cents. This move seems reasonable because if he sells two apples and three oranges he sells five pieces of fruit and receives 10 cents. But now he can charge 2 cents apiece and his arithmetic on each sale is simple.

The dealer is cheating himself. Just to check quickly we shall assume that he has one dozen apples and one dozen oranges for sale. If he sells apples normally at two for 5 cents he receives 30 cents for the dozen apples. If he sells oranges at three for 5 cents he receives 20 cents for the dozen oranges. His total receipts are then 50 cents. However, if he sells the twenty-four pieces at five for 10 cents he will receive 2 cents per article or 48 cents.

The loss is due to poor reasoning on the part of the dealer. He assumed that the average price of the apples and oranges should be 2 cents each. But the average price per apple is $2\frac{1}{2}$ cents and the average price per orange is $1\frac{2}{3}$ cents and the average price of two such items is

$$\frac{2\frac{1}{2}+1\frac{2}{3}}{2} \text{ or } \frac{25}{12}.$$

Thus the average price is $2\frac{1}{12}$ cents per article and not 2 cents.

The foregoing examples involved only simple arithmetic and since we all do such arithmetic rather mechanically we may hardly feel that we have reasoned. We shall therefore undertake a slightly more advanced problem. A farmer interested in land on which to grow crops is offered two parcels. Both are right triangles, one having the dimensions 30, 40, and 50 feet and the other, the dimensions 90, 120, and 150 feet (fig. 1). The dimensions of the second triangle, it will be noted, are three times that of the first. However the price of the second parcel is five times the price of the first.

Since the farmer is naturally interested in buying acreage as cheaply as possible, he must now decide which is the better buy.

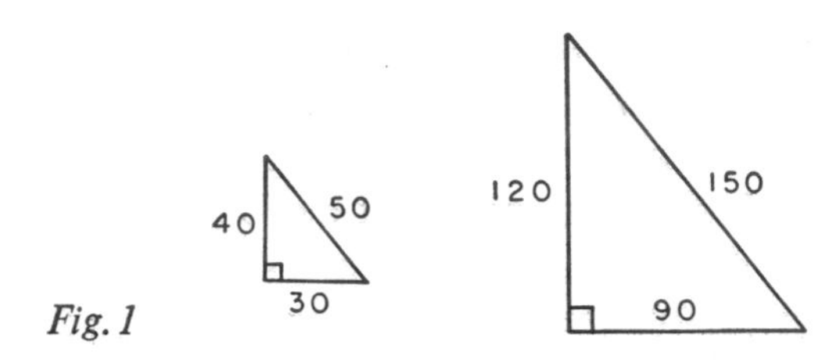

Fig. 1

Offhand it would seem as though the smaller parcel is the better buy because though it presumably contains only one-third the area of the larger parcel it costs only one-fifth as much. Let us assume, now, that the formula for the area of a right triangle, a formula produced of course by mathematical reasoning, is available to us. This formula states that the area of a right triangle is one-half the product of the two sides, or arms. Hence in the first parcel offered to the farmer the area is one-half of 30 by 40 or 600 square feet and in the second, 5400 square feet. Thus the second parcel contains nine times as much area but costs only five times as much. It's a bargain.

Even this last example of the use of mathematical reasoning may not be too impressive to some people. They may claim that the proper visualization of the areas would have sufficed to determine the correct conclusion. Moreover, with continued experience we all develop what we commonly call intuition, a combination of experience, sense impressions, speculation, and crude guessing. Since this faculty helps us to arrive at some conclusions without mathematical reasoning, we had better satisfy ourselves that it is subject to severe limitations.

Let us consider two related problems. First, we have a garden, circular in shape and with a radius of 10 feet. We wish to protect

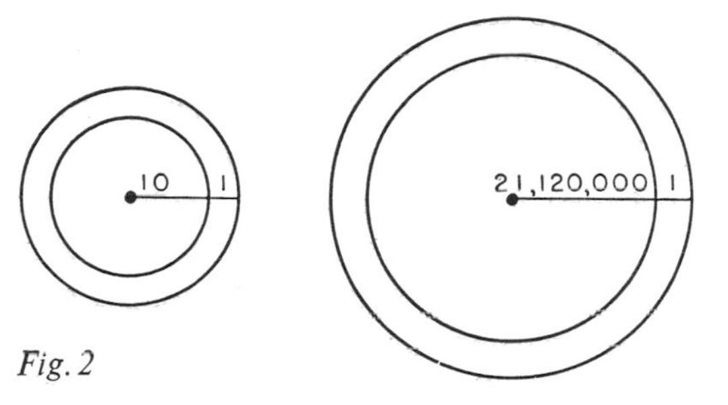

Fig. 2

the garden by a fence that is to be at each point one foot away from the boundary of the garden (fig. 2). How much longer is the fence than the circumference of the garden itself? The answer is readily obtained. The circumference of the garden is given by a formula of geometry; this says that the circumference is 2π times the radius, π being the symbol for a number which is approximately $22/7$. Hence the circumference of the garden is $2\pi \cdot 10$ (the dot standing for "multiplied by"). The condition that the fence be one foot beyond the garden means that the radius of the circular fence is to be 11 feet. Hence the length of the fence is $2\pi \cdot 11$. The difference in these two circumferences is $22\pi - 20\pi$ or 2π. Hence the fence should be 2π feet longer than the circumference of the garden.

We now consider the related problem. Suppose we were to build a roadway around the earth—a trivial task for modern engineers—and the height of the roadway were to be one foot above the surface of the earth all the way round (fig. 2). How much longer than the circumference of the earth would the roadway be? Before calculating this quantity let us use our intuition to at least estimate it. The radius of the earth is about 4000 miles or 21,120,000 feet. Since this radius is roughly 2,000,000 times that of the garden we considered, one might expect that the additional length of the roadway should be about 2,000,000 times the additional length of fence required to enclose the garden. The latter quantity was just 2π feet. Hence an intuitive argument for the additional length of roadway would seemingly lead to the figure of $2{,}000{,}000 \cdot 2\pi$ feet. Whether or not the reader would agree to this argument he would almost certainly estimate that the length of the roadway would be very much greater than the circumference of the earth.

A little mathematics tells the story. To avoid calculation with large numbers let us denote the radius of the earth by r. The circumference of the earth is then $2\pi r$. The circumference or length of the roadway is $2\pi(r + 1)$. But the latter is $2\pi r + 2\pi$. Hence the difference between the length of the roadway and the circumference of the earth is just 2π feet, precisely the same figure that we obtained for the difference between the length of fence and the circumference of the garden, despite the fact that the roadway encircles an enormous earth whereas the fence encircles a small garden. In fact the mathematics tells us even more. Regardless of what the value of r is, the difference, $2\pi(r + 1) - 2\pi r$, is always 2π, and this means that the circumference of the outer circle, if it is at each point one foot away from the inner circle, will always be just 2π feet longer than the circumference of the inner circle.

There are many problems in which intuition can be only of incidental use and wherein mathematical reasoning must carry the entire burden. One of the simplest and yet most impressive examples of what such reasoning can achieve is Eratosthenes' calculation of the circumference of the earth. Eratosthenes (275–194 B.C.) was a famous scholar, poet, historian, astronomer, geographer, and mathematician who lived during the latter period of the ancient Greek civilization when the center of that culture was in Alexandria. In common with most learned Greeks Eratosthenes knew that the earth is spherical; he set out to determine its circumference.

Because he was learned in geography he knew that the city of Alexandria was due north of the city of Syene (fig. 3) and that the measured distance along the surface of the earth from one city to the other was 500 miles. At the summer solstice the noon sun shone directly down into a well at Syene. This means, as Eratosthenes appreciated, that the sun was directly overhead at that time; that is, the direction of the sun was OBS'. At Alexandria at the same instant the direction of the sun was AS, whereas the overhead direction is OAD. Now the sun is so far away that the direction AS is the same as BS', or, mathematically speaking, AS and BS' are parallel lines. Hence it follows from a theorem of geometry, which we may accept for the present, that angles DAS and AOB

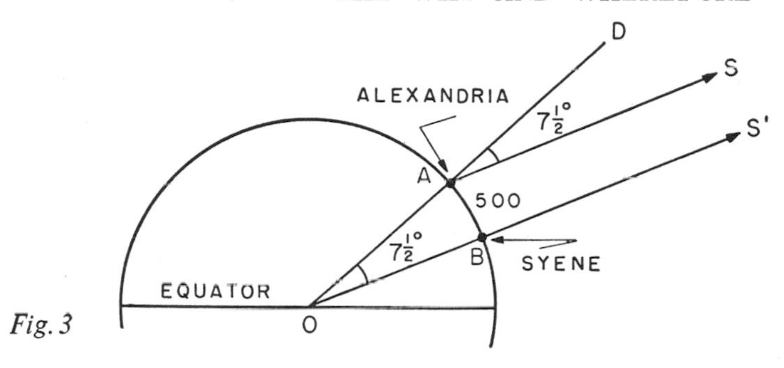

Fig. 3

are equal. Eratosthenes measured angle *DAS* and found it to be $7\frac{1}{2}°$. This, then, is the size of angle *AOB*. But this angle is $7\frac{1}{2}/360$ or $\frac{1}{48}$ of the entire angle at *O*. It follows that arc *AB* is $\frac{1}{48}$ of the entire circumference and since *AB* is 500 miles, the entire circumference is $48 \cdot 500$ or 24,000 miles. Curiously enough, this figure was reduced by later geographers, who used poorer methods, to about 17,000 miles. This figure is the one that reached Columbus; had he known the truer figure of Eratosthenes he might never have undertaken to sail to India because he might have been daunted by the distance to be covered.

It may be clear from even these few simple examples that mathematical reasoning can produce knowledge which guesswork, intuition, and experience cannot produce or can produce only inaccurately. But why should mathematics be studied so extensively and why should it be of interest to all educated people? The primary motivation for the development of mathematics proper and the primary reason for the great importance of the subject is its value in the study of nature. Mathematical concepts and mathematical methods of obtaining knowledge have been most effective in representing and investigating the motions of the heavenly bodies and the motions of objects on and near the surface of the earth, the phenomena of sound, light, heat, electricity, and electromagnetic waves, the structure of matter, the chemical reactions of various substances, the structure of the eye, ear, and other organs of the human body, and dozens of other major scientific phenomena.

But if the main motivation for the development and study of mathematical ideas is the study of nature, why should one study nature? If, for example, one works all his life to earn money, then it is reasonable to ask, what purposes does money serve? Of course the most obvious reasons for the study of nature are the practical advantages that ensue. The farmer studies soil, seeds, methods of cultivation, and the weather to produce good crops. The hunter studies the habits of animals in order to track them down. The sailor studies the currents of large bodies of water, the tides, the weather, and the stars to aid him in navigation. A metallurgist studies minerals in order to produce better iron or steel. Such studies of nature, found even in primitive societies, are obviously useful. In our own civilization the practical advantages derived from the study of nature have reached staggering proportions. We have but to think of our trains, airplanes, and ships, our movies, radio, television, and telephone, our mass production of highly useful equipment for the home, the services performed by electricity and modern medical treatments to appreciate that man has profited by the study of nature.

While the material improvements which have resulted from the study of nature appeal most strongly to people, it is not true that these are really the most compelling reasons for that study. Nor, as a matter of fact, have the practical gains generally been obtained from investigations so motivated. The deeper reason for the study of nature is to satisfy man's intellectual curiosity. Even in the most primitive civilizations, and certainly in ours, there have been people who have sought to answer such questions as, How did the universe come about? Why was man born and what will happen to him after death? What causes and what is light? Is there plan and organization in the solar system? It is true that many people pass their whole lives without being bothered at all by such questions. These people are content with shelter, food, sex, and a few material pleasures. But others are not content to live only on a physical level. They have minds that are aware of the pervasive natural mysteries and are as obsessed to resolve them as any businessman may be to acquire wealth.

The purely intellectual pursuits of the mysteries presented by nature, those "seemingly unprofitable amusements of speculative brains," have produced rewards over and above the satisfaction of curiosity and the improvement in man's physical comforts. Primitive peoples of the past and present have always feared natural phenomena. Thunder, lightning, and hail were regarded as visitations of angry gods and indeed the destructive effects of lightning and hail were certainly some justification for this belief. Primitive peoples also watched with fear the lowering course of the sun in the sky as winter neared. They feared that ultimately the sun would sink so low as to disappear altogether. The day of the winter solstice was a dark one in their hearts because even though many might have remembered that in the past the sun did reverse its course, they had no reason to believe that it would this time. Hence, just before the winter solstice, these peoples would pray for the sun to return to them. When it became evident that the sun was reversing its course in the sky, prayers were replaced by celebrations. These ceremonies, incidentally, are historically the predecessor of the modern Christmas celebration.

Primitive peoples also feared eclipses of the sun and moon. Some believed that such occurrences meant the sun or moon was being swallowed up by a dragon. Hence they tried to appease the dragon by prayers and incantations so that he would return the sun to them, or they made especially loud noises with special instruments to scare the dragon away. Many Hindus today believe that a demon serpent swallows the sun during an eclipse, and they pray for its release. Of course the obvious successes that attended the prayers and rituals served only to convince these people of the correctness of their beliefs.

Such superstitions and fears about the comings and goings of the heavenly bodies have been eliminated by just those people who studied the skies to satisfy their intellectual curiosity. These insatiably curious searchers after truth have replaced the fears and superstitions by modern astronomical doctrines that reveal and proclaim an invariable order in nature, a mathematical pattern to which the heavenly bodies adhere. Scientists have not only studied the paths of the heavenly bodies and know that their movements are regular and periodic, but the mathematical laws they have produced tell us precisely when events such as eclipses are to take place. Unlike primitive people we do not rush out into the open, fall on trembling knees, and pray to the gods to avert a calamity. Instead we go out with stop watches in hand to check the prediction. Because we have the understanding granted by the mathematical laws and because we find century after century that the heavenly bodies continue to follow these marvelously precise mathematical laws, we have the utmost confidence that the sun will reverse its descent in the sky and climb higher after the winter solstice and that the moon will not suddenly stop when it comes between the earth and the sun and eclipse the sun forever. We know that nature is rationally designed and will not be willful or capricious.

We have expelled the demons and the dragons and, some would say, even the delightful fairies; but there are new beauties to compensate for the latter. The mathematical patterns that describe and unite a multiplicity of seemingly diverse, disordered natural phenomena such as the motions of the sun, moon, and planets are joys to some minds. The order that mathematics has revealed in the varieties of nature offers the pleasures that a painting, a piece of music, or a poem offers to others.

Perhaps, then, man has good reasons to reason about nature, and mathematics can be somewhat helpful, but just exactly how is mathematics involved? It is by no means apparent that mathematics should be useful in the study of nature or, for that matter, that any subject should provide the concepts and methods which would uncover the secrets of nature. The intellectually curious person would like to know how a purely human creation, which mathematics is, can give us such insight into and power over nature. How does the mind, so to speak, pull itself up by its own bootstraps or, to put it more elegantly, how is it possible for pure thought to be an instrument for the analysis of nature? As the famous nineteenth-century mathematical physicist James Clerk Maxwell said about any mechanical device he did not understand, "What's the go of that?"

It is not possible to give a simple explanation of how mathematics helps us to investigate and master nature. There are, however, a few facts that may help us to understand just how mathematics goes about this work. Of course it is clear that some simple quantitative facts, such as weights and measurements, and some simple geometrical facts, such as a knowledge of areas, can be helpful to man. Such uses of mathematics are obvious and trivial. Even the ancient Egyptian and Babylonian civilizations employed them. But the values that the Egyptians and Babylonians derived from mathematics may hardly be said to represent even one iota of the accomplishments of mathematics; these people missed the boat insofar as the effective use of mathematics in the study of nature is concerned.

It was the Greeks of the sixth, fifth, and fourth centuries B.C., the people who occupied roughly the area of modern Greece, who really made the great discovery as to how and to what extent mathematical reasoning can be effective. Modern Western civilization learned this secret from the Greeks. The Greek idea is basically simple. One observes nature and finds that certain simple forms such as lines, triangles, and circles occur repeatedly. The heavenly bodies are spheres; light seems to travel in straight lines; the surfaces of lakes are flat; the sides of buildings are rectangles. Number or quantity is also suggested repeatedly by collections and sizes of objects. These concepts of number and geometrical forms, in view of their very prevalence, the Greeks deemed worthy of study.

This fact in itself was a significant discovery, but the next discovery of the Greeks proved to be even more momentous. They observed that certain facts about these concepts are obvious and seemingly basic. Circles are determined by choosing a center and a radius; any two right angles are always equal; equal numbers added to equal numbers or equal lengths added to equal lengths yield equals; and so on. Why not select the most obvious of these facts and see what reasoning could deduce from them? Surely if some new facts could be derived, these facts would apply to *all* those physical objects that possessed the basic properties in the first place. If the area of a circle could be shown by reasoning to be π times the square of the radius, then the area of any circular piece of land should also be π times the square of its radius. Further, perhaps by reasoning one could discover new facts which observation alone would not suggest. These advantages and many more the Greeks expected to derive from reasoning about common concepts on the basis of clearly evident facts. This is the germ of the great Greek discovery, perhaps the greatest discovery man has ever made. Reasoning can produce knowledge that not only covers a multitude of cases in one swoop but may produce physically meaningful information that is entirely unforeseen.

The entire process of starting with some obvious facts, reasoning, and the realization of new knowledge is like going from one place to another by train. The station at which we get on is the physical information we start with. The station at which we get off is the new information we obtain by the reasoning. The train itself is the mathematical train of reasoning that carries us from one place to the other.

Why bother with a train? Sometimes we even have difficulty in getting to the station or to a destination that is not at the train's terminal but some distance from it. Why not walk directly from one place to the other? In the case of an ordinary journey the answer is obvious. But the answer is the same in scientific inquiries. To confine our activity entirely to physical investigations or observations may lead to getting lost in a jungle of physical facts; it may require traversing impassable territory, as in measuring the size of the earth; it may involve climbing mountains so high that the atmosphere becomes too rare to support physical activity; or it may mean the at present impossible task of getting to the moon. But if these obstacles to direct physical investigation do exist, how is it possible for the mathematical engineers to build a "railroad line" from one place to the other? One might say that these engineers are specially equipped for the task, and in a sense this is correct. However, this answer begs the question, and we shall return to it in the last chapter of this book.

To compare the procedures of mathematics with the progress made by a train hardly suggests the power or even the character of mathematical activity. Mathematics is a multistage Roman candle. It is packed with physical gunpowder and set alight by the fire of human minds. It then shoots off in a direction somewhat determined by human thought but not necessarily exactly in the direction planned. Then the candle flares up and successive flashes of light reveal aspects of the physical world not readily illuminated from man's earthbound station. Each successive thrust of the candle into space is followed by flares from higher and higher altitudes that reveal more and more of the constitution of our universe.

Mathematics is commonly regarded as highly abstract and remote from the real world. It is true that as mathematics developed it built more abstract and more complex ideas. An equation involving unknowns is a more abstract and perhaps more complicated notion than that of number, and the reasoning about equations is less immediately subject to physical interpretation than the manipulations of arithmetic, but every abstraction that even the greatest mathematician has introduced is ultimately derived from and is therefore understandable in terms of intuitively meaningful objects or phenomena. The mind does play its part in the creation of mathematical concepts and in determining the direction that the reasoning shall pursue, but the mind does not function independently of the world outside. Indeed, the mathematician who advances concepts that have no physically real or intuitive origins is almost surely talking nonsense. The intimate connection between mathematics and events in the physical world is reassuring, for it means that we not only can hope to understand the mathematics proper but also expect physically meaningful conclusions. From dust thou art to dust returneth may perhaps not be spoken of the soul but it is well spoken of earthborn mathematics.

2: DISCOVERY AND PROOF

Since then reason is divine in comparison with man's whole nature, the life according to reason must be divine in comparison with usual human life.

—ARISTOTLE

MATHEMATICAL reasoning is commonly regarded as different from and even superior to the reasoning other studies utilize. People are wont to describe mathematical reasoning as certain, precise, and exact. Is it actually the case that the mathematician uses a strange brand of reasoning? Does he have knowledge of some secret process that insures his results where other less informed or less gifted people have to fumble? Just what, if anything, is peculiar to the reasoning that mathematics employs?

The mathematician may like to be put on a pedestal, to be extolled, and to be regarded as the superb reasoner among men, but if he is truthful he must admit that he does not know any secrets of reasoning or possess any special powers that are not available to all human beings. Nevertheless, mathematics has been more successful than other branches of human knowledge in its endeavor to erect a reliable and lasting structure of thought. How has this come about? The answer, which at the moment may seem paradoxical, is that mathematics restricts its methods of reasoning more than do other branches of inquiry.

Many methods of reasoning or drawing conclusions are available to all human beings. Of these the most widely used are reasoning by analogy, induction, and deduction. Let us examine these in turn.

A boy who is considering a college career may note that his friend went to college, enjoyed the work, and handled it successfully. He then argues that since he is very much like his friend in physical and mental qualities he, too, should enjoy and succeed in college work. The boy has reasoned by analogy. The essence of such reasoning is to find a similar situation or circumstance and to argue that what is true for the similar case applies to the one in question. Of course one must be able to find a similar situation and one must take the chance that the differences do not matter. One might argue that just because cows eat grass, lions do also. But the similarities between these animals do not extend to their foods. Despite the uncertainties attached to the method of reasoning by analogy, all of us constantly employ it because similar situations are usually readily available.

Now, reasoning by induction. Experimentation would show that iron, copper, brass, oil, and other substances expand when heated and one concludes that all substances expand when heated. The essence of this method is that one observes repeated instances of the *same* phenomenon and concludes that the phenomenon will always occur. Conclusions obtained by induction seem well warranted by the evidence, especially when the number of instances observed is large. Thus, the sun is observed so often to rise in the morning that one is sure it has risen even on those mornings when it is hidden by clouds. Nevertheless, there is much room for error. For example, water, when heated from 0° to 4° centigrade, does not expand; it contracts.

In the two methods of reasoning thus far described, though one starts with some facts that are reliable he obtains a conclusion that is not necessarily reliable. But it is possible to reason in such a way that starting with reliable facts one is able to draw unquestionable conclusions. If we accept as basic facts or premises that honest people return found money and that John is honest, we may conclude unquestionably that John will return money that he finds. Likewise, if we premise that no mathematician is a fool and that John is a mathematician, then we may conclude with certainty that John is not a fool.

These latter examples show that there are ways of combining facts so as to obtain a new but equally sound one. Both examples employ what is called deductive reasoning. A deductive argument consists in combining accepted facts in any way that *compels* acceptance of the conclusion. Of course this characterization of deductive reasoning does not specify just what kinds of combinations of accepted facts yield inescapable conclusions. One might, for example, argue that because the sum of two odd numbers is always even and because the sum of a and b is even, then a and b are odd numbers. This purportedly deductive argument is not correct. We can see the unsoundness of the argument if we substitute 6 and 8 for a and b. In this case the argument says that because the sum of two odd numbers is always even and because the sum of 6 and 8 is even, then 6 and 8 are odd numbers. The *pattern* of the reasoning is the same in the two arguments and the facts we start with are equally true in both cases. Hence the conclusion would have to be true in both cases if correct or valid deductive reasoning had been used.

Apparently, then, there are some ways of combining statements that compel the acceptance of the conclusion and other ways, purportedly deductive, that do not. Who decides which ways are reliable, or, in other words, which forms of deductive reasoning are valid? There is a simple answer to this question. Those people who agree on what is valid deductive reasoning band together and call the others insane. Of course logicians have a little more to add to this matter but, for the present at least, we shall not consult them. The common forms of valid deductive reasoning employed in mathematics carry with them so strong a conviction that we shall not be in doubt as to whether we have employed a valid method. We shall however return to this subject in the last chapter.

Since there are methods of reasoning that yield unquestionable conclusions, why does anyone bother with such unreliable methods as analogy and induction? Not only is the deductively obtained conclusion certain but deductive reasoning is carried on by the human brain and may avoid, for example, expensive experimentations and arduous efforts involved in verifying that some phenomenon repeatedly occurs.

But there are at least two reasons why one does not use deductive reasoning exclusively in nonmathematical investigations. In order to obtain a significant conclusion deductively one must have the appropriate basic facts or premises. Suppose one wished to prove deductively that George Washington was the best President the United States has had. There are no obvious facts on which to base a deductive argument leading to this conclusion. Again, many scientists are now striving to obtain a cure for cancer. Here too the facts prerequisite to a deductive argument are lacking.

There is another equally important reason for resorting to methods of reasoning such as analogy and induction. Let us suppose that a person only slightly familiar with mathematics wished to prove that the sum of the angles of a triangle is 180°. This fact can be established deductively from the basic premises or axioms of Euclidean geometry. But the person who is only slightly familiar with mathematics and yet wishes to establish this conclusion may not be able to find the correct deductive arguments. His alternative might be to measure the angle sums of a number of triangles. He would find that the sum is indeed as close to 180° as accurate measurement can indicate and after measuring perhaps a dozen triangles would conclude *inductively* that the sum of the angles of any triangle is 180.°

In other words, reasoning by induction and by analogy and, where no facts are available, even sheer guesswork and direct trial are necessary in many human endeavors. And even where they are not necessary they often supply a very likely conclusion much more rapidly than the efforts that may be needed to obtain a deductive proof.

Despite the advantages that may accrue from other methods of obtaining knowledge, mathematicians ever since Greek times have limited themselves to conclusions which can be established deductively on the basis of a fixed set of thoroughly reliable premises. Of course the conclusions deduced from such premises are themselves reliable and hence may be used in turn as premises for further deductive reasoning. That is, theorems already established may be used as premises for new proofs. No matter how many successive deductive arguments are involved, each yields certain conclusions. We see therefore why it is that mathematics has attained its reputation for the certainty of its results.

Before discussing the wisdom of the mathematician's insistence on deductive proof we might contrast the method of mathematics with those of the physical and social sciences. The scientist feels free to draw conclusions based on observation, experimentation, and experience. He may reason by analogy as, for example, when he reasons about sound waves by observing water waves or when he reasons about a possible cure for a disease affecting human beings by testing the cure on animals. The scientist may reason inductively; if he observes many times that hydrogen and oxygen combine to form water he will conclude that this combination will always form water. At some stages of his work the scientist may also reason deductively and, in fact, even employ the concepts and methods of mathematics proper. However, he certainly does not confine himself to deductive arguments only.

To contrast the method of mathematics with that of science—and perhaps to illustrate just how stubborn the mathematician can be in his insistence on deductive proof—we might consider a rather famous example. Mathematicians are concerned with whole numbers, or integers, and among these they distinguish the prime numbers. A prime number is a number that has no integral divisors other than itself and 1. Thus 11 is a prime number whereas 12 is not. Now there is a conjecture that every even number can be expressed as the sum of two prime numbers. For example, $2 = 1 + 1$; $4 = 2 + 2$; $6 = 3 + 3$; $8 = 3 + 5$; $10 = 3 + 7$; and so forth. Thousands of even numbers have been tested and the conjecture has been verified in each case. Hence by *inductive* reasoning one could conclude that every even number is the sum of two prime numbers.

But the mathematician does not accept this conclusion as a theorem of mathematics because it has not been proved deductively from acceptable premises. The conjecture, known as Goldbach's hypothesis because it was first suggested by the eighteenth-century mathematician Christian Goldbach, is an unsolved problem of mathematics. However, a scientist would not hesitate to use this inductively well-supported conclusion. The mathematician's insistence on finding a deductive proof even when the truth of the conclusion is undoubted has led to the sarcastic comment that reason is the slow and tortuous method by which those who do not know the truth discover it.

Of course the scientist cannot expect that his conclusions will necessarily stand up, because, as we have seen, induction and analogy do not lead to sure conclusions, but it does seem as though the scientist's procedure is wiser since he can take advantage of any method of reasoning that will help him advance his knowledge. The mathematician by comparison appears to be narrow-minded or shortsighted. He achieves a reputation for precise reasoning but at the price of limiting his results to those that can be established deductively.

But we shall see that far from being shortsighted the mathematicians were infinitely wise. By relying upon and exploiting deductive reasoning, mathematicians have obtained results that would be very difficult or even impossible to obtain by other methods. We have already seen what might be called a trivial example, namely how Eratosthenes calculated the circumference of the earth. Mathematics has accomplished far more. The senses are limited. The eye sees only a small range of light waves and is easily deceived as to sizes and locations of objects. The ear hears only a limited range of sound waves. The sense of touch can reach only to objects accessible to the hand and is not a precise sense at that. On the other hand, man's reason can encompass distances, sizes, sounds, and temperatures beyond the range of the senses. More than that, reason can contemplate phenomena which transcend the senses and even the imagination. Mathematics has thereby been able to create spaces of arbitrary dimension and to predict the existence of the imperceptible radio waves. And because mathematics has confined itself to the soundest methods of reasoning man has, the results of mathematics have endured whereas even some of the most magnificent theories of science have had to be discarded.

The mathematician's insistence on deductive proof has not only limited the conclusions he will accept but it also has meant that the mathematician must confine himself to those areas in which he can find reliable and fruitful facts on which to base his deductive reasoning.

For example, it is possible to prove deductively that at least two people in this world have the same number of hairs on their heads. This fact is of course true for people who have no hair. But it is also true for the others. One can take as premises for deductive reasoning that there are more people in this world than hairs on any person's head, and that there are two billion people in the world. (The precise number does not affect the character of the argument.) If, then, no two people had the same number of hairs, there could be only one person with one hair, only one with two hairs, and so on. Since there are two billion people, we would finally arrive, as we check off these people one by one, at one person with two billion hairs. But this contradicts our premise that there are more people than hairs on any person's head.

Our conclusion has been deductively established and yet it is not a theorem of mathematics. No significant use has been found thus far for the conclusion about the number of hairs on a person's head and even if this particular fact were useful, no further deductions seem to be possible. The subject is sterile. Fortunately, the mathematician has found two subjects in which fruitful as well as reliable premises can be found, namely, the subjects of number and geometric figures. He calls these premises axioms, and to be sure that he uses these precisely as intended and that no other facts are used inadvertently, he states his axioms at the outset of his reasoning.

Because mathematical proof is strictly deductive, mathematics has been described as a deductive science, or as the science that derives conclusions which necessarily follow from the axioms. This description of mathematics is somewhat misleading. For one thing the deductive process requires premises or axioms to start with. These initial premises or axioms, then, cannot be obtained deductively. As a matter of fact, they are derived from observation and experience. But mathematicians must also discover what to prove and how to go about making the proofs. These processes are also part of mathematics and they are not deductive.

How, then, does the mathematician discover what to prove and the deductive arguments that establish the conclusions? The most fertile source of mathematical ideas is nature herself. Mathematics is devoted to the study of the physical world, and this world suggests a multitude of ideas. The process is clearest, though not most striking, in elementary geometry. Once mathematicians decided to devote themselves to geometric forms, there immediately arose such questions as, What are the area, perimeter, and angle sum of common figures? Moreover, it is even possible to see how the precise statements of theorems to be proved can come from direct experience with physical objects. The mathematician might measure the angle sums of various triangles and find that these measurements all yield results close to 180°. Hence the suggestion that the sum of the angles in every triangle is 180° occurs as a possible theorem. Or he might do what Galileo did. To determine the area of a complex figure he made one out of cardboard and compared its weight with a cardboard model of a figure whose area was known. The relative weights correspond, of course, to the relative areas.

After some theorems have been proved, suggestions for others are obtained by the process of generalization. Having found the angle sum of a triangle one can readily ask, What is the sum of the angles in a quadrilateral? a pentagon? and so forth. A less obvious example of a theorem obtained by the method of generalization runs as follows. *AB, CD, EF,* and *GH* are parallel chords of a circle (fig. 4). The centers of these chords lie on the one line. A circle, however, is a special case of an ellipse (fig. 4). Hence it is likely that a similar theorem holds for the ellipse. This generalization can be proved and yields another theorem.

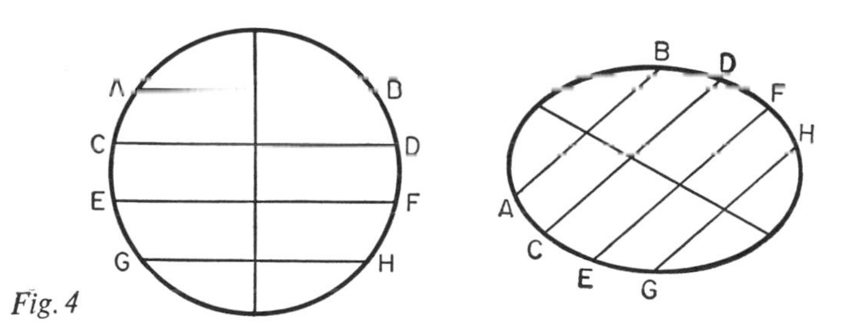

Fig. 4

In the domains of arithmetic and algebra direct calculation with numbers, which is analogous to measurement in geometry, suggests possible theorems. Anyone who has played with integers, for example, has doubtless observed that the sum of the first *two* odd numbers, that is, $1 + 3$, is the square of *two* or 4; the sum of the first *three* odd numbers, that is, $1 + 3 + 5$, is the square of *three* or 9. Similarly for the first four, five, and six odd numbers. Thus, simple calculation suggests a general statement, namely, that the sum of the first n odd numbers, where n is any integer, is the square of n. Of course this possible theorem is not proved by the above calculations. Nor could it ever be proved by such calculations, for mortal man could not make the infinite set of calculations that would be required to establish the conclusion for *every* n. The calculations do, however, give the mathematician something to work on.

Though numerous suggestions for theorems stem directly from

experience—many far more weighty examples will occur in the course of this book—one cannot account for the discovery of what to prove or how to make a deductive proof entirely or perhaps even largely in terms of sheer observation, measurement, or calculations. Let us consider even the trivial example of a quadrilateral in which the opposite sides are equal (fig. 5). To how many people would the idea occur that under these conditions the opposite sides will also be parallel?

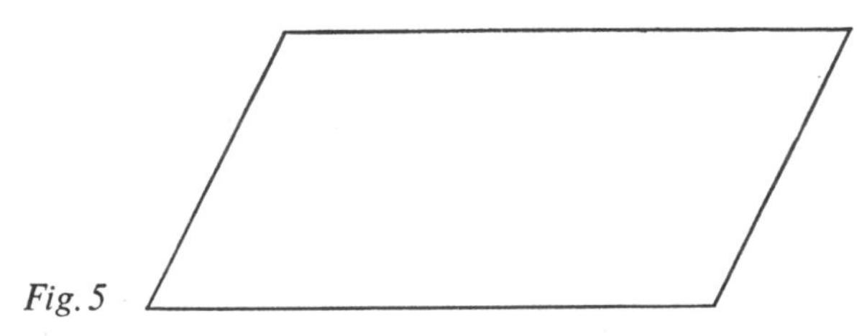

Fig. 5

Consider, also, the matter of proof. One can discover by measurement the likely theorem that the sum of the angles of a triangle is 180°, but how does one go about proving this result deductively? The reader who remembers his elementary geometry will recall that the proof is usually made by drawing a line through one vertex, *A* in figure 6, and parallel to the opposite side. It then turns out that the two angles marked 1 are equal, as are the two angles marked 2.

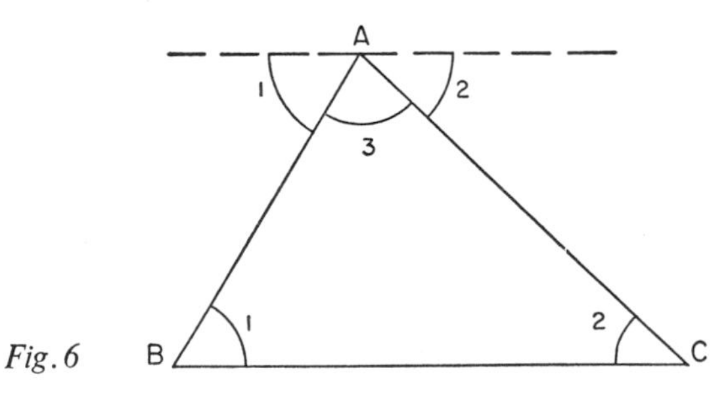

Fig. 6

However, the angles 1, 2, and 3, all with vertex at *A*, do add up to 180° and so the same is true for the angles of the triangle. But this method of proof does not come from experience. Some of the methods of proof seem so ingenious, or so devious and artificial, that they have provoked critical comments. The philosopher Arthur Schopenhauer, who despised mathematics, called Euclid's proof of the Pythagorean theorem a "mean, underhand proof."

Thus, experience and even the more fortuitous sources such as guesswork, trial and error, and pure chance, do not suffice to explain how mathematicians discover the proper theorem or proof. We must include the creative act of the human mind in which imagination, insight, and inspiration are essential. The mental processes that may lead to the correct insight are indeed as difficult to describe as is the creative process in art, literature, or music. One might say of mathematical creation what P. W. Bridgman, the noted modern physicist, has said of scientific method, that it consists of "doing one's damnedest with one's mind, no holds barred."

We have thus far distinguished mathematics from other fields of science and human endeavor by its insistence on deductive proof for the establishment of its conclusions. But there is one other characteristic that, though not peculiar to mathematics, is fundamental there whereas it is only partially involved in other activities. We have said that mathematics is devoted to the study of the physical world. Actually it studies certain abstract concepts, the most elementary of which are derived from properties of physical objects. Physical straight lines have thickness, color, molecular structure, and, often, rigidity. The mathematical straight line has none of these properties. Likewise, other geometrical forms, such as triangles and circles, and the concepts of arithmetic, such as whole numbers and fractions, are abstractions of certain properties of physical objects. On the basis of such abstractions mathematics creates others that are even more remote from anything real. Negative numbers, equations involving unknowns, formulas, derivatives, integrals, and other concepts we shall encounter are abstractions built upon abstractions.

The process of abstraction is far more natural and realistic than appears at first sight. Every young boy envisages the perfect girl, and girls entertain a similar thought about boys. Do these perfect beings exist in the physical world? Almost certainly not. But aside from the pleasures that contemplation of these ideals afford, the thoughts themselves may find application to problems and actions in the world of experience.

Of course other fields of study use some abstractions. The physical concepts of force, mass, and energy are also abstractions from real objects or real phenomena. The concept of wealth in economics is an abstraction from land, buildings, gold, and jewelry, and the concepts of liberty, justice, and democracy are familiar abstractions in political thought. But these fields do attempt to stay close to or deal directly with physical objects and actual events. The study of molecular structure attempts to get at precisely the physical constituents of molecules. The study of wages, income, profits, and sales also deals with concrete phenomena.

The distinction between mathematics and the physical and social sciences with respect to the use of abstract concepts is not a sharp one. Indeed, the influence of mathematics and mathematical ways of thinking on the physical sciences especially has led to ever increasing use of abstract concepts including some, as we shall see, that may have no real counterpart at all, any more than a mathematical formula has a real counterpart.

It would seem that to depart from the real world by concentrating on just a few abstract properties of physical objects, such as the straightness of some physical lengths, would rob mathematics of effectiveness. Yet part of the secret of mathematical power lies in its use of abstract concepts. By this means we free our minds from burdensome and irrelevant detail and are thereby able to accomplish more. For example, if one should study fruits and attempt to encompass in one theory color, shape, structure, nature of skin, relative hardness, nature of pulp, and other properties he might get nowhere because he had tackled too big a problem. However, let him study structure alone and he will soon observe that all have a pulp underneath the skin and a core at the center containing seeds that reproduce the fruit. This is a fundamental fact about fruits that can be used in the study of all species. Similarly, mathematics ignores many properties of the chalk line, the stretched string, and the table's edge to concentrate on just a few properties of lines. The success of abstraction rests on the divide-and-conquer precept.

Mathematics departs further from the study of the physical world in that it idealizes the physical problems it seeks to solve. Suppose a mathematician were considering this problem: An attractive young lady weighing 110 pounds and unable to swim is sitting on the edge of a diving board 100 feet above the water and sunning herself. She is pushed off the board. How long does it take her to strike the water? Insofar as mathematics is concerned the attractive lady is a mass and her mass may, for the purpose of the problem, be regarded as concentrated at one point. She is, in other words, a point mass. The fact that she is pushed off the board means to the mathematician merely that she starts her fall with zero velocity. Since she falls a rather short distance the mathematician will disregard the resistance of the air. Moreover, when air resistance is ignored, all objects fall at the same rate. The young lady's weight is therefore immaterial. But objects falling straight downward under the pull of gravity cover a distance d equal to $16t^2$ feet in t seconds. Hence the mathematician will substitute 100 for d in the formula $d = 16t^2$ and find the corresponding value of t. The entire physical problem is converted into a problem of algebra. Mathematics unquestionably fails to notice some of the beauty in this world in its proclivity for idealizing, yet such are its ways.

Not only young ladies on diving boards but even bodies as big as the earth or the sun may for some purposes be treated as masses concentrated at points. In the study of planetary motion, where mllions of miles separate the bodies involved, the relatively small sizes of the bodies can often be ignored and the entire situation regarded as a problem of points moving along curves. Naturally the idealization of the earth as a point would not do to study motion along the earth's surface. But here another idealization is often used. Even though the earth is not a perfect sphere and is in fact ellipsoidal in shape, it is usually adequate to treat it as a perfect sphere.

The use of mathematical abstractions and idealizations to study the physical world causes many people to reject the knowledge that mathematics offers, but the mathematical process of idealization is not without parallel in other fields where the results are often more highly lauded. A marble statue of a man is not the man, yet it often reveals characteristics of the man that are obscure when we view the living being. Cold marble can reveal features of personality and emotions. The like is true of the mathematical representation of physical phenomena. It must be acknowledged, however, that mathematics is often forced to make idealizations which fall short of the true problem and hence gives only a portion of the knowledge it seeks to embrace. But as Alexander Pope has pointed out, "Who could not win the mistress, must be content to woo the maid."

3: THE SCIENCE OF ARITHMETIC

I had been to school . . . and could say the multiplication table up to 6 × 7 = 35, and I don't reckon I could ever get any further than that if I was to live forever. I don't take no stock in mathematics, anyway.

—HUCKLEBERRY FINN

THE Greeks had a word for it. They called the boring manipulations of arithmetic that are used in commerce, trade, calendar-reckoning, and military problems *logistica*. Most young people would endorse this Greek deprecation of arithmetic for many reasons. But the reason that is pertinent is that we are obliged to learn the subject before we can really understand it. Because the knowledge of counting, adding, subtracting, and the like is regarded as a preparation for "life" we are taught it mechanically from early childhood. The practice takes precedence over the principles. No doubt this introduction to life is not especially cheering. It enables us to handle money efficiently and perhaps serves to arouse our cupidity, but it hardly inspires us.

The Greeks had another word for the *mathematics* of arithmetic. This they called *arithmetica. Arithmetica* was to the Greeks a study of the concept of number, its principles, and its applications to nature. This science of arithmetic was founded by the Greeks. The two preceding civilizations in which a kind of mathematics flourished, the Egyptian and Babylonian, had barely acquired the rudiments of the subject. Let us see what the Greeks meant by the science of arithmetic, for it is their idea that today is the core of the mathematics of number.

The characteristically Greek consideration of the basic concepts of arithmetic, namely the whole numbers and the fractions, was inaugurated by the Pythagoreans, a group or school of mathematicians whose name is derived from their legend-veiled leader,

Pythagoras. He was born on the island of Samos about 569 B.C. After extensive travel in Egypt and India Pythagoras migrated to Croton, a Greek colony in southern Italy, where he founded a school. Pythagoras' teachings were a peculiar combination of mystical and rational doctrines. On the one hand he and his followers were inspired by current Greek religious beliefs such as purification and redemption of the soul from the taint of the physical and the prison of the body. On the other, the members of the community devoted themselves to the study of philosophy, science, and mathematics. Members were pledged to secrecy and required to join up for life. Though membership was restricted to men, women were admitted to lectures, for Pythagoras believed that there was some good in them. The mysticism and secretiveness of the Pythagoreans caused the people of Croton to dislike and suspect them, and they were finally driven out. Pythagoras himself fled to Metapontum in southern Italy and was murdered there. His followers, however, spread to other Greek centers and continued his teachings.

Probably because they had no number symbols (though earlier civilizations did) the Pythagoreans represented whole numbers by pebbles placed on the ground or by dots in sand. They classified these numbers according to the shapes produced by the arrange-

Fig. 7

ments of the dots or pebbles. The numbers we denote by 3, 6, 10, and so on were called triangular numbers because they could be arranged to form triangles (fig. 7). The numbers 4, 9, 16, and so on were called square numbers because they could be arranged as shown in figure 8. In accordance with the Pythagorean terminology we still refer to the latter numbers as square numbers or perfect

Fig. 8

squares, and read 3^2, for example, not as 3 to the second power but as 3 squared.

From even these simple geometrical arrangements some properties of the whole numbers became evident to the Pythagoreans. Thus if we consider the square number 9, and divide the dots by a diagonal line as shown in figure 9, we see that above the diagonal

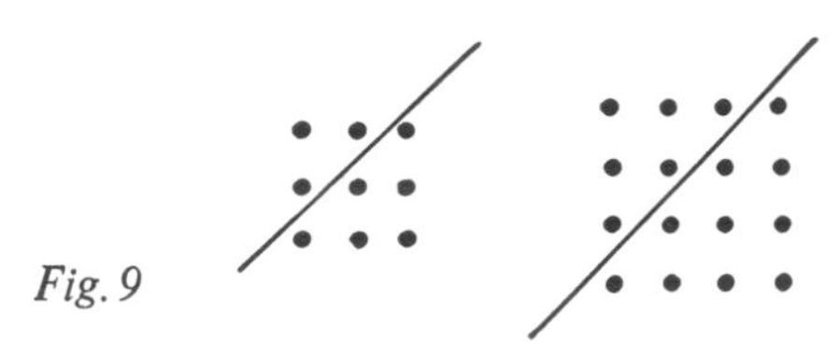
Fig. 9

there is the triangular number 3 and below the diagonal, the triangular number 6. Likewise a diagonal line divides the square number 16 into the triangular numbers 6 and 10. The Pythagoreans concluded that the sum of two triangular numbers is always a square number.

The Pythagoreans also noted that to pass from one square number to another one must add what they called a *gnomon*. The

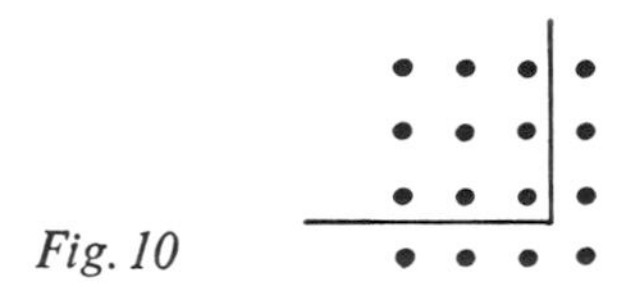
Fig. 10

gnomon is the set of dots to the right of and below the solid lines in figure 10. The gnomon meant a carpenter's square, and it has this shape. If the square number is 9, there are three dots on each side of the square. The gnomon for this number contains two sets of three dots plus the one in the lower right hand corner. Thus, what the Pythagoreans discovered is that

$$3^2 + 2 \cdot 3 + 1 = 4^2.$$

This observation suggested to them that $4^2 + 2 \cdot 4 + 1$ should lead to the next square number or 25, as indeed it does. If the reader will pardon a digression into modern algebraic notation, we may say that what the Pythagoreans discovered is that if n^2 is any square number, then

$$n^2 + 2n + 1 = (n+1)^2.$$

Of course the evidence in hand is not a deductive proof of this general relationship.

To the Pythagoreans the whole numbers were more than symbols for quantities; they were the core of certain concepts and social relationships. The number one they identified with reason, for reason could produce only a consistent whole. Two suggested diversity and so was identified with opinion. Four represented justice because it was the first number that was the product of equals (one was not a number in the full sense because it represented unity rather than quantity to the Pythagoreans). Because these people thought of four as four dots arranged in a square and identified four with justice, the association of the square and justice continues to this day. The square shooter is the man who acts justly. The number five signified marriage because it was the union of the first odd and first even number. Seven was identified with health and eight with love and friendship.

The Pythagoreans made more impressive use of numbers. They saw number and numerical relationships everywhere in nature. In fact they regarded number as the essence of the matter and form of the universe. This doctrine was expressed by Philolaus, a famous fifth-century Pythagorean. "Were it not for number and its nature, nothing that exists would be clear to anybody either in itself or in its relation to other things. . . . You can observe the power of number exercising itself not only in the affairs of demons and gods but in all the acts and the thoughts of men, in all handicrafts and music."

Nature was clearly composed of "fournesses." Thus point, line, surface, and solid were the possible dimensions of all forms. The Pythagoreans and many later philosophers, including Plato and Aristotle, believed that all matter was built up of four elements: earth, air, fire, and water. It was the Pythagoreans themselves who found that harmonious sounds are created when two equally taut strings are plucked, provided that the ratios of the lengths of these strings are formed by simple whole numbers. The ratio of 2 to 1 is still called the octave; the ratio of 3 to 2 is known as the fifth; and the ratio of 4 to 3 is the fourth. Again four numbers are involved.

Four was clearly an important number but the number ten was regarded as the ideal number because it is the sum of 1, 2, 3, and 4, and because the triangle outlined by ten dots contains four members on each edge. Ten, being ideal, must be fundamental in the design of nature. There must, the Pythagoreans thought, be ten moving bodies in the heavens. Unlike most Greeks, the Pythagoreans believed that the earth is in motion. Since they also knew five other planets and observed the sun, moon, and the sphere of stars to be moving, they could account for nine bodies. They asserted that there was a tenth, invisible, body, called the "counter-earth," which together with the earth and the other bodies revolved around a fixed central fire. The ideality of ten also required that the whole universe be describable in terms of ten pairs of categories, such as odd and even, bounded and unbounded, right and left, one and many, male and female, good and evil.

The Pythagoreans saw additional relationships in the motions of the planets. They believed that bodies moving in space produced sounds. They believed, further, that a body moving rapidly gave forth a higher note than one moving slowly. Perhaps this idea was suggested by the swishing sound of an object whirled on the end of a string. Now according to their astronomy, the greater the distance of a planet from the earth the more rapidly it moved. Hence, the sounds produced by the planets varied with their distances from the earth, and these sounds all harmonized. But this "music of the spheres," like all harmony, reduced to no more than number relationships, and hence so did the motions of the planets.

The natural science of the Pythagoreans was of course largely speculation and not at all satisfactory by modern standards. Nevertheless, they made some major contributions. They recognized the importance of number underlying diverse natural phenomena and they made the first steps toward applying that concept to the study of nature. During the twenty-five hundred years since the Pythagoreans flourished the concept of number has been sharpened and extended and efficient schemes for writing numbers and for calculating with them have been created, but the uses we make today of simple numbers in understanding and mastering nature are similar to those made by the Pythagoreans. Indeed, as we shall see in a moment, the whole numbers and fractions enable us to accomplish far more than one would expect with such elementary tools. Let us review first some of the improvements in the science of arithmetic which were made since the time of the Pythagoreans.

These thinkers were wont, as we have seen, to identify numbers with geometrical forms and even physical objects and relationships. But later Greeks, notably Plato, regarded numbers as abstract concepts entirely divorced from geometrical and physical associations, and we have followed their example. Despite the seeming simplicity of this statement most people have great difficulty in appreciating what it means, probably because they think about numbers in connection with applications. It is important that one appreciate just what the abstractness of number implies.

Suppose, for example, a man purchases three pairs of shoes at five dollars a pair. He then reasons that three pairs of shoes times five dollars is fifteen dollars and he hands the salesclerk fifteen dollars. But if this reasoning is sound it should be equally sound to argue that three pairs of shoes times five dollars is fifteen pairs of shoes and perhaps instead of handing the salesclerk fifteen dollars the buyer should walk out with fifteen pairs of shoes.

The difficulty in this example is that the purchaser fails to understand the concept and use of numbers. One cannot multiply pairs of shoes by dollars. In the physical situation in which a man purchases three pairs of shoes at five dollars per pair, he must abstract the 3 and the 5, perform the purely mathematical operation of multiplying 3 by 5, and then apply or interpret the answer in the physical situation. The proper interpretation in the purchase of the shoes demands that the answer be fifteen dollars.

Another example may also help us to appreciate the abstractness of numbers. Mathematically, $2/1$ is equal to $4/2$. But the corresponding physical fact may not be true. One might be willing to accept four half-pies instead of two whole pies, but no woman would accept four half-dresses in place of two whole dresses or four half-shoes in place of a pair of whole shoes.

We see, then, that number is an abstract concept and that when numbers arise in physical situations, one must abstract the quantitative facts, perform the appropriate mathematical operations, and then interpret the mathematical answer. The foregoing examples illustrate one of the major facts about the relationship of mathematics to the physical world. Number is a concept of the mind, no doubt suggested by physical objects but independent of them.

Another of the great abstractions is the number zero, which, peculiarly, is an abstration from the absence of objects. We are of course accustomed to the use of zero and hence do not appreciate that the concept of zero as a number is a remarkable one. Too many people still confuse zero and nothing, and for this reason among others fail to realize just what is meant by zero as a number. It is easy to see, however, that the two concepts must be distinguished. A person's grade in a course he never took is nothing; he may, however, take a course and obtain the grade of zero. If a person has no account in a particular bank, his balance there is not zero but nothing; if he has an account his balance may be zero. To say that zero is a number means that we can operate with zero much as we operate with any other number. Thus, $5 + 0 = 5$ but 5 + nothing is a meaningless expression and yields nothing.

The only restriction on zero as a number is that one cannot divide by zero. Division by zero, so to speak, produces nothing. Because so many false steps in mathematics result from attempted division by zero, we should be clear as to why the operation is not feasible. We say that $6/2 = 3$ because $6 = 2 \cdot 3$, that is, the divisor times the quotient gives the dividend. Hence the answer to $5/0$ should be some number that when multiplied by 0 gives 5. But any number multiplied by 0 gives 0 and not five. Hence $5/0$ has no value or it is nothing. The expression $0/0$ is also meaningless but for a somewhat different reason. The answer to $0/0$ should be some number that when multiplied by 0 yields the 0 dividend. However, any number may then serve as a quotient because any number multiplied by 0 gives 0. But mathematics cannot tolerate such an ambiguous situation. If $0/0$ arises and any number may serve as an answer we shall not know what number to take and hence are not aided. It is as if we asked a person for directions to some place and he replied, Take any direction.

With zero available, mathematicians were finally able to develop our present method of writing whole numbers. First of all we count in units and then represent large quantities in tens, tens of tens, tens of tens of tens, and so on. The use of ten resulted most likely from the fact that man counted on his fingers and when he had used the fingers on his two hands, considered the number arrived at as a larger unit. Thus we represent two hundred and fifty-two by 252. The left-hand 2 means, of course, two tens of tens; the 5 means five times ten; and the right-hand 2, just two units. The use of a zero makes such a system of writing quantity practical because it enables us to distinguish 22 and 202, and moreover to operate with the zero as a number.

Because in our way of writing numbers the position of an integer determines the quantity it represents, the principle involved is called positional notation. Obviously position is important in the description of quantity as it is in other kinds of description. The statement that A loves B is quite different from the statement that B loves A. The system of positional notation we use derives from the Hindus; however, the same scheme was used two milleniums earlier by the Babylonians, but to a more limited extent because they did not have a zero.

Thus, to the abstract concept of number derived from the Greeks, mathematicians added the concept of a zero and an efficient scheme for representing large quantities. Other improvements, which are entirely familiar, are the methods of operating with whole numbers and fractions and the decimal representation of fractions.

But can we do any significant work with the whole numbers and fractions? Of course the usual operations of arithmetic, addition, subtraction, multiplication, and division—or, as Lewis Carroll called them, ambition, distraction, uglification, and derision—are practical. If one wanted to find the total number of cattle in two herds, one of 5280 heads and the other of 10,357 heads, he could, if he chose, go out and count the cattle in the combined herds, but it is slightly more convenient to add the two numbers. Such practical applications and the usual arithmetical operations of commerce are, however, not the really significant ones. Without intending to deprecate the process of making a living or even the use of numbers in such an ancient and honorable occupation as throwing dice we must nevertheless recognize that civilizations that confined their mathematics to such lowly applications did not advance mathematics, science, their understanding of nature, or even their material benefits. Fortunately, even with such simple tools as the whole numbers and fractions one can achieve some important discoveries. For example, the following.

By the time that Galileo became active, in the early seventeenth century, pumps to remove water from flooded mines had been in use for several centuries. A pump was located at the top of a long tube such as is shown in figure 11, and the bottom of the tube was immersed in the water. The pump removed the air at the top of the tube and as it did so the water rushed up the tube. Engineers explained the rise of the water by citing the long-accepted principle

Fig. 11

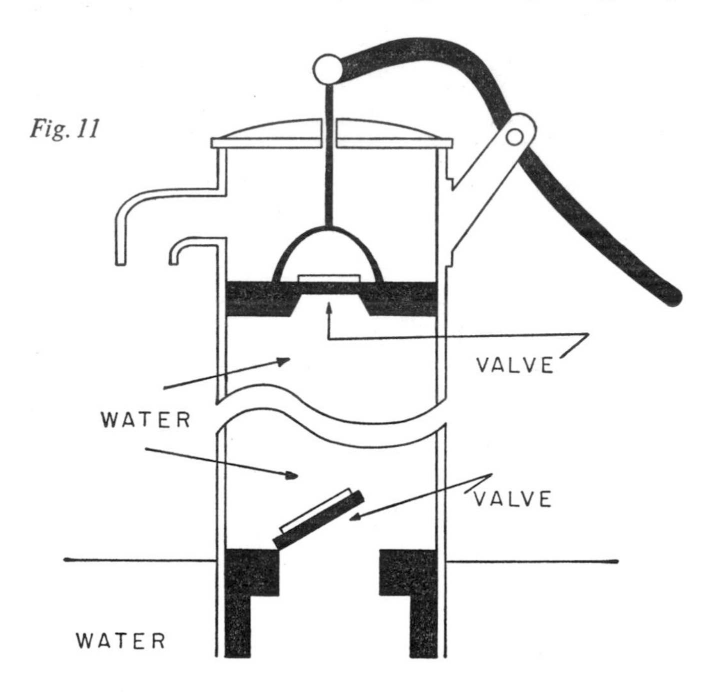

that nature abhors a vacuum; that is, the water rushed up to fill the vacuum created by the pump. This phenomenon was studied by Galileo, about whom we shall learn much more later, and he noted that the water in the tube could be made to rise as much as thirty-four feet above the water level at the bottom but no higher. When told about nature's abhorrence of a vacuum, Galileo skeptically remarked that this was a fine principle but why did nature's abhorrence stop short at thirty-four feet? Galileo's pondering led him to the discovery of the weight of air. The air pressing on the open surface of the water forces the water up the evacuated tube to the height of thirty-four feet, at which point the weight of the column of water exerts just enough pressure to counterbalance the weight of the air pressing on the open water surface.

Numbers were decisive in one of the great achievements of biology. The Englishman William Harvey had become convinced that the blood in our bodies is actually pumped out by the heart to circulate around the body and then return to the heart. However, he had to convince skeptics. He did a little arithmetic, the essence and spirit of which is as follows. Suppose that the left ventricle of the heart can hold about two ounces of blood, a conservative estimate. Suppose further that the heart in contracting forces out at least one-eighth of this, though very likely more. The heart beats about seventy-two times a minute but, said Harvey, suppose that it beats only two thousand times in one half-hour. Then in one half-hour the heart forces into the arteries more than five hundred ounces of blood, which is a quantity greater than all the blood present in the human body. Where does all this blood come from and go to? Obviously the same blood must come back to the heart many times a minute and be pumped out again.

In 1772 Johann Elert Bode (1747–1826) studied the distances from the sun of the planets known at that time and made the following observation. If to each member of the simple sequence of numbers

(1) $$0, 3, 6, 12, 24, 48, 96,$$

each of which comes from doubling the preceding one, the number 4 is added, one obtains the sequence

(2) $$4, 7, 10, 16, 28, 52, 100.$$

If we omit 28 then these numbers are pretty close to the relative distances from the sun of Mercury, Venus, Earth, Mars, Jupiter, and Saturn. Thus, the second sequence says that the distance of Mars from the sun compared to the distance of the earth from the sun is 16 to 10. The actual ratio is 1.52. The astronomers of the late eighteenth century were attracted by the fact that no planet corresponded to the number 28 and they searched for this planet for many years. In 1801 Giuseppe Piazzi (1746–1826) found the asteriod Ceres and since that time over a thousand small asteriods have been found whose orbits are between those of Mars and Jupiter. It is suspected that they are the remains of an exploded planet.

If we now extend Bode's law to the numbers 192 and 384 by further doubling in the first sequence and therefore obtain 196 and 388 in the second sequence, we find that 196 is approximately correct for Uranus but the number 388 is much too large for the next known planet, Neptune, which would correspond to about 300. However, the figure of 388 is approximately correct for Pluto, a planet as big as the earth, lying beyond Neptune. Unfortunately no number in Bode's sequence applies to Neptune and hence his law cannot be accepted as a law of nature. Were a better law available one might dismiss his sequence as a play on numbers or an accidental fit of simple numbers to the relative distances of the planets. However Bode's law was taken seriously for a long time and did lead to the discovery of new planets. Moreover, since we do not have any better law to account for the known relative distances of the planets from the sun, perhaps the wisest position to take is that Bode's law may be a clue or some rough approximation to a more accurate law not fully grasped as yet.

Simple numbers and fractions were sufficient to frame one of the leading ideas of modern chemistry. About one hundred years ago the present theory that all matter is composed of atoms, and that there are different atoms for basically different elements such as hydrogen, oxygen, copper, gold, and mercury, had gained acceptance. Moreover, it was accepted that, while the atoms of any one element have the same weight, atoms of different elements differ in weight.

About 1860 Dmitri Ivanovich Mendeléev (1834–1907) undertook to arrange the known elements in the order of their atomic weights. It turned out that every eighth element among the first sixteen Mendeléev arranged had similar chemical properties to the first one of that set of eight. He found, however, that beyond this point, if he were to continue to arrange elements in the order of increasing atomic weight and yet put elements with similar chemical properties eight positions apart, he had to leave some blank spaces. It seemed a fair inference to Mendeléev that unknown elements belonged in the blank spaces. This reasoning led Mendeléev to look for new elements, and later investigators found three, now called scandium, gallium, and germanium, whose chemical properties Mendeléev could and did predict from a knowledge of the properties of the elements eight positions below. Later discoveries caused some modifications in Mendeléev's periodic law but his arrangement still is the essence of the modern one. Though Mendeléev recognized that he had no physical explanation of why the regularity revealed by his arrangement held, he advocated utilizing the periodicity to discover new elements, to determine their atomic weights, and to predict the chemical properties of these elements, such as their ability to combine with other elements to form molecules.

One of the most startling ideas in modern atomic theory was clearly delineated by arithmetic. From 1910 on famous physicists and chemists, notably J. J. Thomson, Frederick Soddy, and F. W. Aston, had been discovering that most elements formerly regarded as pure are composed of several elements called isotopes, having the same chemical properties but different atomic weights. There is not one substance hydrogen but two substances, one twice as heavy as the other. There are three types of oxygen, of atomic weights 16, 17, and 18. Uranium, very much in the news now, has two isotopes of atomic weights 238 and 235. Actually, the weight of any isotope is not exactly a whole number; but the precise figure is within 1 per cent of the nearest whole number. The fact that these relative atomic weights are so close to whole numbers almost forces the idea that all of these isotopes, and therefore the elements themselves, are made up of multiples of just one fundamental building block. Moreover, if any element is merely a composite of a certain number of building blocks then it should be possible to subtract one or more of these units and change one element into another. Thus, one might change mercury into gold. These conjectures, which of course could be made only after brilliant experimentation supplied the facts, were confirmed by further experiments, and transmutation of elements is now a reality. The fundamental building block is called the proton, and though there are other elements in the various atoms, their contribution to the weight is negligible.

The striking exception to the fact that the atomic weights of isotopes are practically whole numbers is hydrogen. The atomic weight of the lighter isotope is 1.008, which suggests that it contains just one proton. Now the accuracy of the measurements is such that the decimal part in this figure cannot be ignored. Thus, if oxygen were made up of 16 protons its atomic weight should be 16 times 1.008, whereas it is exactly 16. Likewise, when four protons are combined to make helium, the atomic weight proves to 4.0028 rather than 4.032. The difference of 0.03 in atomic weight is lost. If hydrogen, the element of lowest atomic weight, should be the building block (aside from other particles of negligible weight) then something must be lost in the process of fusing hydrogen atoms to form elements of larger atomic weight. This loss of mass, which has been known to be equivalent to energy since Einstein's work of 1905, is called the binding energy. The idea occurred that it might perhaps be utilized. This is now done in the fusion process that takes place when a hydrogen bomb explodes and the extra mass is converted to radiated energy.

One more application of the whole numbers is worthy of mention because it illustrates how an idea that was investigated for many years solely because of its intrinsic interest has proved to be an immensely useful one today. We have mentioned the fact that to express large quantities men counted in tens, then tens of tens, and so forth. We say that ten is a *base* or larger unit of quantity, and using positional notation, write ten as 10. However, other civilizations used other numbers as a base. For example, the Babylonians, for reasons that are obscure, used sixty. Mathematicians of modern times have considered the problem of writing quantity in some base other than ten and in particular have investigated the possible advantages and disadvantages of other bases.

As an example, let us choose base six. To write the quantities from zero to five we would use the symbols 0, 1, 2, 3, 4, 5, as in base ten. The first essential difference comes up when we wish to denote six objects. Since six is to be the base we indicate this larger quantity by the symbols 10, the 1 denoting one times the base, just as in base ten the 1 in 10 denotes one times the base, or the quantity ten. Thus, the symbols 10 can mean different quantities, depending upon the base being employed. To write seven in base six we would write 11, because in base six these symbols mean $1 \cdot 6 + 1$, just as 11 in base ten means $1 \cdot 10 + 1$. Similarly, to denote twenty in base six we write 32 because these symbols now mean $3 \cdot 6 + 2$. To indicate the quantity forty in base six we write 104, because these symbols mean $1 \cdot 6^2 + 0 \cdot 6 + 4$, just as in base ten 104 means $1 \cdot 10^2 + 0 \cdot 10 + 4$ or one hundred and four.

It is clear that we can express quantity in base six. Moreover, we can perform the usual arithmetic operations in this base. We would, however, have to learn new addition and multiplication tables. For example, in base ten $4 + 5 = 9$, but in base six 9 would be written 13. Our new addition table would have it that $4 + 5 = 13$. Likewise, our new multiplication table would have $4 \cdot 5 = 32$. To use base ten we had to memorize the result of adding each number from 0 to 9 to every number from 0 to 9, and the result of multiplying each number from 0 to 9 with every number from 0 to 9. If six were our base we would have to learn to add only numbers from 0 to 5 to the numbers of this set, and proceed similarly for multiplication. Our addition and multiplication tables would therefore be shorter. The disadvantage would be that to represent such a quantity as forty-five we must write 113 and thus use three digits instead of two.

Base twelve has been advocated for daily use because it possesses some advantages that base ten lacks. For example, to express one-third of a unit as a decimal in base ten we must write the nonending decimal 0.3333. . . . To express one-third of a unit as a decimal in base twelve we would merely have to write 0.4, for this would mean $4/12$ or $1/3$. Also, since our English measure of length calls for twelve inches in one foot we could conveniently express 2 feet and 6 inches as 26 inches in base twelve. As far as the operations of addition, subtraction, multiplication, and division are concerned these would be the same as at present except that we would use the addition and multiplication tables belonging to base twelve. After all, base ten has no more to say for itself than that it is the consequence of a physiological detail.

Base two especially impressed the seventeenth-century religious philosopher and mathematician Gottfried Wilhelm Leibniz. He observed that in this base all numbers were written in terms of the symbols 0 and 1 only. Thus eleven, which equals $1 \cdot 2^3 + 0 \cdot 2^2 + 1 \cdot 2 + 1$, would be written 1011 in base two. Leibniz saw in this binary arithmetic the image and proof of creation. Unity was God and zero was the void. God drew all objects from the void just as the unity applied to the zero creates all numbers. This conception, over which the reader would do well not to ponder too long, delighted Leibniz so much that he sent it to Grimaldi, the Jesuit president of the Chinese tribunal for mathematics, to be used as an argument for the conversion of the Chinese emperor to Christianity.

Though Leibniz may have had special reasons for considering base two, neither this base nor any other had until recently been seriously considered as a substitute for base ten. In fact, aside from incidental uses of other bases in higher mathematics to facilitate an occasional proof, the subject of bases other than ten was regarded until recently as an intellectual amusement. But the step from a purely academic, aesthetic, or amusing idea to a weighty scientific application is quickly made when the right application presents itself. Such an opportunity arose in designing the new, very fast, large-scale computing machines. The basic component in these machines is the radio vacuum tube, which can be made to pass current by applying a voltage to it, or can be kept inactive. Two maneuvers are thus possible. But in base two all numbers require only two symbols, the 0 and 1. Hence a number can be recorded by the machine by having a series of tubes, one for the units place, another for multiples of 2, a third for multiples of 2^2, and so forth, "fire" or not depending on whether the digit 1 or 0 occurs in the corresponding place in the number. For example, the number 1101 in base two is recorded by the machine, by activating three tubes while the fourth tube, which records the second place in the number, is kept inactive. The various arithmetic operations are performed by having tubes fire or remain inactive. All large-scale electronic computers perform their arithmetical calculations in base two. The numbers to be worked on are converted from base ten to base two, and the results are reconverted to base ten.

The use of computing machines is growing by leaps and bounds, and computers now furnish the solutions to scientific and engineering problems that were unmanageable previously.

The development of electronic computers that rapidly perform complicated arithmetic operations has led to another study that may prove highly fruitful. According to our present understanding of the human brain and of the nervous system generally the nerve cells respond to electrical impulses much as an electron tube does. Just as an electron tube will "fire" when it receives electrical current beyond a certain minimum value and remain inactive otherwise, so the nerve cells and nerve chains connecting nerve cells will transmit an electrical impulse to whatever organ they may lead when this impulse exceeds a threshold value; otherwise they are inactive. Computing machines also have a memory. That is, they store partial results of calculations and bring up these results when they are needed to combine with other partial results. The result of an addition process may be stored until the result of some multiplication process is obtained, and then the machine will add these two results if it is so directed. Thus, the machine's memory device functions somewhat like the human memory. Since computing machines simulate the actions of nerves and memory, they may give us some clues to the functioning of the human brain and of nerve actions. Though these machines are in speed, accuracy, and endurance superior to the human brain, one should not infer, as many popular writers are now trying to suggest, that computers will ultimately replace brains. Machines do not think. They perform calculations. The machines, to use the words the Greeks used and which we mentioned at the beginning of this chapter, do *logistica* but not *arithmetica*. Nevertheless, we undoubtedly have in the machine a useful model for the study of some functions of the human brain.

It should be clear from these examples—many others could be given—that even with the mathematics of the simple whole numbers important conclusions about inanimate and animate nature can be obtained. We may reasonably expect the applications of mathematics to become more numerous and more weighty as we bring into play new mathematical ideas.

4: THE DEEPER WATERS OF ARITHMETIC

. . . where ignorance is bliss,
'Tis folly to be wise.

—THOMAS GRAY

MATHEMATICS could have been a rather simple and almost innocuous subject if the concept of number had been limited to the whole numbers and fractions. Of course the power of the subject and perhaps the interest in it would have also remained severely limited. Whether or not mathematicians would have preferred this state of affairs and living with peaceful, untroubled minds, fate willed otherwise and it struck an almost mortal blow at that very group of mathematicians who thought they had reached an understanding of the entire universe with just the whole numbers and fractions.

The Pythagoreans, we recall, loved to play with numbers and obtained many pleasing relationships among the whole numbers. Unwittingly they played with fire. One of the relationships that titillated their minds was that the sum of some pairs of square numbers was also a square number. Thus, the sum of 9 and 16 is 25 and the sum of 25 and 144 is 169. The numbers 9, 16, and 25 are, of course, 3^2, 4^2, and 5^2. Numbers such as 3, 4, and 5, whose squares provide the relationship just described, are still called Pythagorean triples. The Pythagoreans liked these combinations so much that they engaged in a search for all possible triples.

These triples also have an interesting geometric interpretation (fig. 12). If two of these numbers are the lengths of the sides, or arms as they are called, of a right triangle, then the third is the length of the hypotenuse or side opposite the right angle. This interpretation suggested to the Pythagoreans the more general fact that the square of the hypotenuse in *any* right triangle equals

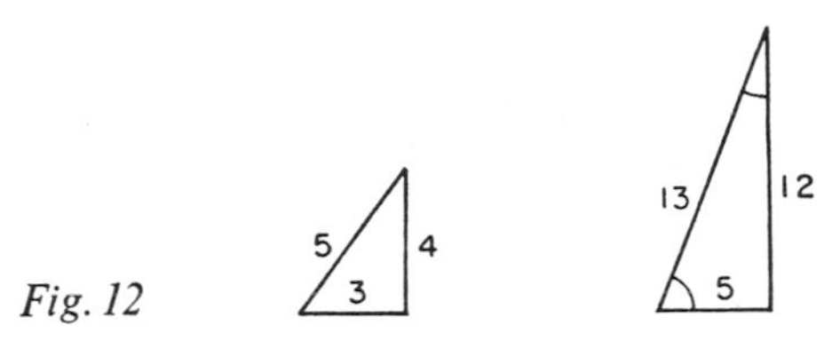

Fig. 12

the sum of the squares of the arms. Whether they knew a satisfactory mathematical proof of this general fact is a much debated question; nevertheless the relationship is still called the Pythagorean theorem. It is said that Pythagoras was so overjoyed with its discovery that he sacrificed an ox in honor of the occasion. But it was this very theorem that proved to be the undoing of the Pythagoreans and caused untold woes to later generations of mathematicians.

It occurred to a Pythagorean one day to examine the seemingly simplest case of the geometric theorem. Suppose a right triangle has arms each 1 unit long. Since the sum of the squares of the arms is the square of the hypotenuse, it follows that the square of the hypotenuse is 2. And since 2 is not a square number, the relationship $1 + 1 = 2$ does not arise from a Pythagorean triple. But the question this Pythagorean raised was, Is the length of the hypotenuse itself, which is obviously not a whole number, at least a fraction, that is, a ratio of whole numbers?

It certainly seemed reasonable that there was some fraction whose square was 2. Even the simple fraction $7/5$, when squared, yields $49/25$, and this square is almost 2. Hence, very likely some ratio of two whole numbers, perhaps large ones, would, when squared, yield 2. However, after many attempts to find a fraction whose square was 2, the Pythagoreans became convinced that there was no such fraction. But of course no real mathematician is content with a very likely surmise; he wants a deductive proof. And the Pythagoreans were mathematicians. Hence one of their school proceeded to investigate this question and proved that $\sqrt{2}$ could not equal a fraction. Let us follow his argument as it is recorded in Euclid's *Elements,* the famous geometry written about two hundred years later.

We shall write the length of the hypotenuse as $\sqrt{2}$, meaning simply by $\sqrt{2}$ a number whose square is 2. Now we suppose that

$\sqrt{2}$ is the fraction a/b, where a and b are whole numbers. Moreover, to make matters simpler, let us reduce the fraction a/b to its lowest terms; this means that if the fraction were $^{14}/_{10}$, say, we would cancel the factor 2 which is common to the numerator and denominator and write it as $^{7}/_{5}$. We have supposed, so far, that

$$\sqrt{2} = \frac{a}{b} \tag{1}$$

and we have reduced the fraction a/b to its lowest terms.

If we now square both sides of (1) we get

$$2 = \frac{a^2}{b^2}. \tag{2}$$

We multiply both sides of equation (2) by b^2 to obtain the form

$$2b^2 = a^2. \tag{3}$$

Since an even integer is one that, by definition, contains 2 as a factor, $2b^2$ is an even integer. Hence, according to (3), a^2 is also an even integer. From the fact that a^2 is even we can conclude that a is even. How do we know this? If a were odd, it would be a number that ends in 1, 3, 5, 7, or 9, and such a number squared must again end in 1, 3, 5, 7, or 9. Hence a^2 would be odd. But it is not. Thus, since a^2 is even, a must be even.

Since a is even, that is, since it contains 2 as a factor, a must be twice some other integer, say d; that is, $a = 2d$. Then by (3),

$$2b^2 = (2d)^2 = (2d)(2d)$$

or

$$2b^2 = 4d^2.$$

Division of both sides by 2 yields

$$b^2 = 2d^2.$$

Hence b^2 is an even integer because it is twice another integer. Since b^2 is an even number, b must be also. The proof is the same as the one we gave in connection with a^2.

At this point we have proved that a and b are even. Both numbers therefore contain 2 as a factor. But this conclusion contradicts the fact pointed out at the beginning of our proof, namely, that the fraction a/b was in its lowest form, for now we find that both a and b contain 2 as a factor. Hence the assumption, stated in (1), that $\sqrt{2}$ is equal to a fraction must be erroneous. This entire argument leaves us with the conclusion that $\sqrt{2}$ cannot be a ratio of two whole numbers.

The Pythagoreans were indeed baffled. Here was a totally new element in the universe that could not be described in terms of whole numbers. Their entire philosophy of nature, which was based on the principle that every phenomenon could be reduced to whole numbers, was threatened. They called this new number *irrational,* which term then meant unmentionable or unknowable though today it means a number not expressible as a ratio of whole numbers. There is a legend that the discovery of $\sqrt{2}$ was made by a member while the entire group of Pythagoreans was on a ship at sea. The member was thrown overboard and the rest of the group pledged to secrecy.

But secrets, particularly a scandal of this magnitude, will out, and mathematicians from Greek times on have not only had to recognize the irrational number but to contend with it. Indeed, as is often the case, once our attention is called to something new we find it all around us. So it was with the irrational number. One can show by the very method which the Pythagoreans used that the square root of any whole number that is not a perfect square, for example $\sqrt{3}$, $\sqrt{5}$, $\sqrt{6}$, the cube root of any whole number that is not a perfect cube, for example $\sqrt[3]{2}$, $\sqrt[3]{3}$, and so forth, are irrational. The number π, which is the ratio of the circumference of any circle to its diameter, is irrational, though this fact was not proved until 1768. Many other examples could be given.

However to recognize the existence of such baffling entities as $\sqrt{2}$, $\sqrt[3]{2}$, and π does not tell what they are. To say that they are irrational, that is, not ratios of whole numbers, is no more informative than the statement that a giraffe is not a cow. What is an irrational? The fact which must be faced squarely is that an irrational is a new kind of number. It is a number because it expresses quantity, for example, the length of the hypotenuse of a right triangle. It is a new kind of number because the Pythagorean proof shows that an irrational number cannot equal a whole number or a fraction.

Our position is simply this. To one who is acquainted with, say, only dogs and cows, an animal means dog or cow. But then he encounters a giraffe, a strange creature indeed. He finds that giraffes have almost all the characteristics of dogs and cows and so he finally realizes that he must enlarge his concept of animal to include giraffes. This is the position mathematics was in when the irrational number was discovered. The Greek mathematicians finally appreciated that they had a new kind of number on their hands and that they must enlarge the concept of number to include these new entities.

But useful numbers are not idle concepts. They must be commingled with other numbers in arithmetic operations so as to lead to new knowledge. The question then arises, How can we add, subtract, multiply, and divide irrational numbers or perform these operations with different types of numbers combined; for example, whole numbers and irrationals? For the moment we shall answer this question by doing just what the mathematicians did when they first explored their discovery of the irrational number; the mathematicians used the operations on whole numbers and fractions as a guide to what should be done with irrationals. Since $\sqrt{2}$ and $\sqrt{4}$ look alike, though the former is an irrational number and the latter just the number 2, it seemed reasonable to be guided by the latter in dealing with the former. Of course, as we have already seen, reasoning by analogy may lead to trouble. Bulls and cows look alike but anyone who tries to pet a bull, thinking it is a cow, will soon discover the essential danger in analogy. Nevertheless, let us proceed in this manner for the present.

Can we combine $\sqrt{2} + \sqrt{3}$ into the simpler $\sqrt{5}$? Let us test this operation on whole numbers expressed as roots. Certainly $\sqrt{9} + \sqrt{16}$ does not equal $\sqrt{9 + 16}$, that is, $3 + 4$ does not equal 5. Hence we should not assert the analogous relation for irrational numbers.

What can we say about the product of irrational numbers? May we write $\sqrt{3} \cdot \sqrt{2}$ as $\sqrt{6}$? Warned perhaps by the case of addition, wherein $\sqrt{3} + \sqrt{2} \neq \sqrt{5}$, we might be tempted to answer, no. However, does $\sqrt{4} \cdot \sqrt{9} = \sqrt{36}$? Since the latter is true because it says merely that $2 \cdot 3 = 6$, we shall operate with irrationals in the same way.

It should be clear that there need be no bewilderment about how to operate with irrationals. We have but to look at operations with whole numbers expressed as roots and see if the operation in question is permissible there.

There is additional justification for what we have said thus far about operations with irrational numbers. While $\sqrt{2}$ is not a fraction it can be approximated as closely as we wish by a fraction. Thus 1.41, which is of course $^{141}/_{100}$, is a good approximation because $(1.41)^2$ is 1.99. Likewise, a good approximation to $\sqrt{3}$ is 1.73. Let us reconsider $\sqrt{2} \cdot \sqrt{3} = \sqrt{6}$. If we replace $\sqrt{2}$ by 1.41 and $\sqrt{3}$ by 1.73 and multiply 1.41 by 1.73 we obtain 2.44, which is a good approximation to $\sqrt{6}$, since $(2.44)^2$ is 5.95. The more accurately we approximate $\sqrt{2}$ and $\sqrt{3}$, the more accurately will the product approximate $\sqrt{6}$. Of course it must be noted that no decimal expression, no matter how many decimal places are used, will ever exactly equal an irrational number; a decimal is a fraction, and an irrational, we saw, cannot equal a fraction.

The fact that one can approximate an irrational by a decimal suggests a question that is often asked and deserves an answer. Why bother with irrationals at all? Why not replace them by decimal approximations? After all, in applications of mathematics one never needs numbers carried to more than some finite number of decimal places because all measurements are accurate only to a few decimal places. Why didn't the Pythagoreans throw the irrational number overboard rather than the member who discovered it?

Even though the irrational number is a bit troublesome to understand and work with one cannot dispense with it. The hypotenuse of a right triangle whose arms are 1 is exactly $\sqrt{2}$ and not some approximation to $\sqrt{2}$. Since mathematics aims to be as exact and as rigorous as human thought can be, to approximate irrationals is to sacrifice the exactness of mathematics.

Practical people will say, however, that mathematicians are too fussy and that a close approximation to one million dollars is as good as the million itself. Perhaps. Suppose that in the course of mathematical work one were led to the expression $(\sqrt{2})^4$, that is $\sqrt{2} \cdot \sqrt{2} \cdot \sqrt{2} \cdot \sqrt{2}$, and let us suppose that four-decimal-place accuracy is required for the particular application. The practical person would then have to set about calculating $(1.4141)^4$. The mathematician can say at once that the value of the expression in question is 4 and spend the next hour gazing contentedly about the world while the practical person carries out and checks his calculation. Moreover the answer 4 obtained by the mathematician is exact, whereas even though 1.4141 is the correct approximation to $\sqrt{2}$ up to four decimal places, the product $(1.4141)^4$ will not be accurate to four decimal places. If the practical man wants his answer correct to four decimal places he will have to start with an approximation to $\sqrt{2}$ that is correct to about seven decimal places and then calculate the product of four such factors.

The irrational number is today a respectable and useful member of the number system but it was not always so. From the time of its discovery by the Pythagoreans until about one hundred years ago it was a highly controversial figure. The reason was simply that mathematicians great and lowly had as much trouble in understanding it and in justifying the rules of operation we discussed above as any young person does when he first encounters it. The gifted Greek mathematicians who followed Pythagoras never did succeed in understanding the irrational number as a number. They evaded the issue by treating all numbers as line segments. As a line segment the irrational number is as innocuous-appearing as a whole number. No one looking at the hypotenuse of a right triangle whose arms are 1 finds the hypotenuse less acceptable as a line segment than either arm.

What did the Greeks do about operations with irrational numbers? For example, how did they obtain the product $\sqrt{2} \cdot \sqrt{3}$? Here, too, they thought geometrically. The product $\sqrt{2} \cdot \sqrt{3}$ was simply the area of a rectangle whose sides were $\sqrt{2}$ and $\sqrt{3}$. But they never said how much the area was numerically; they were content to think of the product as an area. Of course such a "solution" to the problem of irrational numbers is acceptable only if one is content to think about them; it is not useful if one wants answers that can be used in practical work. Where numbers were used—for example, in astronomical calculations—the Greeks used whole numbers and fractions only. Even the greatest Greek algebraist, Diophantus, who lived during the latter part of the Alexandrian Greek civilization (around A.D. 250), rejected irrationals as numbers.

The mathematics and science that developed in Europe after the

Renaissance became much more dependent upon quantitative results and hence upon the use of all types of numbers. The mathematicians had to face once more the problem of the irrational number. But the best they could do was determine the correct operations by analogy with what held for whole numbers and fractions, much as we did above. Nor were they happy about irrational numbers as quantities.

Let us listen to the woes of the greatest German algebraist of the sixteenth century, Michael Stifel. In his *Arithmetica Integra* (1544) he says, "Since, in proving geometrical figures, when rational numbers fail us irrational numbers take their place and prove exactly those things that rational numbers could not prove . . . we are moved and compelled to assert that they truly are numbers, compelled that is, by the results that follow from their use—results that we perceive to be real, certain, and constant. On the other hand, other considerations compel us to deny that irrational numbers are numbers at all. To wit, when we seek to subject them to numeration [when we seek to express them as decimals] . . . we find that they flee away perpetually, so that not one of them can be apprehended precisely in itself. . . . Now that cannot be called a true number which is of such a nature that it lacks precision. . . . Therefore, just as an infinite number is not a number, so an irrational number is not a true number, but lies hidden in a kind of cloud of infinity."

Isaac Barrow, a foremost mathematician of the seventeenth century and the predecessor of Isaac Newton at Cambridge University, maintained that irrational numbers have no meaning independent of geometric lengths. The attacks on irrational numbers continued while mathematicians, finding the numbers more and more convenient and even necessary in the expanding fields of algebra and the calculus, worked with them and defended their steps as we did above. Ultimately, mathematicians overcame their repugnance to irrational numbers and established the proper logical basis for operations with them. We shall have more to say about this matter shortly.

While mathematicians were still looking askance at the Greek gift of the irrational number, the Hindus of India were preparing another brain-teaser, the negative number, which they introduced about A.D. 700. What the Hindus discovered is that they could introduce new numbers that served some purposes as well as the already existing numbers, thereafter called positive, served other purposes. The Hindus saw that when the usual, positive numbers were used to represent assets, it was helpful to have other numbers represent debts. For example, a person with assets of five dollars could represent these assets by the number 5, while a person five dollars in debt could state that his wealth or fortune was -5 dollars.

The Hindus also recognized that one could operate with negative numbers to good advantage if one merely made the rules of operation fit the physical situations. Thus, if a man has assets of five dollars and at the same time debts of three dollars, his net wealth is two dollars. This figure can be obtained by adding 5 and -3 if we define the addition so that $5 + (-3) = 2$; that is, we really subtract the smaller numerical value from the larger one and consider the answer to be positive because the larger of the two numerical values being added belongs to a positive number. If the man's assets were five dollars and his debts eight dollars, his net wealth would be a deficit of three dollars. To get this result mathematically we again subtract the smaller numerical value from the larger one; that is, we subtract 5 from 8 and call the answer -3 because the larger numerical value belongs to the negative number. That is, $5 + (-8) = -3$.

The operation of subtraction with negative numbers can likewise be formulated so as to make it fit physical situations. Suppose a man's net worth is ten dollars and that in reckoning this he had already taken into account that he has a debt of three dollars. Suppose that for some reason the debt is now canceled. What is the man's new net worth? To cancel means in effect to remove or subtract the debt. The man's net worth is now thirteen dollars. Mathematically this means that $10 - (-3) = 13$.

If a man incurs four debts of five dollars each, his total indebtedness is twenty dollars. Symbolically, this means $4 \cdot (-5) = -20$. On the other hand, if a man owes four debts of five dollars each and for some reason these debts are canceled, his net worth is increased by twenty dollars. In symbols, this means $-4 \cdot (-5) = 20$.

Of course negative numbers can be used to represent time before a stated event, where positive numbers are used to represent time after the event. Though we ordinarily write 50 B.C., we could as well write -50. Then to calculate the time elapsed from 50 B.C. to, say, A.D. 30, we should calculate $30 - (-50)$. The result should, of course, be 80 years, and indeed it is that according to the definition of subtraction of negative numbers.

The Hindus saw clearly that if the arithmetic operations of addition, subtraction, and so forth were properly defined for negative numbers, these numbers could be employed to as good advantage as people had previously derived from positive numbers. This contribution of the Hindus seems simple enough to understand once one recognizes that negative numbers are new numbers over and above the numbers previously known. However, it is often more difficult to make a truth known than to discover it. To people to whom the word *number* had always meant positive whole numbers and positive fractions, the very idea of recognizing that there could be other numbers came hard. For many centuries negative numbers were either rejected or treated as second-class citizens.

What was especially difficult for mathematicians to swallow was that negative numbers could be acceptable roots of equations. Though it is not our intention to go into this subject at this point, it may be helpful to note just how negative numbers do occur in equations. The reader will pardon our getting slightly ahead of our story.

Let us consider the following rather trivial problem. A father is 56 years old and his son is 29. When will the father be twice as old as the son? In keeping with the usual elementary algebraic method we let x stand for the number of years from the present time to the time when the father will be twice as old as the son. After x years the father will be $56 + x$ years old and the son $29 + x$ years old. The relation between their ages will then be that

$$2(29 + x) = 56 + x.$$

It happens to be correct algebra to multiply the 2 into both terms in the parentheses and write therefore that

$$58 + 2x = 56 + x.$$

If we now subtract x from both sides of the equation, a step justified by the axiom that equals subtracted from equals give equals, we obtain

$$58 + x = 56.$$

We now subtract 58 from both sides and since $56 - 58 = -2$ we have

$$x = -2.$$

This answer means that the father was twice as old as the son two years ago. Hence, negative numbers answer the question properly. Despite this fact mathematicians who had already accepted negative numbers to represent such quantities as debts or time preceding a given date, or distances opposite to those in a given (positive) direction, rejected negative roots of equations. The famous sixteenth-century algebraist Jerome Cardan called negative roots fictitious, and the founder of modern symbolic algebra, François Viète, discarded negative roots entirely. Descartes, certainly one of the best of mathematicians of all time, also encountered negative numbers as roots of equations, but he called them false on the ground that they represented numbers less than nothing and so were meaningless.

Euler, the greatest eighteenth-century mathematician, believed that negative numbers are greater than infinity. L. N. M. Carnot (1753–1823), a noted French geometer, thought that the use of negative numbers led to erroneous conclusions. As late as 1831 Augustus De Morgan, a distinguished contributor to logic, was uncomfortable about negative numbers occurring as roots of equations. He said that a negative number occurring as the solution of a problem indicated some inconsistency or absurdity. Such a problem, he believed, was incorrectly phrased, or the problem as stated had no solution.

And so the record goes. It was no more than a hundred years ago that mathematicians adopted the correct viewpoint toward these various types of numbers. They recognized first of all that the concept of number could be extended to include negative as well as irrational numbers. These new numbers, suggested by physical uses such as the representation of lengths in the case of irrational numbers or debts in the case of negative numbers, were as legitimate as the whole numbers and fractions. They recognized further that the axioms about number—that is, the premises on which reasoning about number is based—applied as well to the new numbers as to the old ones.

Since we intend to reason about all these types of numbers in later chapters we should note the axioms of number. A few of these are so obviously true that we hardly recognize that we are using them any more than we are usually conscious of breathing. For example, it is certainly correct that $3 + 4 = 4 + 3$ and that $3 \cdot 4 = 4 \cdot 3$. Of course it is correct to interchange the order of the terms in addition and multiplication not just for 3 and 4 but for *any* two numbers. Hence two of the axioms, stated in general terms, are

1. $$a + b = b + a;$$
2. $$a \cdot b = b \cdot a.$$

These two axioms are called the *commutative* axioms of addition and multiplication because the axioms say that we may *commute* the order of the terms.

The next two axioms are as obviously familiar and as acceptable. If one had to calculate $3 + 4 + 5$ he might first add 3 and 4 and then add 5 to the result or he might add 4 and 5 and then add this result to 3. Analogous choices are possible in the calculation of $3 \cdot 4 \cdot 5$. Since the results are the same in both ways of doing the addition, and the same is true in the case of multiplication, we may assume for *any* three numbers that

3. $$(a + b) + c = a + (b + c);$$
4. $$(a \cdot b) \cdot c = a \cdot (b \cdot c).$$

The parentheses are used here to show which two of the three numbers we operate with first. Of course the point of the axioms is that it doesn't matter. These two axioms are called the *associative* axioms of addition and multiplication. Incidentally, the dots used to indicate multiplication are omitted if they are not needed for clarity. Thus, the fourth axiom is usually written as $(ab)c = a(bc)$.

The next axiom is perhaps the only one that requires more attention; it is often incorrectly applied. For the simple numbers 3, 4, and 5 it is evident that

$$3(4 + 5) = 3 \cdot 4 + 3 \cdot 5;$$

that is, $3 \cdot 9 = 3 \cdot 4 + 3 \cdot 5$. Certainly we should expect this fact to be true of any three numbers and so we have

5. $$a(b + c) = ab + ac.$$

This is called the *distributive* axiom of multiplication with respect to addition.

The distributive axiom is often used to pass from the right side of the equality to the left. Thus, we may say that

$$49 \cdot 57 + 49 \cdot 43 = 49(57 + 43).$$

The advantage of using the axiom is clear because from the right side we see at once that the answer is 4900. To describe the transformation of the left side to the right, one often says that he has *factored* the quantity 49 from the two terms on the left side.

The distributive axiom enables us to write that

$$\frac{a + b}{c} = \frac{a}{c} + \frac{b}{c}$$

because the left side means $\frac{1}{c}(a + b)$ and, by virtue of the distributive axiom, this expression becomes

$$\frac{1}{c} \cdot a + \frac{1}{c} \cdot b \text{ or } \frac{a}{c} + \frac{b}{c}.$$

Sometimes mistakes are made by applying the distributive axiom carelessly. Thus in the fraction $(5 + 27)/5$ one is tempted to cancel the fives and obtain $1 + 27$. The correct use of the axiom calls for $5/5 + 27/5$ or $1 + 27/5$. Likewise, one cannot argue that $c/(a + b) = c/a + c/b$. A simple check with numbers will show that the latter is not correct.

We add to the above five axioms the following familiar and certainly acceptable ones.

6. Quantities equal to the same quantity are equal to each other.
7. If equal quantities are added to, subtracted from, multiplied into, or divided into equal quantities the results are equal. The only exception is that one must not divide by zero.

We shall use these seven axioms repeatedly and call attention to them only when the reason for a step is not obvious.

All these axioms must be distinguished from the *definitions* that state just what we mean by addition, subtraction, multiplication, and division of the various types of numbers. For example, we agreed that $5 + (-3) = 2$. This agreement states what we shall mean by the sum of a positive and negative number but it is not an axiom. The distinction between a definition and an axiom may perhaps be seen from an analogy. One may agree as to who shall be considered citizens of a country, for example, those born in the country. Such an agreement constitutes a definition of citizenship. When one specifies, however, that all citizens shall have the right to vote and the duty to bear arms in defense of their country, one is laying down axioms.

This chapter has been so much concerned with details that it may be well to remind ourselves of the main point: With the introduction of the irrational and negative numbers mathematics acquired new tools. One might expect that the power of mathematics to tackle and explore deeper and more significant physical problems would be considerably increased thereby. We shall soon see whether the game is worth the candle.

5: NUMBERS, KNOWN AND UNKNOWN

Algebra begins with the unknown and ends with the unknowable.

—ANONYMOUS

WHEN Karl Friedrich Gauss, one of the greatest mathematicians of all time, was in elementary school, his teacher, harried by endless clerical work, assigned to the class the problem of finding the sum of all the numbers from 1 to 100. The teacher expected that the students would be occupied for quite some time, but Gauss raised his hand immediately and gave the answer. Because the teacher could not believe that Gauss had calculated the answer so rapidly, he reprovingly remarked to the boy that he must have heard this problem before and memorized the answer. The problem happened to be one that the teacher had assigned to many other classes, so he had some basis for his suspicion. But Gauss protested that he calculated it by means of an algebraic relationship. Though this method was certainly not known to students of Gauss's age, the teacher was finally convinced; Gauss's extraordinary ability was recognized; and he was aided by this teacher and others in securing a higher education.

Gauss wanted the sum $1 + 2 + 3 + \ldots + 100$. He used the language of algebra to express the problem in the more general form $1 + 2 + 3 + \ldots n$, where n is any whole number. Being familiar with algebra he knew that this sum could be converted into a more convenient expression. Specifically, he knew that

$$1 + 2 + 3 + \ldots + n = \frac{n}{2}(n + 1).$$

Then, recognizing the generality of this relationship, Gauss had but to substitute 100 for n to see that the sum he wanted was $100(100 + 1)/2$ or $50\ (101)$, which is 5050. Using algebra, Gauss could calculate in a few seconds and with little risk of error what others using mere arithmetic would take perhaps an hour to do and with much greater likelihood of error. These features of algebra that Gauss employed are practically its sum and substance and are therefore worthy of examination—at least to the extent that algebra is worthy of examination.

When we say that algebra is a language we mean that it has its own words and symbols for expressing what might otherwise be expressed in ordinary English. To this extent algebra is as much a language as French or German. But algebra is a special kind of language. In the first place, it is concerned primarily with statements about quantity. Secondly, the language of algebra uses symbols in place of words. To discuss any one of a class of numbers, say the even numbers, a mathematician may say, Let a be any even number. If he wishes to be more impressive he may say let α (the Greek letter alpha) be any even number. Thereafter, in the entire discussion, whenever he wishes to refer to an arbitrary even number, he will use the letter a or α and thus save words and space. Of course he will have to be careful that any statement he makes about a or α applies to *all* even numbers since he introduced the symbols to stand for *any* one.

In addition to using letters to represent various types of numbers, algebra uses special symbols such as $+$ for addition and $\cdot$ or $\times$ for multiplication, symbols such as $=$ for equality, $<$ for less than, and $>$ for greater than, and symbolism such as 3^2 for $3 \cdot 3$ and 3^3 for $3 \cdot 3 \cdot 3$. The use of special symbols enables one to write mathematical statements concisely and precisely. The symbolism also contributes to clarity. The same statements in ordinary words would often be long, complicated sentences, too difficult for the mind to grasp. For instance, to render in words what is represented by $n(n+1)/2$ requires "any positive whole number multiplied into the sum of that whole number and 1 and the product divided by 2," a statement five or six times as long as the symbolic one. In more advanced mathematical work the saving in space and the increase in comprehensibility are enormously greater.

The value of symbolism in algebra, though often dismaying to the uninitiated, cannot be overestimated. Even Jerome K. Jerome's need to resort to algebraic symbolism, though for nonmathematical purposes, reveals clearly enough the usefulness and clarity inherent in such methods. Writing on the subject of love, in *Idle Thoughts of an Idle Fellow,* he said:

> *When a twelfth-century youth fell in love he did not take three paces backward, gaze into her eyes, and tell her she was too beautiful to live. He said he would step outside and see about it. And if, when he got out, he met a man and broke his head—the other man's head, I mean—then that proved that his—the first fellow's—girl was a pretty girl. But if the other fellow broke* his *head—not his own, you know, but the other fellow's—the other fellow to the second fellow, that is, because of course the other fellow would only be the other fellow to him, not the first fellow who—well, if he broke his head, then* his *girl—not the other fellow's, but the fellow who* was *the—Look here, if A broke B's head, then A's girl was a pretty girl; but if B broke A's head, then A's girl wasn't a pretty girl, but B's girl was.*

There is no doubt that the language of algebra is enormously effective in recording and in operating with mathematical concepts. Nevertheless, many people complain about the need to learn this special language. These people would be more justified, however, in complaining that the French people insist on using their own language and the Germans, theirs. Frenchmen and Germans are even foolish enough to translate books written in clearly understandable English into their own languages. After all, everything they can say we can say better in English. Of course neither complaint is warranted. Certainly insofar as the language of algebra is concerned, no substitute serves the purposes so well. Moreover, the language of mathematics is universal.

On the other hand there are some legitimate objections to the language of algebra and of mathematics generally. The symbols do not suggest what they represent, though the English mathematician Robert Recorde (1510–1558), who introduced the equals sign, did say that these suggested parallel lines and he could think of no two things more alike than parallel lines. But generally the symbols are not ideograms, suggesting pictorially what they stand for. The symbols of mathematics, unlike the ideograms of the early Egyptians and Chinese, must be specially remembered.

There is no doubt, too, that mathematicians are generally overzealous about conciseness, and in their passion for brevity indulge in symbols even where these seem no better than a familiar English word or phrase. A faulty judgment has caused mathematicians to equate elegance and conciseness at the cost of intelligibility. Gauss himself wrote elegant, but highly compact, carefully polished papers with no hint of the motivation, meaning, or details of the steps. When criticized, he said that no architect leaves the scaffolding after completing the building. But the fact is that even excellent mathematicians found the reading of Gauss's papers very difficult, and the same is true of the writings of many other mathematicians.

Nor are the symbols very attractive. On the whole, a page of mathematical symbolism is hardly an aesthetic sight. One could not criticize the person who would rather look at a few lines of poetry. The difficulty in remembering the meanings of the symbols and the general unattractiveness of symbolic expressions repel and disturb would-be students of mathematics; the symbols are like hostile standards floating over a seemingly impregnable citadel. Mathematicians who boast about their symbolism are letting their enthusiasm run away with them. (Of course none of these remarks about mathematical symbolism is intended to imply that one who has mastered the language of mathematics and has grasped the ideas and reasoning they convey would fail to be enthusiastic about mathematics itself.)

The unnaturalness of mathematical symbolism is attested to by history. The algebra of the Egyptians, the Babylonians, the Greeks, the Hindus, and the Arabs was what is commonly called rhetorical algebra. These peoples did use minor bits of symbolism aside from number symbols proper, but on the whole they used ordinary rhetoric to describe their mathematical work. Symbolism is a relatively modern invention of the sixteenth and seventeenth centuries, and may be the consequence of the pressure on mathematicians to keep pace with the rapidly expanding and deepening use of mathematics during the rise of modern science. Perhaps, nevertheless, mathematicians should be complimented for making a virtue out of necessity.

The chief innovator of symbolism in algebra was François Viète (1540-1603), the greatest of sixteenth-century French mathematicians. Viète was an amateur in the sense that his professional life was devoted to the law and to service for Henry III and later Henry IV of France. But Viète loved mathematics and spent days completely absorbed in it. He obtained fame by deciphering coded messages that the Spanish, enemies of France at the time, sent to the Spanish-dominated Netherlands. The Spanish thought he had magical powers. Viète is the man who introduced the now general practice of using a letter to denote a whole class of numbers; for example, letting a or b stand for any real number. John Wallis (1616-1703), a noted contemporary of Isaac Newton, says that Viète, in denoting a class of numbers by a letter, followed the custom of lawyers who discussed legal cases by using arbitrary names such as John an-Oakes, John a-Stiles, and John a-Down and later the abbreviations J.O., J.S., and J.D., and still more briefly A, B, and C. Actually, letters had been used occasionally by the Greek Diophantus and by the Hindus. However, in these cases letters were confined to designating a fixed unknown number, powers of that number, and some operations. Viète recognized that a more extensive use of letters and, in particular, the use of letters to denote *classes* of numbers, would permit the development of a new kind of mathematics; this he called *logistica speciosa* in distinction from *logistica numerosa*. Despite Viète's conscious and influential use of letters to expedite quantitative thinking, the growth of symbolism was slow. Even simple ideas take hold slowly. Only in the last few centuries has the use of symbolism become widespread and effective.

The second basic function of algebra is to convert expressions into more useful ones. Gauss's conversion of the sum $1 + 2 + \dots + n$ into the more useful form $n(n + 1)/2$ is a simple example. Algebra is a machine, or more accurately, a collection of machines, There is machinery to factor expressions, to expand expressions, to decompose complicated fractions into simpler ones, and so forth. The object of the conversion in every case is to obtain a form more useful for the problem in hand. For some purposes $a^2 - b^2$ is useful as it stands but for other purposes it may be better to write it in the equivalent form $(a - b)(a + b)$.

The simplest use of the machinery of algebra is in solving equations involving unknown quantities. Just to observe the effectiveness of algebraic processes, let us consider a rather simple physical problem that leads to determining an unknown.

It is known from measurements that sound travels about 1100

feet per second in air. However, it travels about ten times faster in solid bodies such as the earth. If one were listening for the sound of some explosion taking place on the surface of the earth, one would hear much sooner the sound waves that travel through the earth than those that travel through the air. As a matter of fact, the American Indians used to put their ears close to the ground in order to detect the approach of some distant horseback rider. They heard the sound sooner and also more loudly, because sound does not diminish in intensity so fast when it travels through solids. Incidentally, this Indian practice is probably the origin of the modern colloquialism "keeping one's ear to the ground." At any rate, what the American Indian could not do and we can is to calculate the distance to the source of the sound.

Suppose, then, that one hears a sound that comes through the ground and five seconds later he hears the same sound coming through the air. How far away is the source of the sound? Following the established practice in algebra we let x be the unknown distance to the source. Since sound travels at 1100 feet per second in air, the time required for the sound to reach the listener through the air is the distance x divided by the velocity 1100, or $x/1100$. Since sound travels ten times as fast through the ground, the time required for the sound to reach the listener through the ground is $x/11000$. We are told that the difference between these two times is five seconds. Hence

$$\frac{x}{1100} - \frac{x}{11{,}000} = 5. \tag{1}$$

This equation says that the time required for the sound to travel through the air minus the time required for the sound to travel through the ground is five seconds. It expresses the given physical information in the language of algebra.

Now the machinery of algebra comes into play. We wish to find the value of x. The form of the left side of the equation suggests subtraction of fractions. To perform this operation we make the denominators alike by multiplying the numerator and denominator of the first fraction by 10. Thus

$$\frac{10x}{11{,}000} - \frac{x}{11{,}000} = 5.$$

We may now subtract the second fraction from the first and obtain

$$\frac{9x}{11,000} = 5.$$

Multiplication of both sides of this equation by 11,000 gives

$$9x = 55,000;$$

and now division of both sides by 9 reveals that

$$x = \frac{55,000}{9}$$

or

$$x = 6111 \text{ (to the nearest whole number)}.$$

The source of the sound is 6111 feet away.

We see how the processes of algebra can be made to yield information. Of course the example is not important and does not involve much technique. Later, we shall see how algebra is used in the course of more significant investigations.

Before we leave the foregoing problem we might consider a point about the use of mathematics that is of great significance in science and engineering. If we examine the two fractions on the left side of equation (1) we note that the first one is very much larger than the second, in fact, ten times as large. Now an engineer who is interested in the distance to the source of the sound might well argue that the time required for the sound to travel through the earth is negligible compared to the time of travel through the air. He might drop the second fraction and argue that

$$\frac{x}{1100} = 5, \tag{2}$$

and therefore that $x = 5500$. The engineer is of course using all the facts that we used in obtaining the correct solution but, by neglecting the second term of equation (1), he obtains only an approximate answer. In fact his answer is in error by 611 feet. However, he has been able to replace equation (1) by the somewhat simpler equation (2) and still obtain an answer which may be good enough for his purposes.

Of course the advantage gained by making an approximation in the above problem is trivial because it does not involve difficult mathematics even when solved exactly. But in more advanced scientific and engineering work the gain often amounts to the difference between not being able to obtain an answer at all as against obtaining a very useful approximate answer. The general point, then, is that even when it is possible to formulate a physical problem in exact mathematical language it is often practical for an engineer or scientist to neglect some terms in the mathematical formulation in order to expedite the solution. The mathematician would not make this simplification and would persist in attempting to solve the original problem even if he had to hand the problem on to successive generations of mathematicians.

The conclusion one should draw from this comparison between mathematician and scientist or engineer is not that the mathematician is stubborn and impractical. As an individual he may very well be, but in his persistence in attempting to solve the exact problem he is adhering to the spirit and objectives of mathematics. Just how much has been accomplished by this insistence on exact reasoning will be clearer as we get on with our account.

A neat example of the value of the machinery of algebra is furnished by one of the classic stories of the history of mathematics. It concerns the exploits of the greatest mathematician of antiquity, Archimedes. Archimedes lived in the independent Greek city-state of Syracuse, which was in Sicily. King Hiero of Syracuse had ordered a crown made of gold but suspected, when the crown was delivered, that the interior was filled with the baser metal silver. He asked Archimedes to determine, without destroying the crown, whether the crown was made of pure gold. As Archimedes pondered this problem while taking a bath, he suddenly discovered the physical principle that would enable him to solve it. One historical account states that he was so excited by his discovery that he ran outside naked shouting *"Eureka!"* (I have found it).

As he sat down in his filled bathtub Archimedes noted that the amount of water that overflowed was about equal to the amount that his body displaced. The fact that a body immersed in water displaces water equal in volume to the space the body occupies is of course obvious. But just such a casual observation often suggests deeper and more useful thoughts to minds that are active on a problem. Minds in this state seem to take advantage of every idea that occurs. In this case Archimedes saw how he could use the observation he had made.

Bodies of equal weight are not necessarily of equal volume; for example, a pound of silver occupies more volume than a pound of gold. Hence, Archimedes reasoned, a pound of silver would displace more water than a pound of gold. He formed two masses, one entirely of silver and one entirely of gold, but both of the *same weight* as the crown. He then put each into a pail of water filled to the brim. In each case the water overflowed but, as expected, far more overflowed when the mass of silver was immersed. On immersing the crown in the filled pail Archimedes found that it displaced less water than the silver and more than the gold. By this simple quantitative observation Archimedes knew that the crown was some mixture of gold and silver.

But Archimedes wished to determine how much silver and how much gold were in the crown, and here he found that algebra would help him. He supposed that the crown, which, let us say, weighed ten pounds, was made up of w_1 pounds of silver and w_2 pounds of gold. He found that ten pounds of pure silver displaced thirty cubic inches of water. Hence, w_1 pounds of silver would displace $(w_1/10) \cdot 30$ or $3\,w_1$ cubic inches of water. Since ten pounds of pure gold displaced fifteen cubic inches of water, w_2 pounds of gold would displace $(w_2/10) \cdot 15$ or $(3/2)w_2$ cubic inches of water. Hence the crown should displace

$$3w_1 + \frac{3}{2}w_2$$

cubic inches of water. Archimedes measured the volume of water that the crown displaced and found it to be, let us say, twenty cubic inches. Hence he knew that

$$3w_1 + \frac{3}{2}w_2 = 20. \tag{3}$$

He also knew that

$$w_1 + w_2 = 10. \tag{4}$$

Archimedes now had two equations involving two unknowns and he proceeded to apply the machinery of algebra to find them. He multiplied both sides of equation (4) by 3 to obtain

$$3w_1 + 3w_2 = 30. \tag{5}$$

By subtracting the left side of equation (3) from the left side of (5) and, of course, then doing the same with the right sides, he knew, in view of the axiom that equals subtracted from equals give equals, that

$$\frac{3}{2}w_2 = 10.$$

Multiplication of both sides of this equation by ⅔ yielded

$$w_2 = 6\frac{2}{3}.$$

It follows from equation (4) that

$$w_1 = 3\frac{1}{3}.$$

Archimedes could tell King Hiero that the crown contained 3⅓ pounds of silver and 6⅔ pounds of gold.

A good deal of the machinery of elementary algebra is concerned with the solution of equations involving unknowns, and we shall encounter many such uses in the treatment of broader problems. However, we should see that this simple machinery can lead directly to useful results in numerous other types of problems. Let us consider one or two examples.

Suppose a farmer has a fixed amount of fencing, say 100 feet, and he wishes to enclose a rectangular piece of land in order to farm it. What dimensions should he choose? Though restricted to a perimeter of 100 feet and a rectangular shape, the farmer could choose dimensions such as 48 by 2, 40 by 10, 30 by 20, and others. If he is alert, he will notice two things. First, the area enclosed varies considerably with the dimensions chosen despite the fact that the perimeter is always 100 feet. The dimensions 48 by 2 yield an area of 96 square feet; the 40 by 10 rectangle yields 400 square feet; and so on. Second, the more nearly equal the two dimensions are, the larger the area turns out to be. Usually the farmer is interested in enclosing the maximum possible area because he then has more land available for planting. He may then consider the question, What should the dimensions of the rectangle be in order to enclose the *maximum* area and yet keep the perimeter 100 feet? His few choices of dimensions suggest that the largest area would be obtained when the dimensions are equal, that is, when the rectangle is a square. But can he be sure of this fact? Obviously he cannot try all possible dimensions because there is an infinity of possibilities.

A little algebra settles the question quickly. Let us consider any rectangle with a perimeter of 100 feet; the sum of its length and width is 50 feet. Let x be the width of any one of the possible rectangles, then the length is $50 - x$ and the area is $x(50 - x)$. In the particular case where the rectangle is a square each side is 25 feet and the area is 625 square feet. We wish to show that the square has more area than any other rectangle of the same perimeter. In other words, we wish to prove that the area of the square minus the area of any other rectangle is greater than zero. In symbols this statement reads

$$625 - x(50 - x) > 0. \tag{6}$$

We must be careful to note that the letter x in this inequality stands for any possible choice of width. Now, of course, we do not know that the inequality (6) is correct for any choice of x, but let us see if some algebraic manipulations will yield a more penetrable form.

By applying the distributive axiom to the term $x(50 - x)$ we obtain $50x - x^2$. Hence, in view of the first minus sign in (6), we state that

$$625 - 50\,x + x^2 > 0. \tag{7}$$

There is a simple theorem about factoring that tells us we may rewrite (7) as

$$(25 - x)^2 > 0. \tag{8}$$

The inequality (8) tells us what we want to know, for if x is not 25, then $25 - x$ is not zero. Since the square of any nonzero real number is positive, that is, greater than zero, (8) is a correct inequality.

What have we accomplished? The inequality (8) is correct, as we have just seen. We now consider the steps from (8) back to (6), that is, in the reverse order from the way in which we derived (8) from (6). Each step from (8) to (6) is justified by some simple axiom or theorem of algebra and therefore (6) is correct. But (6) says that the area of the square minus the area of any other rectangle with dimensions x and $50 - x$ is greater than zero or that the area of the square is greater than the area of any other rectangle with a perimeter of 100 feet. The simple machinery of algebra has led to an interesting and useful conclusion.

The farmer who faced the above problem considered all possible rectangles with a given perimeter because a rectangular shape is easy to construct. But let us suppose that he were willing to tackle a more involved shape. Could he enclose even more area? He could, of course, try various shapes such as triangles and quadrilaterals, and test them numerically. A little geometry would help at once in eliminating various possibilities. Since our concern here is to see the utility of algebra, let us pass up these various possibilities and consider at once a shape that is at least aesthetically attractive, the circle. The question then becomes, Given a circle and a square of the same perimeter, will the circle contain more area?

Just to show how general the conclusions derived by algebra can be, let us consider not just the square of perimeter 100 feet and side 25 feet but *any* square of perimeter p and let us show that a circle of the same perimeter has more area. We shall borrow from geometry the expression for the perimeter or circumference of a circle. The perimeter of a circle is $2\pi r$ where r is the radius and π is an irrational number whose value is approximately $3\frac{1}{7}$. If the square and the circle are to have the same perimeter, then

$$2\pi r = p$$

or

$$r = \frac{p}{2\pi}. \tag{9}$$

Now the area of any circle is πr^2 and for the value of r given by (9) the area would be

$$\pi \left(\frac{p}{2\pi}\right)^2 = \pi \cdot \frac{p}{2\pi} \cdot \frac{p}{2\pi} = \frac{p^2}{4\pi}.$$

The square of perimeter p has sides of length $p/4$ and area $p^2/16$. Hence we must show that

$$\frac{p^2}{4\pi} > \frac{p^2}{16}. \tag{10}$$

Since π is about $3\frac{1}{7}$ the denominator of the left side is appreciably less than 16. Then the value of the left-hand fraction is larger than the right-hand one, and we have proved that the area of *any* circle is greater than that of a square of the same perimeter.

The problem of enclosing the most land with a given amount of fencing is not a major one; nevertheless the result is more significant and useful than may appear at first sight. Millions of tin cans are made every day in this country, and the manufacturers are concerned with the same type of problem that our farmer was. It is obviously advantageous to use a shape that encloses as much volume as possible with a given amount of tin. Now the volume of a can is the product of the height and the area of the base. Since the circular shape encloses more area than any other of the same perimeter, the circular can will enclose more volume for the same surface area and therefore for the same amount of tin.

These examples of the usefulness of the machinery of algebra must suffice for the present. Elementary algebra as a whole is a huge machine to mechanize thinking. Like the complex machines of modern factories, it enables us to work wonders in almost no time. The mechanization of processes that have to be used repeatedly is of course a great gain, since one does not have to think about them. They become habitual like washing and dressing.

Of course in emphasizing the machine or technique value of algebra we are also stating that in itself elementary algebra is of no great interest. There are people who admire machinery, but, generally speaking, a machine is valuable because it turns out a useful product. The machine may be awe-inspiring but in itself it is no more than a maze of wheels, gears, and rods. So it is with algebra. On the whole the machinery of algebra is not pretty and possesses little intrinsic interest. In itself it has little to say; it serves a subsidiary purpose.

The individual techniques of algebra are like single notes selected at random from large and magnificent musical compositions. When struck in succession these notes are discordant. However, these same techniques, employed in the investigation of more significant undertakings, help to form beautiful patterns of reasoning. They serve, as do the skilled fingers of a pianist, to render the harmonious ideas of great composers. Unfortunately, the usefulness of the techniques of algebra has caused many people to mistake the means for the end and to emphasize these menial techniques to the exclusion of the larger ideas and goals of mathematics. The students who are bored by the processes of algebra are more perceptive than those who have mistakenly identified algebraic processes with mathematics.

The historical associations of the word *algebra* almost substantiate the sordid character of the subject. The word comes from the title of a book written by the ninth-century Arabian mathematician Al-Khowarizmi. In this title, *al-jebr w' almuqabala,* the word *al-jebr* meant transposing a quantity from one side of an equation to another and *muqabala* meant simplification of the resulting expressions. Figuratively, *al-jebr* meant restoring the balance of an equation in transposing terms. When the Moors reached Spain during the Middle Ages, they naturally introduced the Arabic word *al-jebr,* and there, in the form *algebrista,* it came to mean a bonesetter, or a restorer of broken bones, and signs reading *Algebrista y Sangrador* (bonesetter and bloodletter) were found over Spanish barber shops. (Until relatively recent times, barbers performed many of the less skilled medical services.) Thus it might be said that there is a good historical basis for the fact that the word *algebra* stirs up disagreeable thoughts.

Of course generalizations about the dullness of algebraic processes have many exceptions. Occasionally in algebra one encounters simple and yet attractive ideas. Let us consider the very equality that Gauss used, namely,

$$1 + 2 + \ldots + n = \frac{n}{2}(n+1). \tag{11}$$

How does one prove that the left side equals the right side? The method is not obvious and yet a very simple idea does the trick.

Let us call the sum of the first n numbers S and let us write S as equal to this sum and then immediately below it let us write the same sum but with $1 + 2 + \ldots + n$ in reverse order. Thus,

$$S = 1 + 2 + \ldots + (n-1) + n \tag{12}$$

$$S = n + (n-1) + \ldots + 2 + 1. \tag{13}$$

In both equations we have written $n - 1$ for the term next to n since consecutive terms differ by 1. Despite the fact that the proof is now almost complete it still requires one more insight to complete it. We shall add equation (13) to equation (12) but carry out the addition by adding each number of the second line to the one immediately above it. But when we add any member on the right side of the second line to the one immediately above it we always get $n + 1$. Moreover, there are exactly n such sums. Hence, the sum of the two right sides is $n(n+1)$. But according to the left sides this sum is $2S$. Hence,

$$S = \frac{n}{2}(n+1). \tag{14}$$

The ideas in this proof leave some people cold and arouse excitement in others. We shall not dispute about tastes. What is indisputable is that by simple means we have converted the cumbersome expression on the left side of (11) to the simpler and readily computable expression on the right side.

We noted earlier that algebra is a series of techniques. These techniques are powerful because they apply to large classes of numbers and hence are useful for thousands of applications where arithmetic deals with one at a time. Let us pursue this point a bit. We shall have a number of occasions later to find unknowns that satisfy equations such as

$$x^2 - 4 = 0, \tag{15}$$

$$x^2 - 5x + 6 = 0, \tag{16}$$

$$x^2 - 5x + 3 = 0. \tag{17}$$

All of these equations are called second degree equations because x^2 but no higher power of x occurs. Now equation (15) is certainly easy to solve, because from it we have at once

$$x^2 = 4$$

and therefore

$$x = 2 \text{ and } x = -2.$$

The numbers 2 and -2 are called the solutions or roots of the original equation.

Equation (16) is not quite so easy to solve but is still not difficult. One who learns multiplication of algebraic expressions recognizes readily that it is equivalent to

$$(x-3)(x-2) = 0.$$

By inspection we see that when $x = 3$, the first factor is zero and therefore the product of two parentheses is zero. Hence $x = 3$ is one value of x that satisfies the equation, that is, makes the left side equal the right side. Likewise $x = 2$ is a root.

Equation (17) is still harder to solve, but one can find the roots. However, the mathematicians decided several hundred years ago that rather than ponder and struggle with each second degree equation they were likely to encounter they would lick such problems once and for all by attempting to solve the general equation

$$ax^2 + bx + c = 0, \tag{18}$$

wherein *a, b,* and *c* are *any* numbers.

Equation (18) can be solved by a process that is irrelevant at the moment. The solutions—there are two—are

$$x = \frac{-b + \sqrt{b^2 - 4ac}}{2a} \text{ and } x = \frac{-b - \sqrt{b^2 - 4ac}}{2a}\text{*} \tag{19}$$

Thus, in one swoop, the mathematicians solved every second degree equation in x. The device of working with a, b, and c as coefficients in (18) instead of working with particular numbers kills many birds with one stone. It is true that the thought of killing even one bird, let alone many, is repulsive, but there is some comfort in knowing that every equation of the form (18) can be solved.

The mathematician's desire to solve many problems in one general treatment can be ascribed to his laziness. In the long run he has saved effort. But the mathematician is also moved, even goaded, by a passion for generality. Out of sheer curiosity he will tackle the general problem even when he does not have to solve many special cases. This search for generality may pay off unexpectedly.

The solutions (19) establish the fact that every second degree equation in x has two solutions. This result suggests that every third degree equation might have three solutions; every fourth degree equation, four solutions; and so on. Though we shall not bother to prove it here the suggestion is indeed correct. And the mathematician, who loves such harmonious relationships, feels amply rewarded for his daring and efforts in seeking the solution of the general second degree equation.

Of course knowing how many roots there are does not tell us how to find them. The mathematicians could not resist this problem too. The history of their efforts in this direction is both prolonged and exciting. From Babylonian times onward mathematicians had solved a number of particular third degree equations

* For some values of a, b, and c the solutions (19) are complex numbers. But this fact has little to do with the present discussion.

but were unable to find expressions for x in terms of the general coefficients $a, b, c, d,$ of

$$ax^3 + bx^2 + cx + d = 0.$$

This problem was finally solved by the Italian mathematician Nicolo of Brescia, better known as Tartaglia (1499–1557), and published by the famous physician and mathematician Jerome Cardan in his *Ars Magna* of 1545. Not much later Lodovico Ferrari (1522–1565) solved the general fourth degree equation

$$ax^4 + bx^3 + cx^2 + dx + e = 0.$$

It seemed almost certain to the mathematicians that since the general first, second, third, and fourth degree equations can be solved by means of the usual algebraic operations such as addition, subtraction, and roots, then the general fifth degree equation and still higher degree equations could also be solved. For three hundred years this problem was a classic one. Hundreds of mature and expert mathematicians sought the solution, but a little boy found the full answer. The Frenchman Évariste Galois (1811–1832), who refused to conform to school examinations but worked brilliantly and furiously on his own, showed that general equations of degree higher than the fourth cannot be solved by algebraic operations. To establish this result Galois created the theory of groups, a subject that is now at the base of modern abstract algebra and that transformed algebra from a a series of elementary techniques to a broad, abstract, and basic branch of mathematics.

We will not devote ourselves here to the modern developments in algebra, even though many of these have proved useful in the study of nature, because more fundamental creations remain to be discussed. But the history of the theory of equations is instructive as to the general course of mathematics. It shows how the mathematician's enthusiasm for generality has led him from simple and concrete physical problems to new and broad theories, and how mathematics generates its own problems and solves these without regard any longer for the physical world. And then the theory investigated for its own sake finds new applications never intended and certainly not foreseen.

6: THE LAWS OF SPACE AND FORMS

Geometry . . . is the science that it hath pleased God hitherto to bestow on mankind.

—THOMAS HOBBES

THE story is told that the Greek philosopher Aristippus and some friends were shipwrecked on what appeared to be a deserted island near Rhodes. The company was downcast at its ill fortune when Aristippus noticed some geometric diagrams drawn on the beach sand. "Be of good cheer," he told his companions, "I see traces of civilized men."

If man's environment were like that of birds, who constantly encounter changing cloud shapes, or of fish, who see before them just formless water, he would probably never have conceived the study of geometry. Not only are fixed shapes absent in those media but even the notion of a definite locale, a reference point, does not suggest itself. One place is no more readily identifiable than any other. But man lives in a world that contains solid bodies that maintain their shapes and sizes. Like the horse who gets to know his stable, man finds familiar landmarks and constant objects all about him.

The straight line is suggested by the banks of a river, the boundary of a field, and the stretched string. In fact the word *hypotenuse* meant originally, in Greek, "stretched against"; presumably it referred to a string stretched between the ends of the arms of a right triangle. Observation of the heavenly bodies, objects of interest and even great religious significance to most civilizations, suggested the circle and sphere. The surfaces of large bodies of water recommended the plane and the trunks of trees offered the cylinder as geometrical concepts.

How did the creators of geometry get to thinking about perpendicularity, parallel lines, congruent and similar figures, and the like? The answer is that these basic relationships are all suggested by the observation of natural phenomena. A tree grows perpendicularly to the ground. The walls of a house are erected perpendicularly to the ground so that they will not have a tendency to fall in either direction. The banks of a river are generally parallel lines, and the beams to which the walls of a house are nailed are parallel. The geometric notion of congruence is certainly suggested by even everyday experience. We note continually objects that have the same size and shape. In fact, species of animals are identified by size and shape. In our industrial society machines turn out millions of congruent items. The geometric concept of similar figures is likewise physically familiar. A child will draw or build a model of a house without having heard of the word *similar*. And models, which are continually constructed to design buildings, ships, planes, and automobiles, are truly similar to the physical objects they portray.

The figures of geometry are suggested by physical objects, but, as in the case of number, primitive civilizations did not abstract these figures. Some progress toward abstraction was made by the Egyptians and the Babylonians, who recognized that geometrical forms were worthy of study in themselves. But it was the Greeks who took the decisive step of concentrating on idealized concepts such as point, line, triangle, square, and circle, and undertook an extensive study of the properties of these idealizations.

We have already noted that mathematics studies abstract concepts and not physical objects. But this point seems to warrant special emphasis in connection with geometry. Because geometrical concepts are more readily visualized than the concepts of arithmetic and algebra and are in fact often represented by figures drawn with chalk or pencil, there is a tendency to identify the concepts with the pictures that we draw to illustrate them. But the pictures are not mathematical concepts. Chalk marks and penciled lines possess color, molecular structure, and width, and these properties must be stripped away to attain the mathematical idealization called the straight line.

Though the concepts of geometry are not apparently more useful than those of number, the first extensive development of deductive mathematics took place in geometry. There are several reasons for this sequence of events. The concepts of geometry are intuitively more appealing and their properties more readily suggested by observation, experience, and even measurement. Secondly, the Greeks of the classical age (600–300 B.C.), who conceived abstract, deductive mathematics, were attracted by the geometrical forms and motions they observed in the heavens. Therein they saw the grandeur and mystery of nature and phenomena of high aesthetic value. Being speculative philosophers rather than practical men, they were content with a qualitative account of what they saw, that is, a description in terms of shapes, and they did not pursue numerical details that might have been useful in navigation or precise calendar-reckoning.

Thirdly, though the Greeks were intellectually among the most powerful of people, they failed to understand the more complicated creations of arithmetic such as the irrational number. They could understand $\sqrt{2}$ as the length of a hypotenuse of a right triangle but not as an addition to the concept of number. Because the Greeks were interested in a firm logical approach to mathematics and could see their way clearly in geometry, they favored that subject. Hence, geometry was developed systematically much earlier than arithmetic and algebra, and it was regarded for centuries as the prime example of mathematical reasoning.

The development of geometry with which most of us are familiar is the one framed by the Greek mathematician Euclid, who lived in Alexandria about 300 B.C. Unfortunately we know very little about this man other than the fact that he was the author of a work taught to hapless beginners. His famous creation was the *Elements,* a text on plane and solid geometry. Divided into thirteen books, it is an organization of the separate discoveries of numerous Greek mathematicians from Asia Minor to Sicily who preceded him over a period of about three centuries. Euclid was the great master who arranged the scattered conclusions of his predecessors so that they all followed by deduction from the one set of axioms he selected. Modern geometry texts are variations on Euclid's presentation and usually contain a brief selection from that vast work of 467 theorems and many corollaries.

Euclid began his treatise with some definitions of the basic geometrical concepts. He then introduced ten axioms that tell us precisely what facts or properties we may utilize about these concepts. These axioms are largely familiar to us. Given two points, there is one and only one line passing through them. The whole is greater than any of its parts. A circle may be drawn with a given center and radius. All right angles are equal. Having defined parallel lines as two lines in a plane that have no point in common, Euclid included an axiom about them. The formulation of this axiom which is most familiar to us and which is equivalent to Euclid's reads: Given a point and line, one and only one line may be drawn that passes through the given point and is parallel to the given line. With these axioms Euclid was prepared to deduce assertions about the concepts.

It is important to realize that the choice of axioms about points, lines, angles, circles, and the like is not unique. Modern textbooks do not always give the very same axioms that Euclid used, but the variations are not unreasonable. What matters in the choice of axioms is that they should be so unquestionably true about the concepts that we are willing to accept them as a basis for further reasoning, and that they should lead by valid deductive reasoning to the establishment of conclusions that are not quite so obvious or are not at all obvious. The extraordinary merit of Euclid's axioms is that though they are simple and unquestionable in themselves, they lead to profound consequences.

Of course the reader who has examined the theorems of Euclidean geometry may wonder why anyone bothers to prove some of them. Isn't it obvious that if two sides and the included angle of one triangle equal respectively two sides and the included angle of another, the triangles must be congruent, that is, identical except for position? Even some of the Greeks themselves objected to the simpler Euclidean theorems as too obvious. For example, the Epicureans, a school of Greek philosophers, picked on Euclid's theorem that the sum of two sides of a triangle is greater than the third side. Said the Epicureans, "Any ass knows this theorem. If fodder is placed at one point and the ass is at another, the ass does not traverse two sides of a triangle to reach the fodder but goes directly to it." Every layman recognizes at once the truth of some of the Euclidean theorems. If a swimmer wants to reach shore as quickly as possible he heads along the perpendicular to the shore line. (As a matter of fact, even an animal will behave as wisely.)

Many geometrical theorems indeed seem obvious, but if they can be deduced from axioms, then the theorems are utterly beyond question. We are more secure in our reliance upon them than if we had merely accepted them intuitively. Many a statement that seems obviously true may not be so, for our senses are notoriously deceptive. As we shall see, some of the most significant developments of mathematics and science have resulted from investigating the obvious.

The major value of proof is that it establishes beyond doubt conclusions that are not at all obvious. Consider the theorem which states that the line joining the mid-points of two sides of a triangle is parallel to the third side and equal in length to one-half of this side (fig. 13). Surely, this result is not obvious and yet by a simple proof it can be unquestionably established. The certainty and permanence of mathematical results are secured by proof.

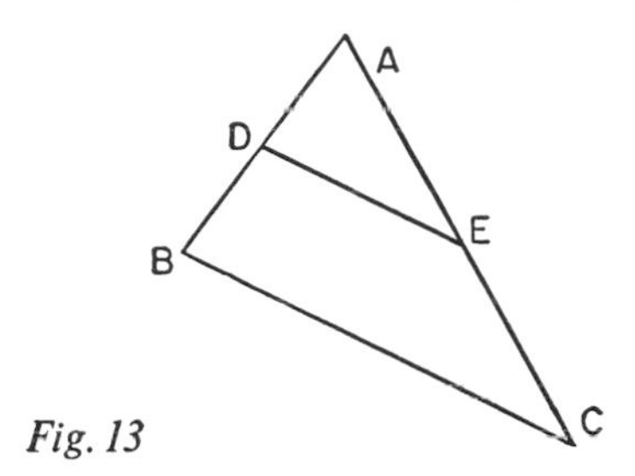

Fig. 13

But the key question in many people's minds is not whether proof does or does not guarantee the conclusion; rather, it is, Why do we want to prove any theorems at all? One answer—and one especially true of geometry—is that many proofs and results are delightful. The satisfaction of proving what is intuitively foreseen or of checking intuitively what a theorem reveals is especially enjoyable in geometry. The methods of proof themselves often contain an element of ingenuity or surprise that pleases some people at least. However, the most important reason for investigating geometrical problems is the same as for algebraic ones, namely, to shed light on natural phenomena. Let us consider first a minor and already familiar situation.

In the preceding chapter we considerately aided a farmer by answering the question, Which rectangle of all those with the same perimeter contains the most area? Numerical considerations suggested, we may recall, that the square would do best, and we then proved this fact by the use of a little algebra. Just to show that there are often many ways of establishing the same result, we shall consider a geometric proof of the same fact.

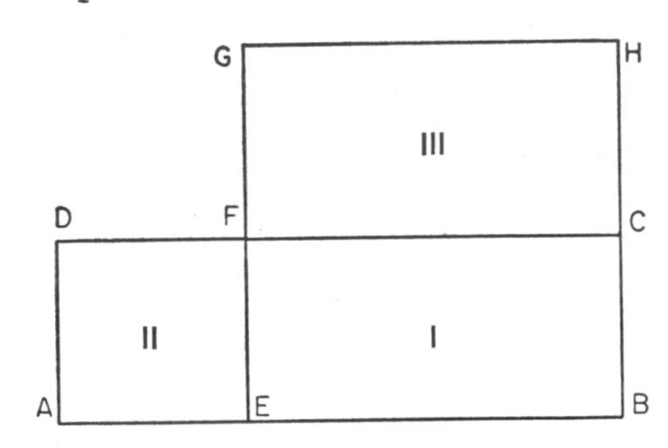

Fig. 14

Let $ABCD$ (fig. 14) be any rectangle with perimeter p. If this rectangle is not a square then one side, AB, is larger than the other. On the larger side let us construct a square, $EBHG$, with perimeter p. Now, $AE + EB + BC$ is half the perimeter of the rectangle. Hence

$$AE + EB + BC = \frac{p}{2}. \tag{1}$$

But since the square has the same perimeter as the rectangle, half of its perimeter is also $p/2$; in symbols,

$$EB + BC + CH = \frac{p}{2}. \tag{2}$$

If we subtract equation (2) from equation (1) we obtain

$$AE - CH = 0. \tag{3}$$

Addition of CH to both sides of this equation yields

$$AE = CH. \tag{4}$$

Now the smaller side, AD, of the rectangle must be less than any side of the square because the square has the same perimeter as the rectangle. Hence

$$AD < GH. \tag{5}$$

The area of rectangle II in figure 14 is $AE \cdot AD$ and the area of rectangle III is $CH \cdot GH$. But equation (4) tells us that $AE = CH$, while equation (5) tells us that $AD < GH$. Hence the area of rectangle II is less than the area of rectangle III.

Since the area of the entire rectangle $ABCD$ is the sum of the areas of I and II and the area of the square is the sum of the areas of I and III, the square has greater area than the rectangle.

We have proved precisely the same result that was established in the preceding chapter. It is a matter of taste as to whether one prefers the geometric proof to the algebraic one.

As is so often the case in mathematics, a theorem created to solve one problem also solves a number of other problems. A builder of modestly priced homes uses the above theorem whenever the size of the lot permits it. The perimeter of the floor area of the house determines the number of feet of wall that the house requires and hence the cost of the walls. For this cost, the builder would wish to enclose as much area as possible because the house would then offer the most living space. Our theorem tells the builder that the floor shape should be a square.

The farmer or builder might be content with the answer to his problem and forget mathematics thereafter, but the mathematician, having obtained such a neat result, would deliberately alter the problem to see if he could obtain another just as exciting. As a variation, he might consider the following problem. Speaking in concrete terms merely to interest those who use mathematics, he might say, Let us suppose that the farmer's land lies alongside a

Fig. 15

river and that the farmer wishes to enclose a rectangular piece of land along the river bank with a given amount of fencing. No fencing is required along the river bank itself. What shape should the rectangle take if its area is to be a maximum? In terms of figure 15 the problem is to determine the dimensions of rectangle $ABCD$ under the conditions that fencing is required only along DA, AB, and BC; the total amount of fencing is fixed in advance, and the area is to be as large as possible.

The problem could indeed be tackled as a new and independent problem of geometry but the mathematician with his characteristic laziness always seeks to use any previously established conclusion that may help him to answer the question immediately. He knows that the square is the best shape if all *four* sides of the rectangle require fencing. Can the present problem be solved by utilizing the solution of the earlier one? It can be, but, as is usual in mathematical work, an idea is required.

In the present case the idea is to complicate the problem somewhat. The mathematician would propose the following question. Suppose that the farmer wished to fence off a rectangular area lying on *both* sides of CD and symmetrical with respect to CD. Also, if the farmer had p feet of fencing at his disposal, let us now suppose that he has $2p$ feet of fencing available. What shape

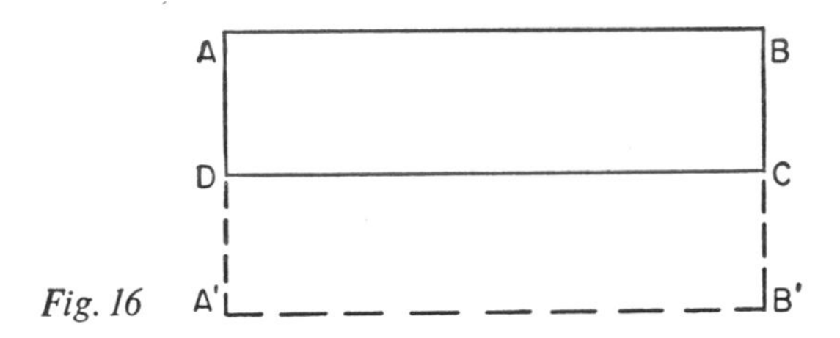

Fig. 16

should he choose for the rectangle $A'B'BA$ (fig. 16) if the perimeter is to be the fixed quantity $2p$ and the area is to be a maximum?

The answer to this problem is, of course, to make $A'B'BA$ a square with side $2p/4$ or $p/2$. However, when the area $A'B'BA$ is a maximum, the area $ABCD$ is a maximum because the latter area is always one-half of the former. Moreover, since the perimeter of $A'B'BA$ is $2p$, the perimeter of $ABCD$ will be p and thus meet the original condition. Hence the answer to our original problem is to choose the dimensions of rectangle $ABCD$ so that AB equals $p/2$ and AD, which is one-half of AA', is equal to $p/4$.

The major result—namely, that of all rectangles with a given perimeter the square has maximum area—thus leads to pretty thinking and even to practical applications. Such values are more than enough to keep the mathematician engrossed in the subject. His next thought would be to generalize the question and seek the answer. However, there are several directions in which he can generalize. Instead of asking, Of all rectangles with given perimeter, which has the maximum area? he might consider the question, Of all *quadrilaterals* of given perimeter, which has the maximum area? The answer, which we shall not establish here, is still the square. The mathematician might then consider pentagons, and ask, Of all pentagons with a given perimeter, which has the maximum area? Now the square is a regular quadrilateral; that is, its sides are all equal to each other and the angles are all equal. Hence, a reasonable conjecture in the case of pentagons is that the regular pentagon would furnish maximum area. And this is the correct result. In fact, it is not hard to prove the general theorem that of all polygons with the same number of sides and the same perimeter, the regular polygon has maximum area.

It is possible to raise and answer still other questions on the same theme. Instead of comparing polygons with the same number of sides, suppose one compared two polygons with different numbers of sides. For instance, of a square and regular pentagon with the same perimeter, which has the greater area? The answer is the pentagon, and a more general result is that of two regular polygons with the same perimeter the one with the larger number of sides has the greater area. But this result suggests a very exciting possibility. Even though the perimeter is kept constant one can find polygons of more and more sides. These polygons approach the circle in shape. Hence, it seems likely that the circle has greater area than any polygon with the same perimeter. This conclusion can also be proved deductively. In the preceding chapter we showed merely that a circle has greater area than a square of the same perimeter, but it is not too hard to prove the more general result. In all this we see how the mathematician can raise a series of interesting questions, suggested by one seemingly minor theorem, and obtain both useful and attractive results. Apropos of the conclusion that a circle bounds more area than any other figure with the same perimeter, one ancient Greek writer commented that it was a good thing to know this fact because the real estate men of his time assumed that a plot of ground with larger perimeter had more area.

Man is naturally interested in maximizing quantities such as area. It means a saving of money in fencing or in tin for cans. But nature is equally interested, and one of the truly deep connections between mathematics and nature has come from the recognition that nature seeks to maximize or minimize various quantities, presumably in the interest of effectiveness and economy. This theme will be a recurring one and we shall have many examples of nature's keen mathematical insight. Let us see here how geometry reveals this wisdom of nature.

From Greek times onward mathematicians and scientists have continually pursued the subject of light. The intense interest in this phenomenon is understandable: man depends upon light for his very existence and of course finds the phenomenon striking and pervasive. The rosy-fingered dawn may be just a pleasant theme for the poet but to both primitive and civilized man it promises light for vision, the generally welcome arrival of heat, nourishment for crops, and the banishment of the terrors and fears of the night. To the mathematically minded scientist light offers another domain of inquiry in his search for laws of nature. Despite its pervasiveness light is a most baffling subject; and even today it is still one of the great mysteries. Though progress in the discovery of the mathematical laws of light has been slow, some remarkable laws have nevertheless been obtained, and the depth of these laws has kept pace with the progress in mathematics proper.

One of the simplest and most readily observable phenomena of light is the fact that it is reflected from brightly polished surfaces. Is there any regularity and mathematically describable property in the process of reflection? There is, of course, and it was made known to us by Euclid in his *Optics*. As the Greeks had observed, light travels in ordinary air along straight lines. When regarded as the paths of light these lines are called rays. A ray that strikes a mirror is reflected and goes off in some different direction. Euclid records for us a simple relationship between the directions of the incident and reflected rays.

He tells us that the light ray AP (fig. 17) that strikes the mirror at P, will be reflected in the direction PB so that angle $1 =$ angle 2. This law of reflection can be observed experimentally by sending a beam of light along AP and noting the direction of the reflected beam. Angle 1 is called the angle of incidence and angle 2, the angle of reflection. (Most books call angle 3 the angle of incidence and angle 4 the angle of reflection, but the equality of angles 1 and 2 imples the equality of angles 3 and 4 and conversely. We shall feel free to use either pair of angles.) Nature,

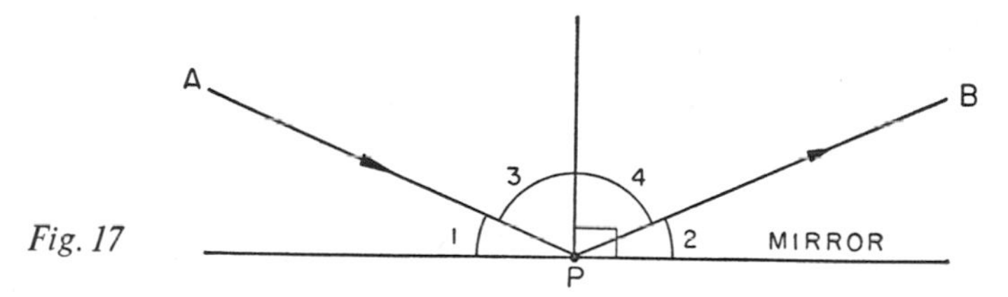

Fig. 17

we see, has obligingly allowed light rays to be reflected in accordance with a very simple mathematical law. This law was the opening wedge in the mathematical attack on the theory of light.

Let us now see if with this fact and the usual facts of geometry we can deduce any new knowledge about the behavior of light. We must note first the following trivial consequence of the law of reflection. Suppose a person at B (fig. 18) looks into a mirror to see

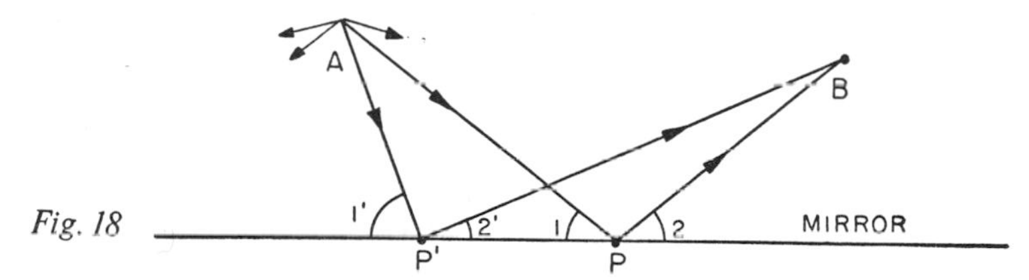

Fig. 18

the reflection of an object at A. Now light rays spread out from A in all directions and strike the mirror at various points. Each of these rays will be reflected in accordance with the law of reflection. However, only one of the rays issuing from A will be reflected to B, namely, the ray AP of figure 18 for which angle $1 =$ angle 2. Were any other ray, such as AP', reflected to B, than angle $2'$ would have to be the angle of reflection. But angle $1'$, as an exterior angle of triangle $AP'P$, must be greater than angle 1. Likewise, angle 2, as an exterior angle of triangle $P'BP$ is greater than angle $2'$. Since angle $1 = 2$, it follows that angle $1'$ is greater than angle $2'$ and therefore that $P'B$ cannot be the reflected ray corresponding to the incident ray AP'.

The Greek mathematician Heron (first century A.D.) noted the fact we have just discussed, namely, that just one ray from A would reach the point B, and he discovered another important property of the path from A to P to B. This path is the shortest possible one that anything can take in going from A to the mirror and then to B. Thus Heron asserted that the path $AP + PB$ is shorter than any

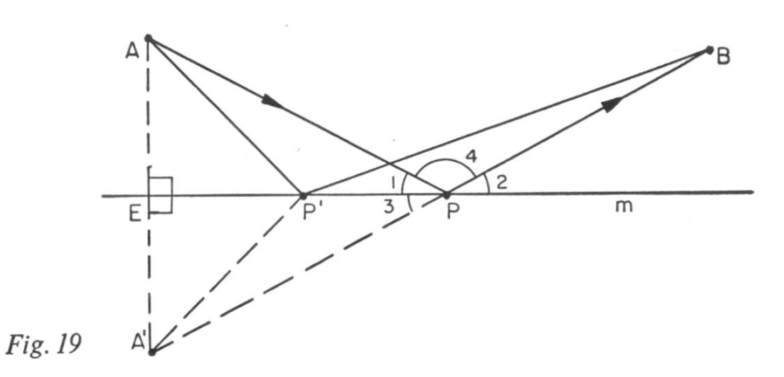

Fig. 19

other path such as $AP' + P'B$ with P' on the mirror (fig. 19). Let us examine Heron's proof.

Like all proofs of any consequence, this one requires an idea. Nature herself supplies it. When the person at B looks into the mirror along BP he sees the image of A at A', and this image appears to be as far behind the mirror as A is in front. Moreover the line AA' is perpendicular to the mirror. Hence, said Heron, let m be the line representing the mirror and let A' be that point on the perpendicular AE to the line m for which $A'E = AE$. It follows at once that the triangles AEP' and $A'EP'$ are congruent, so that $AP' = A'P'$. Likewise the triangles AEP and $A'EP$ are congruent, so $AP = A'P$. Instead, therefore, of comparing $AP' + P'B$ with $AP + PB$, let us compare $A'P' + P'B$ and $A'P + PB$.

Since the eye at B looks along BA' when it sees A', this fact suggests that $A'PB$ is a straight line, and if it is, then $A'P + PB$ is shorter than $A'P' + P'B$. Hence, we can establish the desired conclusion if we can prove that $A'PB$ is a straight line. This fact would follow from the knowledge that angle 4 + angle 1 + angle 3 is a straight angle. But certainly angle 4 + angle 1 + angle 2 is a straight angle, becaue m is a straight line. Also, angle 1 = angle 3 because triangles APE and $A'PE$ are congruent, and angle 1 = angle 2 because AP and PB are incident and reflected rays; then angle 3 = angle 2. Hence, angle 4 + angle 1 + angle 3 is a straight angle; $A'PB$ is a straight line; and our theorem is proved.

The theorem concerns the behavior of light rays, but we should note that it has a purely mathematical content which is independent of the physical phenomenon of light. The mathematical theorem says that of all broken-line paths from A to any point of a line m and then to B, the path APB, for which AP and PB make equal angles with m, is the shortest. This abstract, geometrical theorem about lines, angles, and shortest paths is likely to have other interpretations and applications. And it does.

Let us suppose that m in figure 19 is the shore of a river and that A and B are two inland towns. A trucking company that delivers merchandise from boats on the river to towns A and B wishes to build a pier on the river so as to serve both towns. About the same amount of merchandise is trucked to both towns. The company asks this question: Where should it build a pier on the shore m so that the total trucking distance from the pier to A and from the pier to B is least? The geometric theorem we proved above supplies the answer. The pier should be located at that point P on m for which AP and PB make equal angles with m.

The same geometric theorem has many other applications. Let us forget commerce for a moment and play billiards. A billiard ball that hits the side of the table behaves just as a light ray does in striking a mirror; that is, it bounces off at the same angle at which it strikes. Hence, if a billiard player wishes to direct a ball at A (fig. 20) so as to hit the side m of the table and then hit a ball at B, he consciously or unconsciously aims it at the point P, for which

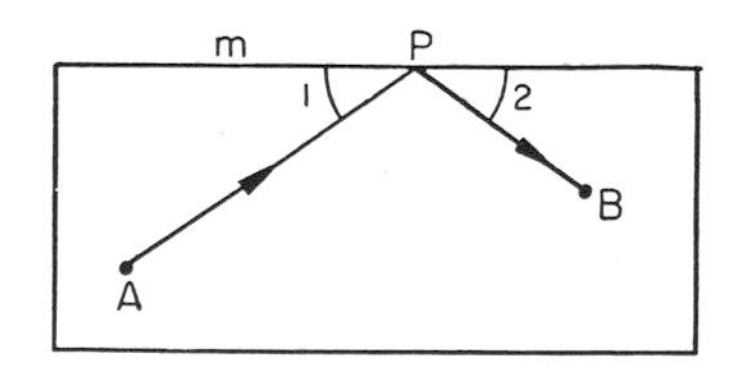

Fig. 20

AP and PB make equal angles with m. This choice of direction guarantees that the ball at A will hit the ball at B. But what is clear in addition is that the billiard player has selected the shortest path from A to m to B. Of course this discussion does not prove that a knowledge of mathematics makes a good billiard player.

Though our interest at the moment lies primarily in the mathematical behavior of light, we should pause to note how one geometrical theorem, obtained during the course of the study of light, applies to many different physical situations. The geometrical theorem proper concerns abstract concepts such as points and lines, but the very abstractness of it proves to be of advantage in making different applications. We therefore see how even in a simple subject such as geometry the abstractness of mathematics pays unexpected dividends.

We have learned a few mathematical facts about light, among them the not insignificant one that under certain conditions of reflection light knows enough to take the shortest path. Let us now test the value of this knowledge for other practical problems. Let us investigate what might be accomplished by successive reflection in two mirrors.

Suppose a person stands at the point P between two mirrors, m and n of figure 21, which intersect at the point O. Could the person see in mirror n the side of himself that faces mirror m? Should this be possible, a light ray would have to issue from P, strike the mirror m at some point, say X, be reflected according to the law of reflection, strike the mirror n, at Y, say, and be reflected to P. Now we know from our study of reflection of light in a mirror that the light ray that starts at P and is reflected at X will appear to an observer at Q to come from P', the mirror image of P in m. Likewise the light ray that actually goes from Y to P but can be thought of as going from P to Y would after reflection at Y appear at Q to come from P'', the mirror image of P in n. Hence the line from X to Y will have to pass through P' and P''. Let us therefore first locate P' and P'' as the mirror images of P in m and n respec-

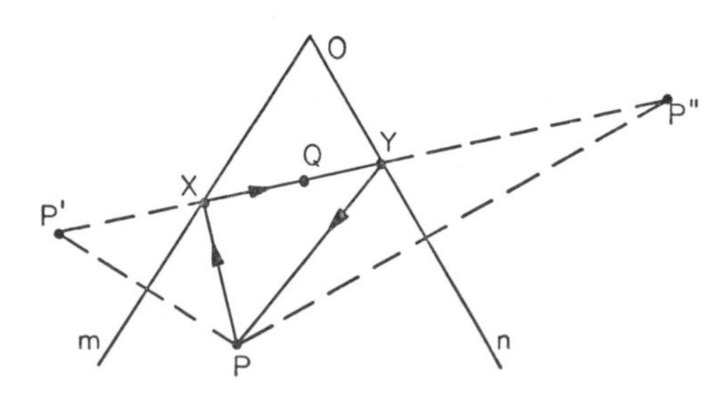

Fig. 21

tively. We now draw $P'P''$, and where this line cuts m we mark the point X and where it cuts n we mark the point Y. Then the light ray that takes the path PX, XY, YP satisfies the law of reflection at X and at Y and returns to P. The person at P who looks in the direction PY will see the light that issues from P in the direction PX. Of course the other light rays that leave P and undergo successive reflection at mirrors m and n will not return to P.

The combination of mirrors we have just analyzed is often found in clothing stores. A customer can view the fit of a suit in the back by looking into one of the pair of mirrors. Often the two mirrors are hinged at O so that the customer can change the angle there and see different parts of his back. If one should play with such a pair of mirrors he would find that when the angle XOY becomes 90°, no part of his side or back is any longer visible. Let us see what happens to rays issuing from P in this case.

Let PX be *any* ray that starts from P and strikes the mirror m

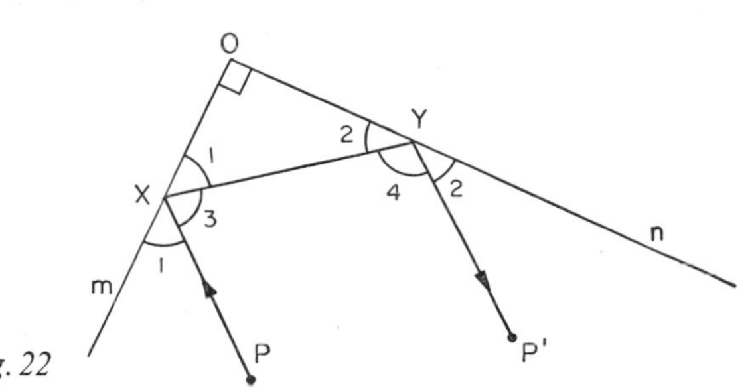

Fig. 22

at X (fig. 22). This ray is reflected at X, hits n at Y, say, and is reflected from n. What can we say about the direction that the ray now takes? Since light rays obey the law that the angle of incidence equals the angle of reflection, the two angles marked 1 in the figure are equal, as are the two angles marked 2; since a straight angle contains 180° we may then say that (using the symbol $\measuredangle$ to mean *angle*)

(6) $$2\measuredangle 1 + 2\measuredangle 2 + \measuredangle 3 + \measuredangle 4 = 360°.$$

But, since XOY is a right triangle,

(7) $$\measuredangle 1 + \measuredangle 2 = 90°.$$

If we use this fact in (6) we obtain

$$180° + \measuredangle 3 + \measuredangle 4 = 360°$$

and therefore

$$\measuredangle 3 + \measuredangle 4 = 180°.$$

This last fact implies that the lines PX and $P'Y$ are parallel because the interior angles 3 and 4 on the same side of the line XY are supplementary. Now PX was *any* ray entering the space between the two mirrors. Hence, any such ray is reflected parallel to itself.

This fact, interesting in itself, can be put to use. Suppose we place three mirrors at right angles to each other so that they form a corner at O (fig. 23). We can show by a slight extension of the

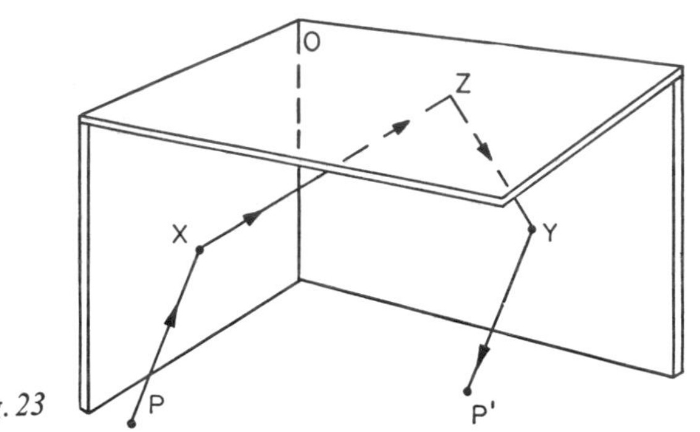

Fig. 23

above reasoning that any ray PX striking one mirror will emerge from the corner, after being reflected successively by all three mirrors, in a direction parallel to the incident ray.

Since radio waves behave like light waves (see chapter 21), a radio wave sent out from a radio system on the ground and reflected by the combination of the three mirrors will travel back in the direction of the radio set and be detected by the set. Because many rays of the radio beam that enters the corner are reflected back to the transmitting station, a strong signal is reflected.

Such corner reflectors are suspended from freely drifting balloons; the direction of the balloon is followed by having the radio set constantly send out signals in various directions and by noting the direction from which the reflected signal comes. Because the balloon drifts with the wind, its change in direction together with other information, namely, the distance of the balloon from the radio set, can be used to calculate the velocity of the wind.

We could continue to demonstrate how Euclidean geometry can be applied to the study of light and other physical phenomena. The vast range of these applications may be gleaned from the fact that the Greeks alone proved over a thousand theorems, and mathematicians since Greek times have added a considerable number of new ones. All of these theorems offer information about geometrical figures directly and about geometrical properties of physical happenings. However, we should not lose ourselves in details but rather recognize two pre-eminent values in this branch of mathematics. The first is the evidence it supplies of the power of human reason to derive new knowledge by deductive reasoning. The thousand and more theorems all follow from ten simple axioms. Recognition of this amazing power of the mind has inspired scientists, logicians, philosophers, theologians, and all seekers of truth. Euclidean geometry is still taught as the prime example of how mathematical reasoning proceeds and how much it can accomplish. Of course many people deplore and deprecate the study of Euclidean geometry, their argument being that one can learn reasoning and the power of reasoning by studying logic. Nevertheless, anyone who seeks to exhibit a significant example of reasoning almost inevitably turns to mathematics.

The second supreme value of Euclidean geometry is the evidence it proffers of the rational and indeed mathematical design of nature. A theorem deduced by hundreds of steps of reasoning from the basic axioms applies to the sizes and shapes of physical objects as accurately as the axioms themselves. The bridges and buildings designed in accordance with such theorems hold up. The obvious implication is that nature is designed in accordance with the rational pattern uncovered by Euclidean geometry. When in addition a pervasive phenomenon such as light is found to behave in accordance with geometrical principles, and deductive arguments based on these principles are found to lead to new truths about light, the argument for the mathematical design of nature becomes immensely stronger. Historically, it was Euclidean geometry that, developed to a large extent as a votive offering to the God of Reason, opened men's eyes to the existence of design and to the possibility of uncovering it by the pursuit of mathematics.

7: THE DIMENSIONS OF THE HEAVENLY SPHERES

On he passed, far beyond the flaming walls of the world, and traversed in mind and spirit the immeasurable universe.

—LUCRETIUS

ACCORDING to history, Euclidean geometry arose from man's efforts to measure areas of land, and the literal meaning of the word *geometry,* namely, earth-measure, supports history. Arithmetic and algebra, on the other hand, arose from man's efforts to count and manage property: cattle, crops, money, and other possessions. But the combination of simple algebra and simple geometry enabled man to pass from a knowledge of the world at his feet to knowledge of the world far above him. With just this elementary mathematics he determined the sizes and distances of the heavenly bodies and constructed systematic and rational accounts of the seemingly irregular and puzzling motions of the mysterious wanderers in the sky. In turn this knowledge about the heavens has taught man far more about his worldly lot than the most accurate accounts of his material possessions.

Interest in the heavenly bodies existed even in the most primitive societies. The light and heat of the sun, the spectacular colors that the sun and moon often assume, the bright lights of the planets that appear and disappear at various times of the year, the amazing panorama of lights from the Milky Way, and eclipses have excited wonder, admiration, speculation, and, in some cases, terror. However, any accurate knowledge about these phenomena was limited in pre-Greek times to the periods of revolution of the sun and moon and to the times of appearance and disappearance of some planets and stars. This information was unfortunately inadequate to yield any estimates of the sizes and distances of these bodies and still less adequate to furnish any physical or mathematical picture of their relative motions and of the structure of the heavens.

Even the Greeks of the classical period had only crude and erroneous ideas as to these sizes and distances. The Greek philosopher Anaximander, who lived from about 611 to 545 B.C., believed that of the heavenly bodies the stars are nearest to us and the sun, most distant. The sun's diameter he estimated to be about twenty-seven or twenty-eight times that of the earth. Leucippus (about 450 B.C.) believed that the sun was most distant, the moon nearest, and the planets in between. Plato's order of the heavenly bodies was Moon, Sun, Venus, Mercury, Mars, Jupiter, and Saturn. This order was accepted by Aristotle. Some, like Democritus, who flourished about 420 B.C., thought that the sun and moon were smaller than the earth. Even Eudoxus, the foremost mathematician of the classical Greek period and the originator of the first great astronomical theory, believed that the sun is just nine times as distant from the earth as the moon is.

That these estimates should have been crude or grossly inaccurate is understandable. Two facilities were lacking to these philosophers and mathematicians. They had no instruments worthy of the name with which to make angular measurements, and they did not have the mathematical methods that would have permitted them to draw some conclusions from these measurements. It was also true of many of the philosophers that they preferred to draw conclusions from mystical, poetical, and aesthetically pleasing principles rather than to set about creating the necessary instruments and mathematical methods.

Interest in quantitative knowledge and the willingness to accumulate this knowledge finally arose in the second great Greek period, when the center of that civilization was moved to Alexandria by the man after whom the city is named. It is perhaps irrelevant to trace the changes in the character of the Greek civilization that took place in the Alexandrian period. What is important, however, is that in that city the Greeks came into close contact with Egyptians and Babylonians, and the wealth of astronomical observations acquired by the Egyptians and Babylonians over several millenniums became more accessible. Equally pertinent is the fact that those successors of Alexander who ruled the Egyptian empire, the Ptolemys, built a great home for scholars, called the Museum, and spent lavishly to equip a famous library. They also provided funds to construct carefully graduated instruments which were put to use to make far more accurate measurements of the angular bearings of the heavenly bodies and of the angles these bodies subtend at points of observation on the earth.

During the Alexandrian period Eratosthenes, Apollonius, Aristarchus, Hipparchus, Ptolemy, and dozens of other luminaries applied themselves to the study of geography and astronomy. The latter subject in particular became a science, with observations and mathematical reasoning replacing vague doctrines, preconceived schemes, and mystical and aesthetic preferences. Of the men who advanced this science two were outstanding. They improved the astronomical data, they created the mathematics that enabled them to utilize these data, and they constructed the astronomical theory that was to predominate for over fifteen hundred years.

Hipparchus, the older man, lived most of his life in Rhodes. At the time he was active, about 150 B.C., Rhodes was a flourishing commercial and intellectual Greek state and a rival of Alexandria. Hipparchus knew fully what was going on at Alexandria. He knew, for example, Eratosthenes' *Geographica,* and wrote a criticism of it. He had possession of older Babylonian observations and those made at Alexandria from about 300 to 150 B.C. Of course he made many observations of his own. He is famous also for having discovered through the study of these observations the precession of the equinoxes, that is, the slow change in the times of occurrence of the spring and fall equinoxes and the summer and winter solstices.

Very little of what Hipparchus wrote is known to us directly but fortunately his work was taken up and expanded by the more celebrated astronomer, Claudius Ptolemy. Ptolemy lived and worked in Alexandria about A.D. 150. Though even less is known about his personal life than about Hipparchus', his comprehensive treatise, the *Almagest,* not only survived but became the standard work in astronomy up to the time of Copernicus.

The mathematical basis that Hipparchus and Ptolemy created for their work is today known as trigonometry. The new idea here is remarkable not only for its simplicity but also for its wide applicability. Indeed the trigonometric methods of obtaining such seemingly inaccessible facts as the distance to the sun and the size of the sun are far simpler than the geometric proof that the area of a circle is πr^2. In pursuing the idea of Hipparchus and Ptolemy we shall not adhere to their formulation, methods of proof, and notation; we shall use a modern version. However, the changes are minor.

Let us consider two right triangles (fig. 24) for which we suppose that angle A = angle A'. In addition, we know that the right angle C must equal the right angle C'. In view of the fact that the

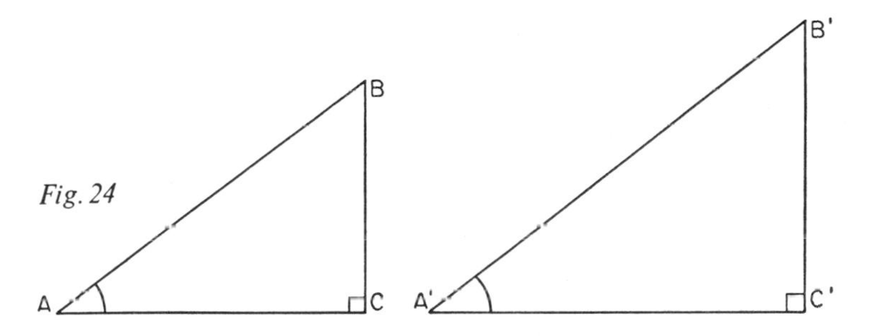

Fig. 24

sum of the angles in each triangle is 180°, it follows that angle B = angle B'. Hence, if one acute angle of a right triangle, angle A in our discussion, equals an acute angle of another, all three angles of the first triangle equal the corresponding angles of the other.

If two triangles are such that the three angles of one equal the corresponding angles of the other, the triangles are called similar. Hence, just because angle A = angle A' and the triangles ABC and $A'B'C'$ are right triangles, these two triangles are similar.

It is a theorem of Euclidean geometry that if two triangles are similar, the ratio of any pair of sides in one equals the ratio of the corresponding sides in the other. As applied to the right triangles of figure 24 this theorem says, for example, that $BC/AB = B'C'/A'B'$. But $A'B'C'$ is *any* right triangle provided only that the angle A' equals angle A. Hence, in any right triangle containing an angle equal to angle A the ratio of any two sides of that triangle will equal the corresponding ratio in triangle ABC. We need consider therefore only the ratios that can be formed in triangle ABC.

Since these ratios are quite useful, it is customary to give them names. However, since they vary with the size of angle A, we must speak of these ratios in connection with the angle. The three important ratios are

$$\text{sine } A = \frac{BC}{AB} = \frac{\textit{side opposite } \measuredangle A}{\text{hypotenuse}},$$

$$\text{cosine } A = \frac{AC}{AB} = \frac{\textit{side adjacent to } \measuredangle A}{\text{hypotenuse}},$$

$$\text{tangent } A = \frac{BC}{AC} = \frac{\textit{side opposite } \measuredangle A}{\text{side adjacent to } \measuredangle \text{A}}.$$

The names sine, cosine, and tangent for these particular ratios arose for historical reasons that are detailed and not important; one could use any names just to avoid the lengthier descriptions given above. As a matter of fact, the mathematician, with his usual niggardliness with pen and ink, doesn't even bother to write out these full names but refers to them as sin A, cos A, and tan A respectively.

Thus far we have merely introduced a new thought, namely, that the ratio of two sides of a right triangle, to be specific, say sin A, is the same in all right triangles having an acute angle equal to angle A. If, however, we could calculate this ratio for one such triangle we could record it and perhaps use it. Though we shall not dally with the details of calculating these ratios, let us see what we can expect. The hypotenuse of a right triangle is the side opposite the 90° angle, whereas each arm is opposite an acute angle. A theorem of Euclidean geometry says that in any triangle the greater side lies opposite the greater angle; hence the hypotenuse is always larger than either arm. Since the sine of an angle is the ratio of the side opposite to the hypotenuse, the sine of any acute angle is always less than 1. When the angle A is very small (fig. 25), the sine of that angle is likewise small; in fact the closer the angle is to 0°, the closer its sine value is to 0. On the other hand, when the angle A is close to 90° in size (fig. 26), the side opposite becomes almost as large as the hypotenuse and the sine will be close to 1 in value. The angle A cannot become 90° and still be an angle of a right triangle, because the sum of all three angles of the triangle must be 180°. However, because as angle A approaches 90° in size, the sine value approaches 1; the sine of 90° is taken to be 1 by an ex-

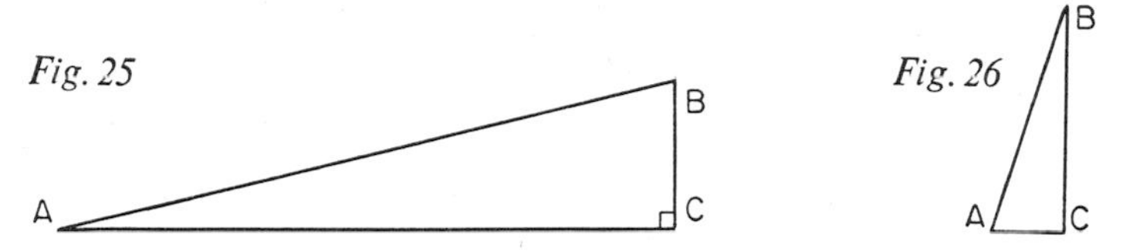

plicit definition for this case. Clearly, we could make a similar analysis of the possible ranges of cosine and tangent values as angle A varies from $0°$ to $90°$.

Except for a few angles, the calculation of the ratios is not readily made. Moreover, though there are a number of well-established relationships among these ratios that can be used for this purpose, these relationships and the calculations are not especially interesting in themselves. Fortunately, the chore of proving the relevant trigonometrical identities, as they are called, and of calculating the ratios, was performed by Hipparchus and Ptolemy and the results compiled in tables. The trigonometric tables found in modern texts are essentially what these men built up. Let us express our gratitude to Hipparchus and Ptolemy for doing this dirty work for us, and see what can be accomplished with the ratios.

Before we wander out into the vast realms of outer space to apply our new mathematical tool to the measurement of the heavenly bodies, let us test its strength on solid earth. The Greek mathematician and engineer Heron startled his colleagues with a little display of the power of trigonometry. He showed them how to dig a tunnel under a mountain by starting at both ends at once and having the two borings meet each other.

Let us suppose that the tunnel is to be dug under the mountain shown in figure 27. Of course the digger at either end cannot sight the opposite one and thereby work toward him. Heron overcame the difficulty by advancing essentially the following simple plan. He suggested choosing a point B on one side and point A on the other, such that both are visible from a point C to the right, with C chosen so that angle ACB is a right angle. He then measured AC and BC. Now

$$\tan A = \frac{BC}{AC} .$$

Since Heron knew BC and AC he knew $\tan A$. By looking up tables

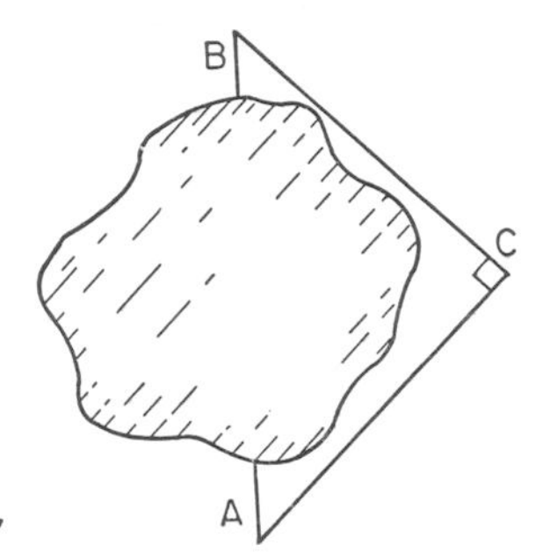

Fig. 27

he could find which angle A had the tangent value of BC/AC. Hence the digger at A had to make certain only that he followed the path that made the required angle A with AC. Angle B was similarly determined and the course fixed from this end. By continuing along these paths the diggers would meet somewhere under the mountain. Moreover, by using the Pythagorean theorem Heron calculated the length AB, and could tell the diggers exactly how far they would have to go in order to meet.

Of course Heron's method utilized the preliminary mathematical work that Hipparchus before him had done to calculate the values of tan A for various angles A. It should now be clear that this boring and fatiguing mathematical work has saved millions of hours of boring and fatiguing physical labor in digging tunnels and in thousands of other applications.

Our example of the use of the trigonometric ratios, though practical and demonstrative of the effectiveness of a mathematical idea, is dwarfed in significance by the tasks that Hipparchus and Ptolemy performed in measuring the sizes and distances of the heavenly bodies. We have already related how the Alexandrian scholar Eratosthenes measured the circumference of the earth. Let us take his result for granted and see how Hipparchus and Ptolemy advanced beyond this point. We shall simplify their methods slightly but the principles involved are the very ones used by these masters.

Let us find the distance from the earth to the moon. Suppose that the moon M in figure 28 is directly over the point Q on the earth's surface. The statement that the moon is directly over Q means geometrically that M lies on the line from the center E

of the earth through Q. Let P be another point on the earth's surface from which M is just visible when it is directly over Q. The statement that M is just visible at P means geometrically that the line from M to P is tangent to the earth's surface at P. It follows from a theorem of Euclidean geometry that the radius EP is perpendicular to PM, so that EPM is a right triangle. We seek the distance EM, that is, the distance from the center of the earth to the moon, the latter being regarded as a point in space.

For simplicity, let us suppose that P and Q are on the equator, which would be the case at certain times of each month. Then, since EP is the radius of the earth, we know from Eratosthenes'

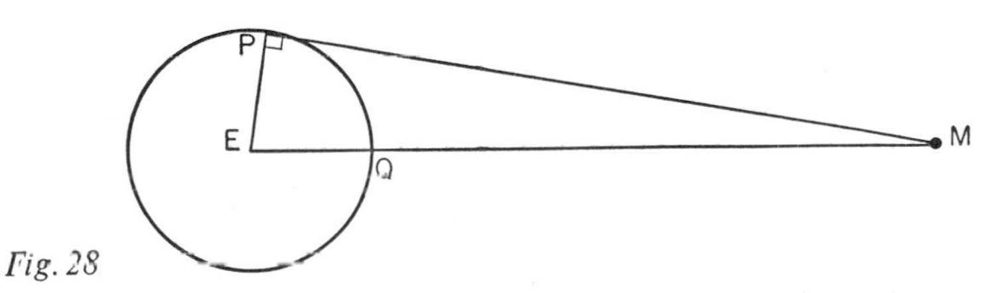

Fig. 28

work that this length is about 4000 miles. The angle E is determined by the arc one would traverse in going from P to Q. Whatever fraction this arc is of the entire circumference, the angle E is of 360°. Hence, by measuring PQ, we can determine angle E. Alternatively, angle E is the difference in longitude between P and Q and hence knowledge of the longitudes of these places gives angle E. (Incidentally, Hipparchus himself introduced the system of locating points on the surface of the earth by means of latitude and longitude.) At any rate the angle E can be determined and its size is 89°4′12″.

At this point the trigonometric ratios become useful, for

$$\cos E = \frac{PE}{EM} = \frac{4000}{EM} \cdot$$

From Hipparchus' tables one can find that cos 89°4′12″ is 0.0163. Hence

$$.0163 = \frac{4000}{EM} \cdot$$

By multiplying both sides of this equation by EM, a step justified by the axiom that equals multiplied by equals give equals, we obtain

$$.0163\ EM = 4000.$$

If we now divide both sides of this equation by 0.0163 and carry out the division on the right side we obtain

$$EM = 245{,}000.$$

This value is only approximately correct because we used 4000 miles instead of 3960 for the radius of the earth and we used only a four-place decimal value for the cosine. Our result also includes the distance EQ or the radius of the earth. Hipparchus, incidentally, gave the figure of about 280,000 miles.

Precisely the same method can be used to find the distance to the sun; the only difference is that angle $E = 89° 89' 81.2''$. The result is that the distance to the sun is approximately 93,000,000 miles. This distance varies, of course, with the time of the year. As is almost evident from the fact that angle E must be known to fractions of a second, Hipparchus could not have done well in computing the distance to the sun. His estimate was 10,000,000 miles.

Let us consider next the slightly different problem of calculating the radius of the moon. An observer at point P on the earth's surface (fig. 29) can determine the size of angle A, the angle between the line PM to the center of the moon and the line PQ, which is

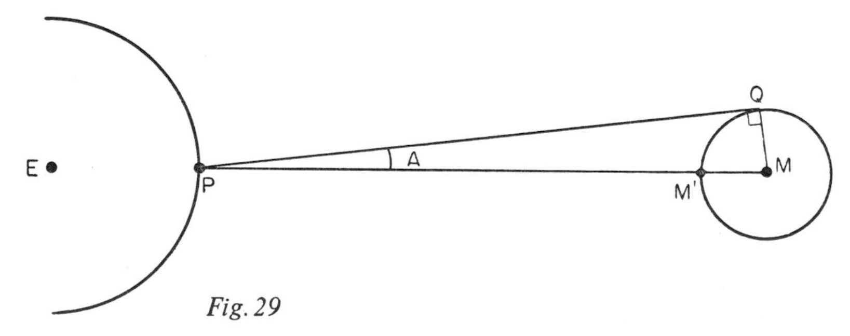

Fig. 29

just tangent to the moon's surface. This angle is about 15′. The distance PM is the distance from the *surface* of the earth to the center of the moon and is therefore 4000 miles less than the distance obtained above; that is, PM is about 240,000 miles. (This number does not include the distance MM' but, as we shall soon see, this omission introduces very little error.) We note that QM, the radius of the moon, is perpendicular to the tangent PQ. With these preliminaries taken care of we may now state that

$$\sin A = \frac{QM}{PM}$$

or

$$\sin 15' = \frac{QM}{240{,}000}.$$

Hence

$$QM = 240{,}000 \sin 15'.$$

But sin 15′, from Hipparchus' tables, is 0.0044. Hence

$$QM = 1056 \text{ miles}.$$

A more accurate value is 1080 miles. We see now that the neglect of the distance MM', an error of 1080 in 240,000, would make very little difference in the result.

By this very same method we can obtain the radius of the sun, which proves to be 432,000 miles. Thus the sun's radius is about 400 times that of the moon. It follows from a theorem to the effect that the volumes of two spheres are to each other as the cubes of their radii that the sun's size is $(400)^3$ or 64,000,000 times the size of the moon.

By making observations on the surface of the earth we were able to calculate readily the earth's distance from the sun. But now suppose that we wished to calculate the distance of Venus from the sun. We are not in a position as yet to make observations from the surface of Venus. How, then, shall we determine this distance? The substitute for the somewhat difficult task of getting to Venus and withstanding the heat there while making the necessary observations is just a little mathematics.

First note a mathematical detail. Let S, E, and V (fig. 30a, p. 100) be the positions of the sun, earth, and Venus at any time. We may regard S, E, and V as the vertices of a triangle. If we assume that the paths of the earth and Venus around the sun are circular then we have the condition that ES and VS are fixed lengths. The question at the moment is, For what positions of V and E is angle E largest? Let us compare this figure with the one in which SV is perpendicular to EV (fig. 30b). In the latter figure

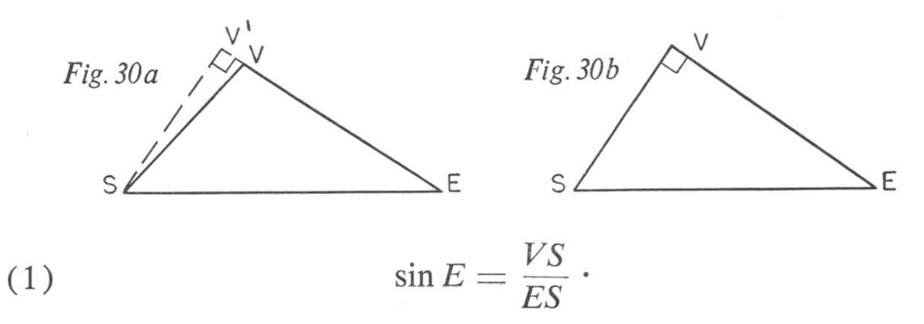

Fig. 30a Fig. 30b

(1) $$\sin E = \frac{VS}{ES} \cdot$$

If in the first figure we drop a perpendicular SV' from S to EV, we see that in this first figure

(2) $$\sin E = \frac{V'S}{ES} \cdot$$

But $V'S$ is less than VS because $V'S$ is the perpendicular from S to the line EV. Comparison of (1) and (2) shows then that angle E of the second figure is larger. In other words, angle E is largest when SV, the line from the sun to Venus, is perpendicular to EV, the line from the earth to Venus. Conversely, when angle E is maximum, SV must be perpendicular to EV.

By measuring the values of angle E as the earth and Venus move

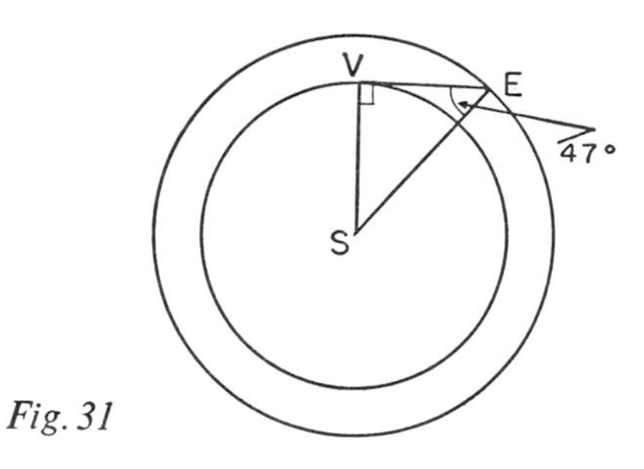

Fig. 31

around the sun in their respective paths (fig. 31) one finds that the maximum value of angle E is 47°. But just at that value, SV must be perpendicular to EV. Hence

$$\sin 47^\circ = \frac{VS}{ES},$$

or

$$VS = ES \sin 47^\circ.$$

Since sin 47° is 0.7314 and ES is the distance of the earth from the sun,

$$VS = 93{,}000{,}000\ (.7314),$$

or

$$VS = 68{,}000{,}000.$$

All of the values we have obtained are, of course, modern values. We note from the data that the angle subtended by the moon at the surface of the earth hardly exceeds 30′. The comparable value for Venus is 1′6″, for Jupiter 50″, for Mars 26″, all measured at their least distance from the earth. These angles are all very small and could not be accurately measured by Hipparchus and Ptolemy since they did not have accurate instruments, and, of course, they did not have telescopes. Because the angles involved in calculating the distances to the stars are even smaller, these distances were not known at all to the Greeks. We now know that the light from some nebulae takes millions of years to reach us. The Periclean court of fifth-century Athens was startled by Anaxagoras' conjecture that the sun might be as large as the mainland of Greece. We can imagine, then, how much even the approximate figures revealed by the Alexandrian astronomers must have staggered their contemporaries.

The difficulty in measuring the sizes and distances of the stars continued to balk astronomers up to modern times. During the seventeenth and eighteenth centuries the main method was to try to determine a difference in the direction of a star from one position of the earth in its orbit as against the position six months later. A British professor of astronomy at Oxford, James Bradley (1693–1762), who had worked on this very problem, was disturbed to find that the direction of a star at any one time was not what he expected it to be. While riding in a sailboat on the Thames one day in the year 1728 Bradley observed that the direction of the wind appeared to change when the boat changed its course. Bradley realized that it was not the wind which shifted but that the direction of the wind relative to the boat had changed. This thought suggested that the motion of the earth in its orbit relative to the speed of light acted somewhat as the sailboat did relative to the speed of the wind. Instead of pursuing the problem of the change in direction of a star due to the earth's changed position in space, on which problem he had made no progress, Bradley recognized that he had a new method for determining the velocity of light, a very difficult task in view of the immense speed involved.

The principle Bradley used is readily understood. If one is traveling in a train while rain is falling outside, one is likely to observe that the rain, which is actually falling straight down, appears to fall along a path that slopes toward the rear of the train. Of course one soon realizes that this backward motion of the drops is due to the forward motion of the train. But suppose that one wished to keep in view one drop in the air that would hit the ground directly alongside the observer. The observer at *A* (fig. 32) would have to choose a drop, *O* say, such that as he moves from *A* to *B* the drop will fall in the same time from *O* to *B*. Moreover, to keep the drop in view the observer will *always have to look in the direction parallel to AO*. Thus when at *A′* he will be looking in the

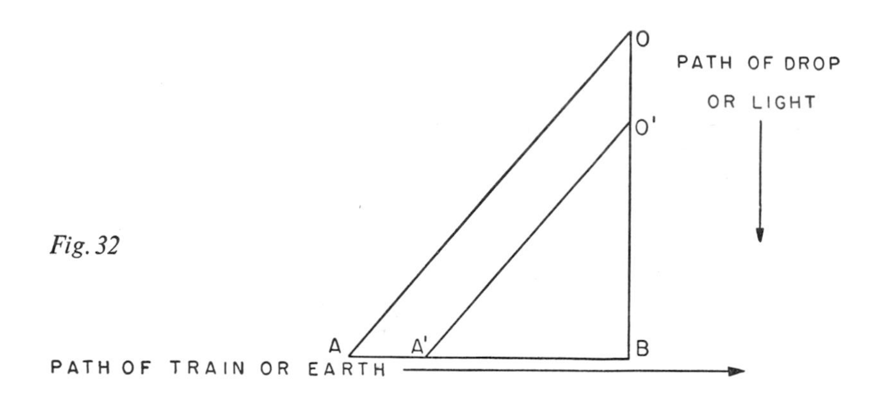

Fig. 32

direction *A′O′*. Since *AB* is the distance the observer moves in the same time that the drop falls the distance *OB*, the ratio *OB*/*AB* is nothing else than the ratio of the velocities of the two motions.

Now suppose an astronomer on the earth, which is at the point *A* in space, wishes to observe a star at *O*. He must point his telescope along the direction *AO* so that as the earth moves to *B* the telescope keeps the moving light in view and so that when the earth reaches *B* the light will enter the telescope. The angle *OAB* (fig. 32) is determined by the relative velocities of light and of the earth as it moves in its orbit. (The rotation of the earth can be neglected because it is so much slower than the velocity of revolution around the sun.) In other words, in the astronomical case, *OB* is the distance the light travels from the star as the earth moves from *A* to *B*. The angle at *A* can be measured and a typical value obtained by Bradley was 89° 59′ 40″. Moreover, the orbital velocity of the earth is readily calculated because we know the radius of the earth's path around the sun to be 93,000,000 miles and the circumference (assuming for simplicity that the path is circular) is 2π times the radius. Since the time taken by the earth to traverse this circumference is one year, the orbital velocity of the earth is about 18.5 miles per second. Hence, Bradley, argued,

$$\tan 89^\circ\ 59'\ 40'' = \frac{c}{18.5},$$

where c is the velocity of light. The value of tan 89° 59′ 40″ is about 10,000. Hence $c = 185{,}000$ miles per second. The value accepted today is about 186,000 miles per second. Of course this is the velocity of light in a vacuum. The velocity of light in other media—water, for example—is less.

In addition to determining the velocity of light Bradley called attention to an important astronomical principle now called aberration. The observer at *A* who wishes to see the star at *O* must point his telescope in the direction *AO*. He will not actually see the star until he reaches *B*, but as he moves from *A* to *B* he must keep his telescope pointed along the direction parallel to *AO*. Hence when he finally receives the light from the star, his telescope is pointed in a direction that, in our example, is 20″ off from the true direction of the fixed star. Thus the star that is directly over *B* will appear to be 20″ to the right. The general principle that Bradley noted is that the motion of the earth introduces an error in our observations of the directions of the heavenly bodies.

Still another phenomenon of light, troublesome to astronomers ever since Ptolemy's day, proves to be the key to many puzzling features found close to home. The Alexandrian Greeks had become familiar with the bending or refraction of light when it passed from one medium to another. For example, a light ray that passes from air to water changes direction when it enters the water so that if i is the angle the ray makes with the perpendicular to the surface while traveling in air (fig. 33), and r is the angle it makes with the perpendicular in water, then r is less than i. Over many centuries physicists and mathematicians tried to find the correct mathematical relation between i and r, which are called the angles of incidence and refraction respectively. Of course, for different values of i, the corresponding values of r differ. There is, moreover, the further complication that for any given i, the value of r depends

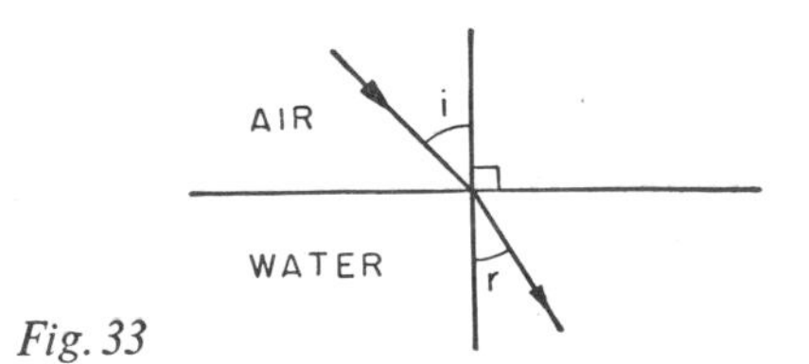

Fig. 33

upon the two media involved. Thus for the same i the value of r for glass is different from that of the r for water.

The correct law was found in the early seventeenth century by Willebrord Snell and René Descartes; it involves the velocity of light in the two media. Let v_1 be the velocity of light in the medium from which the light is incident and v_2, the velocity in the refracting medium. Then the correct law is

$$\frac{\sin i}{\sin r} = \frac{v_1}{v_2}.$$

Peculiarly, the ratio of the sines of the angles is the ratio of the velocities. Again the trigonometric ratios permit us to express a law of nature.

We need not pursue Snell's and Descartes's reasoning that led to this law in order to use it. As a matter of fact Descartes's argument was wrong even though he obtained the correct result. Let us consider the case of light going from air to water. The ratio of v_1 to v_2 is then $4/3$. Hence for these media

$$\frac{\sin i}{\sin r} = \frac{4}{3}. \tag{3}$$

To familiarize ourselves with the implications of this law let us suppose light from the sun, when it happens to be practically overhead, strikes the surface of a body of water. Then angle i is close to 0°. Then sin i is likewise close to 0. Since the ratio of sin i to sin r must be $4/3$, sin r will be even closer to 0 in value and hence r will be even smaller than i.

Now let us suppose that the sun is near the horizon when its light strikes the surface of a body of water. The rays enter the water at an angle of incidence of nearly 90° (fig. 34). The sine of an angle nearly 90° is practically 1. The question we raise is, How big is angle r? In view of equation (3), if sin i is equal to 1, sin r must equal $3/4$. With the aid of trigonometric tables we find that angle $r = 49°$. In other words, the sun's rays penetrate the water at a rather moderate angle.

Now let us consider what a person who is at point P *in the water* sees. Since the light rays come toward him along the direction OP he will conclude (if he knows no mathematics) that the sun lies in the direction PO. In other words, the sun will appear to be rather high in the sky and not at the horizon.

More than that, we have just seen that when light passes from air

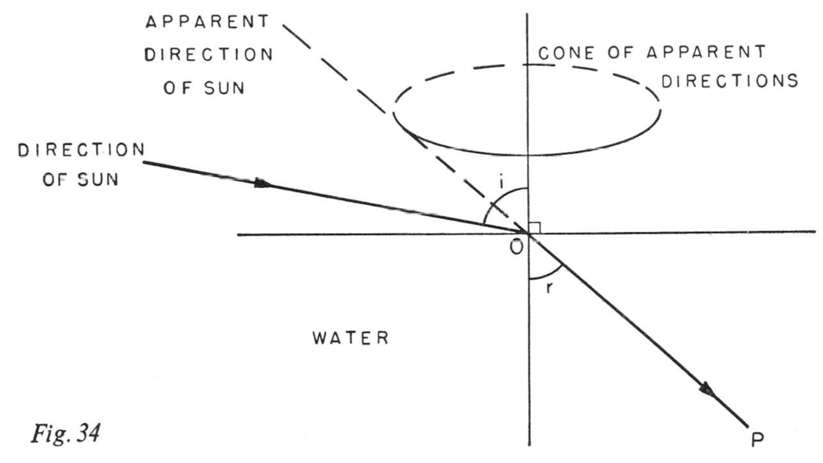

Fig. 34

to water the angles of refraction can vary only between 0 and 49° as the angles of incidence vary from 0 to 90°. Hence a person under water will believe that everything above water lies somewhere within a 49° angle of the perpendicular to the surface. Since, as far as we know, a fish knows no mathematics or physics, he must believe that the world above the surface lies in this cone about the perpendicular. This distorted picture of the world above, as seen from positions in the water, is called the fish-eye view of the world. Of course a man in a submarine or a swimmer under water would obtain the same view.

Let us consider what happens when light passes from water to air. Suppose that a flashlight beam is sent in the direction shown in figure 35. Now we must regard the angle of incidence i to be the

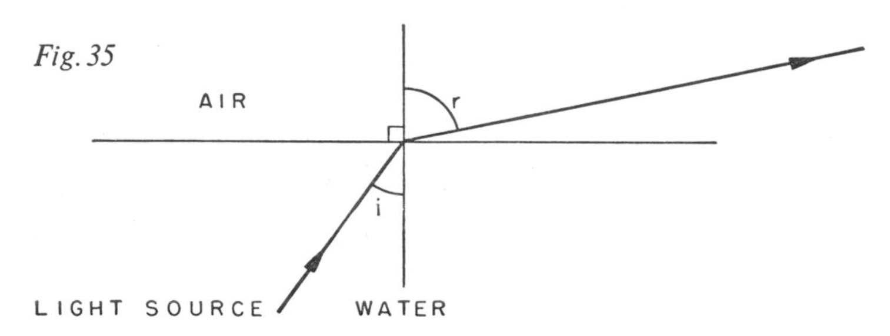

angle made by the incident ray with the perpendicular in the water and the angle of refraction to be in the upper medium. Moreover the ratio v_1 to v_2 is now 3 to 4. Hence

$$(4) \qquad \frac{\sin i}{\sin r} = \frac{3}{4} \cdot$$

Thus angle r must be larger than angle i. When $i = 49°$, $\sin i$, as we found above, is ¾. Hence, from (4), $\sin r = 1$ and $r = 90°$. The rays will therefore emerge and travel parallel to the surface of the water.

But suppose now that i is greater than 49°. Then $\sin i$ is greater than ¾. Since equation (4) can be written

$$\sin r = \frac{4}{3} (\sin i) ,$$

when $\sin i$ is greater than ¾, $\sin r$ must be greater than 1. How large is angle r? There is no angle whose sine value is greater than 1 because the side opposite the acute angle of a right triangle cannot be greater than the hypotenuse of that triangle.

Here's a pretty mess, indeed! Light in the water hitting the surface at an angle of incidence greater than 49° cannot emerge from the water because mathematics says that there is no angle of refraction for it. What should these baffled light rays do? Well,

Fig. 36

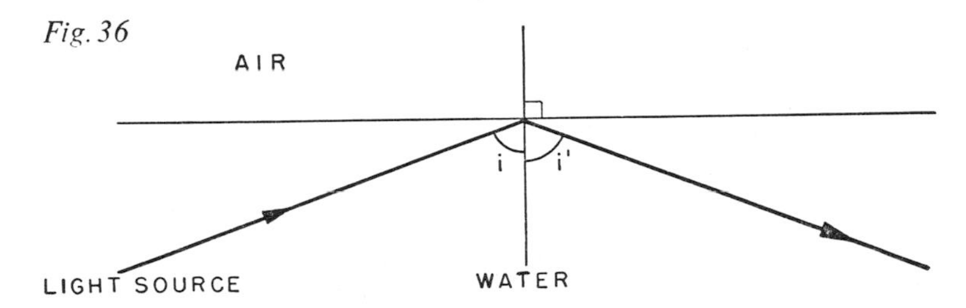

rather than defy the laws of mathematics the light rays stay under water and allow themselves to be *reflected* at an angle i' equal to angle i (fig. 36).

Thus, by means of mathematics, we have predicted a rather amazing phenomenon. When light passes from one medium to another in which the velocity of light is greater, there is an angle of incidence called the critical angle—49° in the above example—beyond which all the light is reflected. For angles of incidence beyond the critical angle the surface between the two media acts like a perfect reflector and the light is reflected according to the law of reflection. This phenomenon is called total reflection. The word *total* is used because when the angle of incidence is below the critical value a small percentage of the light is reflected, though most of it is refracted; but above the critical value, all the light is reflected.

Like most phenomena discovered in the general study of nature this one has many practical applications. Let ABC of figure 37 be a piece of prism-shaped glass such that ABC is an isosceles right

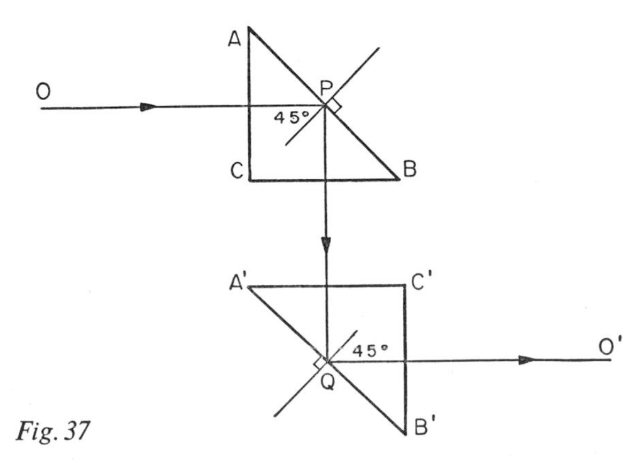

Fig. 37

triangle. Suppose light enters from the left and perpendicular to face *AC*. Since the angle of incidence is zero the angle of refraction is likewise zero, and the light goes through unchanged in direction. Now let us suppose that the prism is made of flint glass, for which the ratio of the velocity of light in air to that in the glass is 1.65. The critical angle for this glass is 37°. The light that has entered the glass strikes the back face *AB* at an angle of 45° to the perpendicular, is totally reflected at the same angle, travels to a second prism-shaped piece, *A′B′C′*, enters at *A′C′*, is totally reflected at *Q*, and then emerges from *B′C′* horizontally. The faces *AB* and *A′B′* act like perfect mirrors, though unsilvered. They are in fact preferable to mirrors because they do not deteriorate from tarnishing. The combination of the two prisms serves to translate the incident light from *OP* to *QO′*. That is why this combination is useful in periscopes and in binoculars.

We have seen in this chapter how the trigonometric ratios based on the concept of similar right triangles can be used to triangulate the heavens, determine the velocity of light, and express the precise law of refraction for light when it passes from one medium to another. We can see now what Plato meant when he declared in his famous dialogue, the *Timaeus,* that the world can be made intelligible in terms of right triangles.

8: THE REVOLUTIONS OF THE HEAVENLY SPHERES

This terror then and darkness of mind must be dispelled not by the rays of the sun and glittering shafts of light, but by the aspect and law of nature.

—LUCRETIUS

INSTEAD of being daunted by the immense sizes and distances of the heavenly bodies, which a little trigonometry had disclosed, the Greeks were emboldened by this victory and tackled a far more ambitious undertaking. They sought to discern the design governing the motions of the sun, moon, planets, and stars. Even the tion of the heavens. But it was Plato, to whom the conquest of early Greeks had made some bold conjectures about the organiza-Mount Everest would have seemed an insignificant achievement, who launched the Greeks on the task of discovering the precise geometrical plan. He pressed the problem upon his students, and one of them, Eudoxus, the foremost mathematician of the classical Greek period, succeeded in erecting a mathematical scheme that described the grosser features of the motions of the sun, moon, and planets. However, it was deficient in many respects, and the problem continued to occupy the minds of the Greek mathematicians for many centuries.

A much more laudable effort was made by the Alexandrians Hipparchus and Ptolemy. As Ptolemy said, they set out to base astronomy "on the incontrovertible ways of arithmetic and geometry," and they established a theory that lasted fifteen hundred years. The climactic step, however, was made by Copernicus and Kepler. Their contribution not only provided the most successful astronomical theory, demonstrated the power of mathematics to master nature's ways, and altered the intellectual life of Europe, but initiated a new pattern of scientific thinking in which mathematics was to play a far more fundamental role than it had in all the previous centuries.

To appreciate just what these men accomplished and how it led to major creations in modern mathematics and science, we must go back a bit. The algebra, the geometry, and the trigonometry examined thus far have been presented much as they would be in any modern school or text. We have also described some modern applications of these mathematical subjects. Nevertheless, the ideas and the bulk of the theorems date back to the Greeks. In the sixteenth century the Europeans began to take a big hand in mathematics and in the mathematical investigation of nature. Why did we not hear of them before? What happened to the Greeks? And how did it come about that the center of mathematical life shifted to Europe?

The Greek civilization was gradually destroyed by a combination of forces too detailed to relate here. It was wiped out by the Mohammedan conquest of Alexandria in A.D. 640, though an independent but stultified Greek center did continue to exist in the Eastern Roman, or Byzantine, Empire centered around Constantinople. During the thousand or so years that the Greek civilization flourished nothing significant appears to have occurred in Europe. Actually very little is known about the history of Europe during the centuries preceding the Christian era and even during the first few centuries of this era. But we do know that the Germanic tribes, the ancestors of most Americans, occupied that region and that these peoples were barbarians. They had no written language, no arts, and no science or mathematics. Like most primitive peoples they had lots of mythology.

The Germanic tribes signalized their entry into recorded history by raiding and pillaging regions of the Roman empire. Through such contacts with the Romans, and through Roman military occupation of France and Britain, the barbarians learned something of civilization. But the major civilizing influence, especially after the fall of the Roman Empire, was the Christian Church. This influence was certainly beneficial in that it produced a stable, settled, and more peaceful state of affairs in Europe—the barbarians learned to live in peace for long stretches of time and to concentrate their killing during short intervals—but from the standpoint of the continued development of mathematics and the study of nature the improvements in Europe were of no help. The Christian leaders insisted that knowledge of man and nature could be derived entirely from the Bible, and they held to a strict interpretation of Biblical statements. The little Greek and Roman knowledge that filtered into medieval Europe was opposed as pagan.

Unfortunately, the Biblical account of creation and the structure of the universe is of Babylonian origin and contains the most naïve and fanciful views. The earth was the center of the universe. This particular doctrine happened to be basic in the highly mathematical geocentric theory of Hipparchus and Ptolemy but the agreement on this point was entirely coincidental, and here the resemblance of Greek and early medieval European thought ended. Though every educated Greek after the time of Pythagoras believed in the sphericity of the earth—Aristotle, for one, gave excellent arguments in support of this belief—medieval Europe believed the earth was flat. Whereas the Greeks had amassed extensive and remarkably accurate geographical knowledge, which Ptolemy recorded in his *Geography*, the Europeans' geographical knowledge gleaned from the Scriptures and limited travel was highly erroneous. Medieval Europe accepted without question the existence of the most deformed types of men and animals in various corners of the world and of Hell beneath the earth and Heaven above.

Knowledge of the Greek mathematical and scientific works finally reached Europe. These works had been in the hands of the Arabs, who, conquering areas along the Mediterranean, including Spain, Sicily, and southern Italy, came into contact with the Europeans. European intellectuals, princes, and some church leaders eagerly sought copies of the Greek works and either personally undertook or financed translations from the Arabic or directly from the Greek into Latin. Plato, Aristotle, Euclid, Apollonius, Archimedes, Ptolemy, and Diophantus were read avidly. The transformation of Europe from a purely Christian culture to a fusion of the Greek and Christian cultures had begun, and the way was prepared for revolutions in medieval thought.

The first of these was initiated by Nicolaus Copernicus (1473–1543), a Pole. After studying mathematics and science at the University of Cracow he decided to go to Italy because that country had become the great center for the newly recovered Greek learning. He enrolled as a student at the University of Bologna in 1497, learned Greek, and, in particular, studied under a foremost "Pythagorean," Domenico Maria da Novara. Though there is some question as to how much Copernicus was influenced by Novara, the latter, a practical astronomer, did teach Copernicus how to make astronomical observations. After further studies in law and medicine at various centers in Italy he undertook the duties of canon of the Frauenberg cathedral in East Prussia. These duties were administrative and fortunately light enough to permit him to devote years to further astronomical observations and to the construction of his new astronomical theory.

The fruit of these years is the classic *On the Revolutions of the Heavenly Spheres,* published in 1543, the year of his death. To a world that had satisfied its egotism and scientific needs by drawing upon the geocentric theory of Hipparchus and Ptolemy, Copernicus presented a startlingly new and fully developed heliocentric theory of the heavenly motions.

To follow Copernicus' innovations let us note first the essential idea in the scheme of Hipparchus and Ptolemy. Under their plan a planet, P in figure 38, moved on a circle called an epicycle whose center S, the sun, itself moved on a larger circle, called the deferent.

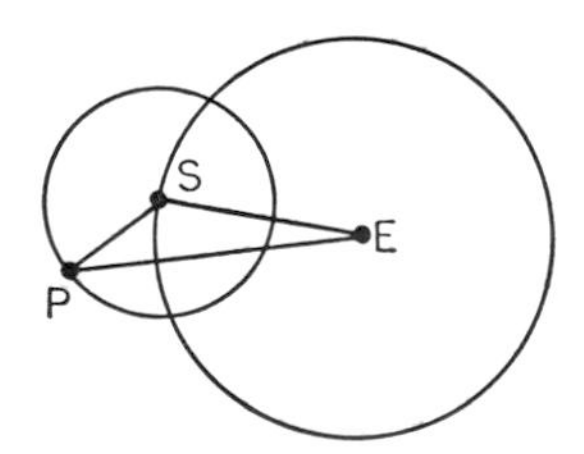

Fig. 38

The center of the deferent was the earth, E. The radii of the two circles, epicycle and deferent, as well as the velocity with which P moved around S and S around E were fixed by Hipparchus and Ptolemy to make the positions of a planet fit the observations made on the earth. Actually the Ptolemaic scheme was more complicated. For each heavenly body there were several epicycles, each of whose centers moved on the next one and the final one on a deferent. Moreover, each such system had to be specially constructed to fit the motion of the heavenly body it represented. But to understand Copernicus' changes we shall confine ourselves to the essential principle just described.

Copernicus observed that by having the planet *P* revolve about the sun *S* (fig. 39), and by having the earth *E* also revolve about the sun, the positions of *P* as observed from *E* would still be the same. Hence the motion of the planet *P* is described by the one circle whereas the geocentric view calls for two circles. Of course

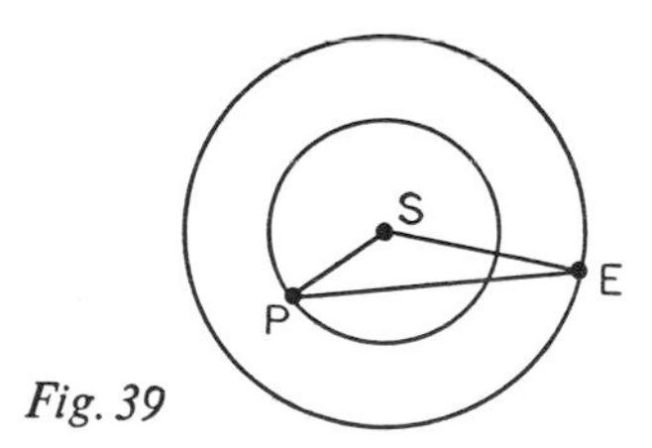

Fig. 39

the motion of a planet around the sun is not strictly circular and Copernicus added epicycles to the circles shown in figure 39 in order to describe the motions of *P* and *E* more accurately. Nevertheless, he was able to reduce the number of circles required in his time to "explain the whole construction of the world and the whole dance of the planets" from seventy-seven to thirty-four. Thus the heliocentric view permitted a considerable simplification in the description of the planetary motions.

It is of interest to note in connection with the storm that later broke over the works of Copernicus and Kepler (whose work we shall discuss shortly) that about 1530 Copernicus circulated a short account of his new ideas in a little tract called the *Commentariolus* and that Cardinal Nicolaus von Schönberg, the Archbishop of Capua, wrote to Copernicus urging him to make the full work known and begging him for a copy to be made at the Cardinal's expense. However, Copernicus dreaded the furor he knew his work would generate and for years shrank from publication. He entrusted his manuscript to Tiedemann Giese, Bishop of Kulm, who, with the aid of Georg Joachim Rheticus, a professor at the University of Wittenberg, got the book published. A Lutheran theologian, Andreas Osiander, who assisted in the printing and feared trouble, added an unsigned preface of his own. Osiander stated that the new work was a hypothesis that allowed computations of the heavenly motions on geometrical principles, adding that this hypothesis was not the real situation. Whoever takes for truth what was designed for different purposes, he added, will leave the science of astronomy a greater fool than when he approached it. Of course Osiander did not reflect Copernicus' views, for Copernicus believed that the motion of the earth was a physical reality.

The Copernican hypothesis of a stationary sun considerably simplified astronomical theory and calculations but otherwise it was not impressively accurate. Copernicus fell far short of even 10′ of accuracy in predicting the angular positions of planets. He therefore tried variations on the basic plan of deferent and epicycle, with the sun, of course, always stationary and either at or near the center of the deferent. Though these variations were not much more successful the failures did not diminish his enthusiasm for the heliocentric view.

The definitive improvement was made some fifty years later by that almost incredible mystic, rationalist, and empiricist, Johann Kepler, a German who combined wonderful imaginative power and emotional exuberance with infinite patience in acquiring data and meticulous adherence to the dictates of facts. The personal life of Kepler contrasts sharply with that of Copernicus. The latter had obtained an excellent education as a youth and lived a retired and secure life, which he was able to devote almost uninterruptedly to his theorizing. Kepler, born in 1571 with delicate health, was neglected by his parents and received a rather poor education. Like most boys of his time who showed some interest in learning he was expected to study for the ministry. In 1589 he enrolled at the German University of Tübingen where he learned astronomy from an enthusiastic Copernican, Michael Mästlin. Kepler was impressed by the new theory but the superiors of the Lutheran Church were not and questioned Kepler's devoutness. Since Kepler objected to the narrowness of the current Lutheran thought, he abandoned a ministerial career and accepted the position of Professor of Mathematics and Morals at the University of Graz in Styria, Austria, where he lectured on mathematics, astronomy, rhetoric, and Vergil. He was also called upon to make astrological predictions, which at the time he seems to have believed in. He set out to master the art and practiced by checking predictions of his own fortunes. Later in life he became less credulous and used to warn his clients, "What I say will or will not come to pass."

At Graz Kepler introduced the new calendar advocated by Pope Gregory XIII. The Protestants rejected it because they preferred to be at variance with the sun rather than in accordance with the Pope. Unfortunately, the liberal Catholic ruler of Styria was succeeded by an intolerant one and Kepler found life uncomfortable. Though he was protected by the Jesuits for a while, and could have stayed by professing Catholicism, he refused to do so and finally left Graz.

He secured a position as assistant to the famous astronomical observer Tycho Brahe, who had been making the first large revision of astronomical data since Greek times. Kepler was to benefit enormously from the new information that Brahe placed at his disposal. He referred to these data as the one honor Providence bestowed on him. On Brahe's death in 1601 Kepler succeeded him as "Imperial Mathematician" to Emperor Rudolph II of Bohemia. This employer, too, expected Kepler to cast horoscopes for members of the court. Kepler resigned himself to this duty with the philosophical view that nature provided all animals with a means of existence. He was wont to refer to astrology as the daughter of astronomy who nursed her own mother.

About ten years after Kepler joined him the Emperor Rudolph began to experience political troubles and could not afford to pay Kepler's salary. Kepler had to find another job. In 1612 he accepted the position of Provincial Mathematician at Linz. But other difficulties still plagued him. While he was at Prague his wife and one son died. Kepler remarried but at Linz two more of his children died; the Protestants did not accept him; money was tight; and Kepler struggled for existence. In 1620 Linz was conquered by the Catholic Duke Maximilian of Bavaria and Kepler suffered stronger persecution. Ill health began to plague him. The last few years of his life (he died in 1630) were spent in trying to secure publication of more books, collecting salary owed to him, and searching for a new position. On his tombstone Kepler's friends inscribed a verse Kepler himself left for this use:

Once I measured the skies,
Now I measure the earth's shadow;
Of heavenly birth was the measuring mind,
In the shadow remains only the body.

Kepler began his work on astronomical theory with the firm conviction that nature was designed in accordance with mathematical principles, and he cast about for some beautiful and neat mathematical principle that would fit the obvious requirements. Though a modest and humble man, his approach was that which a God, seeking a pleasing design for a new world, might take. But the schemes he tried to use, notably one employing the five regular solids favored by the Greeks, did not fit the data on the paths of the planets nor on their distances from the sun. Thus far Kepler's work was subject to the criticism that Aristotle made of the Pythagoreans: "They do not with regard to the phenomena seek for their reasons and causes but forcibly make the phenomena fit their opinions and preconceived notions and attempt to construct the universe." But the enlightened Kepler had too much regard for facts to persist with theories that failed to agree with observations and that could not yield accurate predictions.

It was after Kepler had acquired Brahe's data and had made additional observations of his own that he became convinced he must discard the astronomical patterns his predecessors and he had conceived. The search for laws that would fit these new data culminated in two famous results. These Kepler presented to the world in his book *On the Motion of the Planet Mars,* published in 1609.

The first of these laws broke all traditions and introduced the ellipse in astronomy. This curve had already been studied intensively by the Greeks about two thousand years earlier. Its mathematical properties were therefore known. Whereas the circle is defined as the set of all points that are at a constant distance, the radius, from a fixed point, the ellipse can be defined as the set of all points the *sum* of whose distances from two fixed points is constant. Thus, if F_1 and F_2 are the fixed points (fig. 40) and P is a typical point on the ellipse, the sum $PF_1 + PF_2$ is the same no matter where P may be on the ellipse. The two fixed points F_1 and F_2 are called the foci. Kepler's first law states that each planet moves along an ellipse and that the sun is at one of the foci. The other focus is just a mathematical point at which nothing physical exists. The motion of *P* around the ellipse represents the motion of a planet around the sun. Of course each planet moves on its own ellipse, of which one focus, the one at which the sun exists, is the same for all the planets. Thus, the upshot of fifteen hundred years

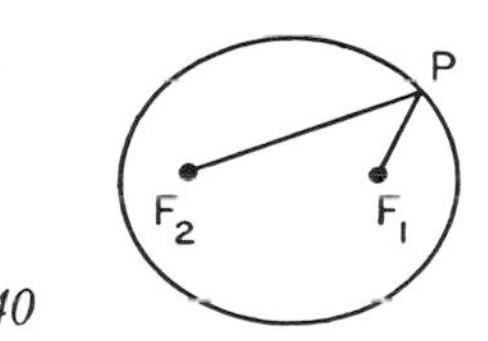

Fig. 40

of attempts to use combinations of circles to represent the motion of each planet was the replacement of each combination by a simple ellipse.

Kepler's first law tells us the path pursued by a planet but it does not tell us how fast the planet moves along this path; if we should observe a planet's position at any particular time, we would still not know when it will be at any other point on that course. One might expect that each planet would move at a constant velocity along its path but observations, the final authority, convinced Kepler that this was not the case. Kepler's second discovery states that the area swept out by the line from the sun to the planet is constant. That is, if the planet moves from *P* to *Q* (fig. 41) in one

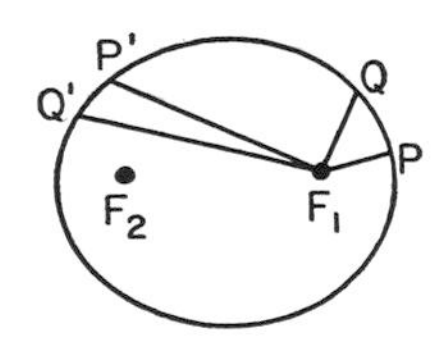

Fig. 41

month, say, and from P' to Q' in the same time, then the areas F_1PQ and $F_1P'Q'$ are the same. Kepler was overjoyed to find that there was a readily statable mathematical law of planetary velocities. Apparently God preferred constant area to constant speed.

Kepler's third law was the culmination of his long search for a relationship among the distances of the planets from the sun. Since each planet's distance from the sun varies during its revolution around the sun, Kepler searched for a harmonious principle based on the greatest and least distances. When the study of these distances failed to reveal God's design Kepler turned to a related quantity, the angular velocity of the planets. His firm conviction that the mathematical laws of nature expressed a harmony incorporated by God now led him into one of the strangest flights of reasoning in the whole history of science. Harmony meant to Kepler literally some musical harmony. He therefore associated a note with the daily angular velocity of a planet and since the angular velocity of a planet varied from a least to a greatest value during the period of revolution, this range of velocities corresponded to a musical interval. The musical intervals of the several planets, he then argued, must belong in different registers that should depend upon the planets' mean or average distances from the sun (fig. 42). His next

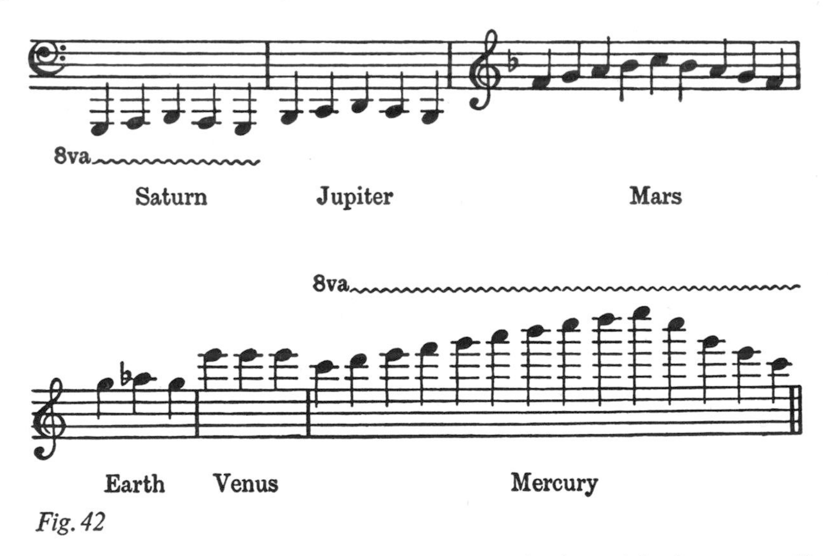

Fig. 42

thought was to relate the mean angular velocity with the mean distance from the sun. By substituting the period of revolution for the mean angular velocity Kepler finally found a law that worked and allowed him to show how the music of the planets produced a concert. Of all this work science has chosen to retain only the purely mathematical law.

Kepler's third law states, in symbols, that $T^2/D^3 = K$, where T is the period of revolution of a planet, D is its mean distance, and

K is the same constant for all the planets. Since the earth's mean distance from the sun is 93,000,000 miles and the time of revolution is one year, we can substitute these values of D and T in the law and determine K. Then the law obviously can be used to compute a planet's mean distance from the sun if one knows its period of revolution, or the other way around.

No doubt Kepler would have preferred to find some relationship among the distances themselves of the planets, but the result he did obtain overjoyed him so much that after stating it in his book *The Harmony of the World* (1619), he broke forth into a paean to God:

> *The wisdom of the Lord is infinite; so also are His glory and His power. Ye heavens, sing His praises! Sun, moon, and planets glorify Him in your ineffable language! Celestial harmonies, all ye who comprehend His marvelous works, praise Him. And thou, my soul, praise thy Creator! It is by Him and in Him that all exists. That which we know best is comprised in Him, as well as in our vain science. To Him be praise, honor, and glory throughout eternity.*

Because we are today taught to accept the heliocentric theory and Kepler's laws we no longer appreciate the full significance of Copernicus' and Kepler's achievements. It will profit us to go back for a moment to survey the setting in which these men worked and to see what their mathematics really accomplished.

We should recall that Copernicus and Kepler worked in the sixteenth and early seventeenth centuries. The geocentric view had been in force since the days of Ptolemy and fitted in very neatly with the thoroughly entrenched religious doctrines. The earth was at the center of the universe and man was the central figure on the earth. He was the apple of God's eye, and for him the sun, moon, and stars had been specially created. The heliocentric theory denied this basic dogma. It implied that man was an insignificant speck of dust on one of many whirling globes. Was it likely, then, that he was the chief object of God's ministrations? The new astronomy also destroyed Heaven and Hell, which had very reasonable geographical locations in the geocentric view.

Early in the sixteenth century the Copernican theory was accepted even by some religious leaders as true in philosophy though false in theology, a distinction such as modern man might make between a convenient hypothesis and truth. However, alarmed by the Protestant revolution and sensing the threat of any doctrine that conflicted with its teachings, the Church insisted once more on the literal interpretation of the Scriptures, as it had in the medieval period, and denounced the heliocentric heresy. Nor were the Protestants more receptive to the new astronomy. They had based their revolt on the to them unwarranted Catholic deviations from the Scriptures, and they now objected to Copernicanism as contrary to the holy writings. Luther said of Copernicus, "The fool will upset the whole science of astronomy, but as the Holy Scripture shows, it was the sun and not the Earth which Joshua ordered to stand still." Many other passages in the Scriptures were cited against Copernicus. In 1616 the Copernican doctrine was declared heretical by the Roman Inquisition; in 1620 this organization forbade all publications teaching the doctrine. In 1633 Galileo's famous book, *Dialogue on the Great World Systems,* was also put on the *Index of Prohibited Books.* The Church's ban on such writings and teachings was not lifted until 1822.

Though Copernicus and Kepler were deeply religious men they were not tied to literal interpretations of holy writings. They believed their new doctrines to be part of God's design of the universe and therefore wholly acceptable. Copernicus made clear his willingness to battle narrow-minded theologians and professors. In a letter to Pope Paul III he wrote, "If perhaps there are babblers who, although completely ignorant of mathematics, nevertheless take it upon themselves to pass judgment on mathematical questions and, improperly distorting some passages of the Scriptures to their purpose, dare to find fault with my system and censure it, I disregard them even to the extent of despising their judgment as uninformed." Kepler, too, ignored the conservative religionists to the extent of finding himself out of favor with Catholics and Protestants.

Copernicus and Kepler had to contend with more than conservative religionists. There were serious scientific objections to the new theory. One of these was a basic astronomical argument. If the earth was in motion, then the direction of the stars, which seemed to be fixed in the heavens, should be different from different parts of the earth's orbit. But, as we noted earlier in connection with Bradley's work, this difference was not observed in the sixteenth and seventeenth centuries. Hence the reality of the motion of the earth was indeed questionable. Copernicus disposed of this question by stating that the distance of the stars is immense in comparison with the orbit of the earth. This answer did not appeal to his contemporaries; they argued that the distances necessary to make the change in direction insensible were inconsistent with the fact that the stars were clearly observable.

Copernicus' explanation, of course, proved to be the correct one, though even he would have been astonished to hear some of the modern figures for the distances of the stars from the earth. The change in direction of the stars when viewed from one position on the earth's orbit as opposed to another was first measured by the mathematician Bessel in 1838 and proved to be 0.76″ for the nearest star.

The astronomical argument was taken seriously by only a few specialists, but there were other valid scientific objections to a moving earth that every layman appreciated. Neither Copernicus nor Kepler could explain how the heavy matter of the earth could be gotten into and kept in motion. That the other planets were also in motion even under the geocentric view did not disturb people who believed the heavenly bodies were of light stuff and therefore readily moved. About the best answer that Copernicus could give was that it was natural for any sphere to move. Equally troublesome was the objection, why did not the earth's rotation cause objects on it to fly off into space, just as an object whirled at the end of a string tends to fly off? Indeed, Ptolemy had rejected the rotation of the earth for this very reason. Further, why did not the earth itself fly apart? To the latter question Copernicus replied that since the motion was natural it could not have the effect of destroying the body; he also gave some vague arguments about gravity. Then he countered by asking why the skies did not fall apart under the very rapid daily motion presumed by the geocentric hypothesis. Entirely unanswered was the objection that if the earth rotated from west to east, an object thrown up into the air should fall back to the west of its original position. Again, if, as practically all scientists since Greek times believed, the motion of an object was proportional to its weight, why did not the earth leave behind objects of lesser weight? Even the air surrounding the earth should have been left behind. Though he could not account for the fact that all objects on the earth moved with it, Copernicus "explained" the continued presence of the atmosphere by arguing that the air was earthy and so rotated in sympathy with the earth. All of the foregoing scientific objections to the new heliocentric theory were genuine and arose out of the fact that the age still accepted Aristotelian physics. What was needed, and was subsequently created by Galileo and Newton, was a totally new theory of motion.

Every layman had another and most reasonable argument against the new heliocentric theory in that he did not feel either the rotation or the revolution of the earth, despite the fact that the theory had him rotating through about $3/10$ of a mile per second and revolving around the sun at the rate of about 18 miles per second. On the other hand, the observer apparently did see the motion of the sun. To the famous astronomer Tycho Brahe, these and other arguments were conclusive proof that the earth must be stationary.

In view of the variety of weighty objections offered to the new ideas, Copernicus' willingness to pursue them is one of the enigmas of history. Almost every major intellectual creation is preceded by decades and even centuries of ground-breaking work that, in retrospect at least, makes the decisive step appear to be natural. But Copernicus had no immediate scientific predecessors, and his sudden adoption of the heliocentric view, despite the unquestioned acceptance for fifteen hundred years of the geocentric view, today seems decidedly unnatural. In the company of the other astronomers of the sixteenth century Copernicus stands forth as a colossus.

It is true that Copernicus had read the Greek works and noted that many of them advanced the idea that the earth was in motion, but none had attempted to work out a mathematical theory on this basis, whereas the geocentric theory had been intensively developed. Nor did Copernicus' observations suggest that something radically new was called for. His instruments were as crude as those of his predecessors and his observations no better. He was disturbed by the complexity of Ptolemaic theory, which by his time had become entangled in many more epicycles to make the theory fit Arabian and European observations. In the magnificent dedication of his book to Pope Paul III, Copernicus remarks that he was first induced to seek a new theory when he found that mathematicians were arguing among themselves as to the soundness of Ptolemaic theory. Nevertheless, historically, the appearance of his work is as surprising as a mountain suddenly rising from a calm sea.

One can understand the motivation and deeds of Copernicus and Kepler only by comparing them with missionaries who, fired by religious zeal, invade unknown lands to convert hostile and savage tribes. Indeed, religious convictions of a special sort account for the direction of their work. Copernicus and Kepler were convinced that the universe was a systematic, harmonious structure framed on the basis of mathematical principles. Their religious beliefs also assured them that God so designed the universe. The glimpse of a new possibility that might reveal the greater grandeur of God was sufficient to arouse them and to fire their thoughts. The results of their efforts satisfied their expectations of harmony, symmetry, and design in the divine workshop. The very simplicity of the new theory was proof that God would have used it in preference to a more complicated design. The appeal of this symmetry and simplicity is clear in Copernicus' own statement: "We find, therefore, under this orderly arrangement, a wonderful symmetry in the universe, and a definite relation of harmony in the motion and magnitude of the orbs, of a kind it is not possible to obtain in any other way." It so happens that Ptolemy had asserted, as do modern scientists, that in explaining phenomena it is necessary to adopt the simplest hypothesis that will coordinate the facts. Copernicus turned this very argument against Ptolemaic theory when he introduced the simpler hypothesis; but Copernicus, because he believed the universe to be the work of God, interpreted the simplicity he had found to be the true design. Since Kepler improved on the work of Copernicus, he had all the more reason to believe that he had found the very laws that God had incorporated in the construction of the universe.

The work of Copernicus and Kepler is the work of men searching the universe for the harmony their commingled religious and scientific beliefs assured them must exist, and in aesthetically satisfactory mathematical form. To men who were convinced that an omnipotent being, designing a mathematical universe, would certainly prefer the superior features of the new theory, there could be only one conclusion—the heliocentric theory was true. Kepler said of the theory, "I have attested it as true in my deepest soul and I contemplate its beauty with incredible and ravishing delight."

There was also a mystic element in their thinking that now seems anomalous in great scientists. The inspiration to conceive and carry out a heliocentric theory may have come from some vague and even primitive response to the power of the sun. Copernicus wrote, "The earth conceives from the sun and the sun rules the family of stars." He reinforced this argument by the statement, "For who could in this most beautiful temple place this lamp in another or better place than that from which at the same time it can illuminate the whole." Copernicus also argued that the motion of the heavenly bodies must be uniform and circular, or a composition of circular motions, since only a circle could bring a body back to its original position. Moreover, any variation in the velocity of the motion could be caused only by a change in the motive power (God) or by a variation in the body moved, both of which assumptions were absurd.

Kepler, too, placed great emphasis on the sun as the heart of the universe and tried to devise means whereby the sun, which he regarded as a living force, moved the planets around. He imagined lines of force stretching out from the sun to each planet. As the sun rotated it supposedly moved these lines around as though they were spokes attached to the hub of a wheel, thereby forcing the planets to move. The different resistances of the planets accounted for their different velocities. Shortly after he heard of William Gilbert's experimental demonstration that the earth is a magnet, Kepler tried to account for the planetary motions by supposing the sun to be a magnet that attracted each of the planetary magnets.

But despite the presence of mystic, poetic, and religious influences, Copernicus and Kepler were thoroughly rational in rejecting any speculations or conjectures that did not agree with observations. What distinguishes their work from medieval vaporizings is not only the mathematical framework of their theories but their insistence of *making the mathematics fit reality*. In addition, the preference they showed for a simpler mathematical theory is a thoroughly modern scientific attitude.

One must also admire the intellectual breadth and courage each exhibited. The Greek astronomers had to break with the prevailing religious anthropomorphism of gods who ordered the affairs of the universe from day to day according to their own human-like wills and desires, and who controlled the destinies of the heavenly bodies, thunder, lightning, the winds over the seas, and the lives of men. Instead, the Greek scientists substituted a purely rational and impersonal pattern, which, though instituted by God, remained inviolable thereafter. It was as if these early thinkers lived in a different culture. In the case of Copernicus and Kepler, the idea that the planets followed an invariable pattern that could be mathematically described was no longer new, but their willingness to think in terms of, and indeed their preference for, a heliocentric theory, despite the arguments we have cited against the reasonableness of such a view in their time, marks them as heroes in the intellectual battles man has fought to attain a secure dwelling place for science.

Despite the weighty scientific arguments against a moving earth, despite the religious and philosophic conservatism, and despite the affront to common sense, the new theory gradually won acceptance. Mathematicians and astronomers were impressed, especially after Kepler's work, with the simplicity of the new theory. Also, it was far more convenient for navigational computations and calendar reckoning, and hence many geographers and astronomers who were not convinced of its truth began to use it nevertheless.

Equally helpful were the observations Galileo made with the newly invented telescope. He noted four moons revolving around Jupiter (twelve are now known). This observation was enormously significant for it showed that planets other than the earth had bodies revolving around them in space; it followed that it was equally likely for the earth to be moving and yet have a moon revolving around it. Evidently, too, there were more than seven moving bodies in the heavens, a number that had been defended as inviolable on Scriptural grounds. Galileo also saw mountains on the moon, spots on the sun, and a bulge around the equator of Saturn, which we now call the rings of Saturn. These observations destroyed the belief that the heavenly bodies, unlike the earth, were perfect bodies. The telescope revealed that the Milky Way was a swarm of stars that gave off light like our sun. Hence the belief in a unique world—our own system of sun, moon, and planets revolving around the earth—became questionable. All of these observations did not prove that the heliocentric theory was correct but they served to challenge and ultimately destroy older convictions. Galileo was himself sufficiently impressed with Copernicus' work—he ignored Kepler's for reasons that are not clear—to argue most effectively for it in his *Dialogue on the Great World Systems,* in which he gave numerous arguments in favor of a moving earth.

Since the acceptance of the heliocentric theory was gradual, it is difficult to say which arguments and forces were most decisive. Perhaps the view expressed by the modern physicist Max Planck is most applicable here. In referring to the inability of people to cast off old ideas in favor of new ones, Planck said that new ideas win acceptance because the opponents gradually die off and young people, attracted by and more receptive to strange theories, investigate and adopt them. The same thought was expressed more sarcastically by a Harvard professor: "People now accept the undulatory theory of light because all those who formerly accepted the corpuscular theory are now dead."

The gains that resulted from the introduction of the heliocentric theory were great. The practical value of the new theory in more readily locating the positions of ships at sea by means of the positions of the sun, moon, and planets and its value in calendar-reckoning have already been noted. But these values are dwarfed by the larger understanding it has given us of nature's behavior. Though Ptolemaic theory had already afforded to the world considerable evidence for the regularity, lawfulness, and intelligibility of nature's behavior, the heliocentric theory was far more impressive. The necessity to complicate Ptolemaic theory by the addition of epicycles in order to make the theory fit the increasing and increasingly accurate astronomical observations had cast some cloud over the light this theory had shed. The simplicity and ready intelligibility of Kepler's work was a much more convincing demonstration of nature's mathematical design. The ability of modern scientists to predict astronomical events, such as eclipses, to the fraction of a second forcefully drives this conviction home to us.

The belief in the lawfulness and intelligibility of nature's behavior has implications far beyond the motions of the heavenly bodies. Medieval Europe had subscribed to a number of superstitions. Sickness was a punishment of God. Plagues and earthquakes were expressions of God's anger. People could be infected by the Devil and become witches, and such demons had magical powers or could exercise evil influences on others. Kepler's own mother was accused of sorcery and thrown into prison. She was suspect because she had never been seen shedding tears and would not look other people in the face. Under the threat of torture she was asked to confess deeds that, if admitted, would have led to burning at the stake. In her defense Kepler dared not attack the belief in witchcraft but asserted only that she was innocent.

Even in modern times one finds similar superstitions in areas not yet penetrated by science. During an eclipse of the sun that was visible over Ceylon on June 20, 1955, many Hindus drank a medicine specially prepared for the occasion, believing it would grant them good health, beauty, and wisdom. Some six thousand of them ended up in hospitals, but the medicine will doubtless be taken again during the next eclipse.

Most primitive societies have associated the planets with their gods and assumed that these gods determined the daily events on earth. Hence, the study of the planets to fathom and predict the intentions of gods has long been an institution. Within the medieval European world this "science" of astrology flourished. Its principles were as seriously accepted and applied as any modern scientific doctrine. Every medieval and Renaissance court had a royal astrologer who advised the duke or prince he served. The astrologers foretold for these men the outcomes of battles and campaigns. Men such as Roger Bacon, who even in the thirteenth century was a clear and outspoken champion of the experimental method in science, and Jerome Cardan, one of the foremost mathematicians and physicians of the sixteenth century, subscribed to astrology. Kepler himself was not only required to cast horoscopes but, in his youth at least, seemed to believe in them. Even the mathematical books that Renaissance Europe acquired from the Arabs contained applications to astrology. Since the gods caused and cured illnesses, doctors were required to "read the stars." Mathematics and medicine were in fact closely connected through astrology. During the twelfth century the University of Bologna had a school of Mathematics and Medicine, and Galileo lectured on astrology to medical students. It is significant that during the seventeenth century laws were passed in England and France forbidding the insertion of astrological prophecies into almanacs.

The heliocentric theory utterly destroyed the supposed scientific basis for astrology and the superstitions associated with astronomical phenomena. The earth itself was one of the planets. Clearly, then, the powers assigned to the other planets and stars were fictitious. The new theory also destroyed the doctrine, taught by Aristotle and accepted and embellished by the medieval world, that all objects below the moon were imperfect, heavy, and subject to decay whereas the heavenly bodies were perfect, composed of some light ethereal substance, and unalterable. The whole material universe was now seen to behave in the same way. Galileo, in his *Dialogue on the Great World Systems,* presented precisely this doctrine of the uniformity of the workings of nature throughout the universe.

No less significant was the influence of the work of Copernicus and Kepler on subsequent scientific thought. A new science of motion was needed to account for objects staying on a moving earth. The study of motion became the central one in the seventeenth century, and we shall see how Galileo and Newton succeeded in re-establishing it on a totally new basis.

But there were larger implications for science. Because Copernicus and Kepler believed in the reality of the earth's motion and therefore in the truth of the heliocentric theory, they asked man to accept a theory that violated his sense impressions. We do not feel the rotation and revolution of the earth and yet the new theory asks us to reject the evidence of our senses and accept the motions because one can thereby obtain a more satisfactory mathematical theory. What Copernicus and Kepler were asking, then, was to let reason and mathematics be the determining factor in accepting what is true of nature. Modern science follows the behest of Copernicus and Kepler in that it constructs vast theories that are largely mathematical and make no appeal to the senses or are unintelligible in terms of sensory or intuitive perceptions. Thus, the theory of relativity describes the universe of space and time in terms of a four-dimensional non-Euclidean geometry that we cannot visualize at all. Modern atomic theory describes fundamental particles, such as the electron, as a something being both wavelike and yet localized as a solid bit of matter, both of which classes of properties are unified and treated in a mathematical theory.

Of course these theories have something to say about our sense perceptions. They are based to start with on principles derived from observation and experimentation, and they make predictions that can likewise be checked by observation or experimentation. But between the fundamental principles and the predictions lie extensive mathematical theories that dictate the interpretation of the phenomena involved. So it is with the heliocentric theory. Observations made on the earth, which for this purpose can be regarded as stationary, furnish the starting point. However, the mathematical theory built on these data requires us to think in terms of the earth and all the other planets moving around the sun. Mathematical reasoning about the positions and speeds of heavenly bodies then tells us where we should look from our "stationary" earth to detect an eclipse, say. The prediction is confirmed by direct sensory evidence.

There is one more implication that modern science has perceived in the work of Copernicus and Kepler. The same observational data that Hipparchus and Ptolemy organized in their geocentric theory of deferent and epicycle can also be organized under the heliocentric theory of Copernicus and Kepler. Despite the latters' belief that the new theory was true, the modern view is that either theory will do and there is no need to adopt the heliocentric hypothesis except to gain mathematical simplicity. Reality seems far less knowable than Copernicus and Kepler believed, and today scientific theories are regarded as inventions of man. Modern astronomers might agree with Kepler that the heavens declare the glory of God and the firmament showeth His handiwork, but they now recognize that the mathematical interpretations of the works of God are their own creations.

9: THE SCIENTIFIC REVOLUTION

Beside the mathematical arts there is no infallible knowledge, except it be borrowed from them.

—ROBERT RECORDE

THE mathematics that has been created since about the year 1600 is enormously greater than that which even the Greek geniuses produced. Moreover, this newer mathematics, in intimate collaboration with science, has affected, one might say molded, the character of our modern civilization so markedly that we now recognize we live in a scientific age. Our ability to construct bridges and skyscrapers; our success in harnessing the power of water, electricity, coal, and the atom; our effective employment of light, radio waves, sound, and the many natural elements; and our understanding of the inner structure of matter and the vast reaches of the universe have already reached spectacular proportions, and the future holds forth the prospect of dwarfing the past in such attainments. To understand the reasons for the prolific mathematical activity of our era and to appreciate just what it aims to accomplish, we must examine the society that arose in Europe during the Renaissance. Insofar as mathematics and science are concerned, the fifteenth and sixteenth centuries were crucial.

It is not an exaggeration to say that a series of revolutions took place during the Renaissance that overturned the medieval civilization and initiated a totally new one. The revolution in astronomical thought and its momentous consequences have already been discussed. Others affected the social order, political organizations, religion, methods of warfare, world outlook, scientific activity, art, and indeed almost all phases of European civilization and culture.

Revolutions have causes. About 1200 the Greek treatises on mathematics, science, literature, and philosophy began to become known in Europe. In succeeding centuries these works attained a wider circulation and hence exerted more influence. The influx of Greek works was broadened when the Turks captured Constantinople, the last remaining center of Greek culture. The scholars of the Byzantine Empire sought refuge in Europe and brought with them Greek manuscripts as yet unknown to the scholar in Europe proper.

These works, avidly read by the Europeans, described a strange way of life. The very fact that another people could view nature differently and worship other gods provoked questions about the Biblical account of creation, the will of the Christian God, and the meaning and purpose of man's life on this earth. Strange ideas are likely to breed healthy doubts about one's own beliefs.

There were also rather pointed contrasts between the Greek and medieval thought and values. The Greek writers emphasized the power of reason to determine the nature of the universe and man's role in it. Medieval leaders relied upon the teachings of the Scriptures and on interpretations made by ecclesiastical authorities. The Greeks stressed freedom of the mind; the Church demanded and received conformity to the established doctrines. Deviations were labeled heresies and suppressed. The Greeks wished to understand the structure of the physical world. The Church regarded natural phenomena as unimportant and life on earth as a mere prelude to and preparation for life in heaven. The Greeks enjoyed physical life, food, sports, the development of the body, and beauty. The Church railed against these pagan physical pleasures as injurious to the soul and distracting from the preparation for salvation. The body was but a temporary home for the soul and tainted it. Clearly the Greek outlook and values challenged the long established insular order of medieval Europe.

Another signal development of the Renaissance was the growth of large-scale commerce and industry. It was the desire of the merchants to expand their trading interests that caused them to promote the geographical explorations of the fifteenth and sixteenth centuries. The ensuing discovery of America and of a route to China around Africa enlarged man's home and brought much knowledge to Europe of foreign lands, their beliefs and customs and ways of life. Catholics met Hindus, Chinese, and the American

Indians. Knowledge of new forms of plant and animal life and of new climates stimulated men's imaginations.

The spread of old and new knowledge was expedited immensely by the introduction of two improvements in the art of bookmaking. Paper, made of cotton and later of rags, replaced costly parchment. The invention of printing replaced the laborious copying of manuscripts, and thousands of copies of books could be made in less time than a dozen handwritten ones. The Greek works that had been flowing into Europe and the books written by contemporary men were spread wide. The importance of these inventions may well be gauged if we stop to think of our present-day dependence on books for formal education and the general dissemination of information.

Another happening of the Renaissance, the introduction of gunpowder in the thirteenth century to propel cannon balls, revolutionized methods of warfare. The subsequent development of muskets aided in granting the common man an importance formerly possessed only by the armored noble.

Equally important for its indirect effects on the course of history was the introduction of the compass. The Europeans learned of the existence of this instrument and began to use it for navigation. It made possible, or at least encouraged men to attempt long sea voyages and so was instrumental in bringing to Europe the knowledge that resulted from these explorations. The phenomenon of magnetism, which the compass exhibits, excited the greatest minds and was regarded as one of the three wonders of the new age, the other two being printing and gunpowder.

The power of glass lenses to magnify objects was discovered by the Europeans in the thirteenth century. Roger Bacon was so impressed by this phenomenon that he immediately conceived of some grand projects. He believed that by means of lenses the sun and moon could be made to appear over the heads of an enemy and thus frighten him into flight. He also believed that Julius Caesar used lenses to see from Gaul what was going on in England. The more practically minded lens grinders designed lenses for eyeglasses. More important, however, was the discovery made by these same lens grinders that a proper combination of lenses could form a telescope, a potent instrument, as Galileo immediately demonstrated, for the investigation of the heavens. At about the same time that the telescope was put to such dramatic use by Galileo—that is, early in the seventeenth century—the microscope, whose historical origin is obscure, brought within range the minute details of animal and plant life. Another domain of nature was thus opened up to investigation.

One more major event, the Protestant revolution, shattered the existing order in Europe. Doctrinal arguments, which had gone on throughout the history of the Church, now developed into serious intellectual schisms. The Protestant revolution was supported by the merchants, who, seeking to enjoy life on this earth, resented the asceticism preached by the Church and the Church's condemnation of usury. Many secular princes supported the Protestants, hoping to break the political and military power of the Church so that they might rule unfettered.

The Reformation as such did not liberalize thought or free men's minds. Nevertheless, when religious leaders such as Luther, Calvin, and Zwingli dared to challenge the papacy and Catholic doctrines, the common person felt encouraged to do likewise. The Protestants, as revolters, were necessarily more tolerant and so gave protection to thinkers whom the Catholic Church might have suppressed. In fact the Protestants were forced at times to maintain that variations of belief are the necessary consequences of free inquiry. Moreover, though reason was the Devil's harlot to Luther, the Protestants, who advocated a return to the literal teachings of the Scriptures, used reason to combat Catholic interpretations. Reluctantly, the Protestant leaders altered their ethical codes in favor of the rising merchant class in order to gain support. Whereas the Church opposed the practice of usury, the Protestants favored it. Luther also sided with many princes in their fight to gain power over the papacy—a dangerous step for one who merely wished to transfer power to his own brand of religion. Since people were called upon to choose between Catholic and Protestant claims, independent thinking was unintentionally encouraged.

These challenges to the existing social and intellectual order were so stirring and so grave that they were bound to produce fundamental changes. Slowly but surely the intellectual outlook was refashioned. A major positive doctrine of the Renaissance proclaimed "back to nature." Scientists began to abandon their endless rationalizing on the basis of dogmatic principles, vague in meaning and unrelated to experience, and turned to nature herself as the true source of knowledge. This appeal to nature and to observation had been urged long before by Roger Bacon, Cardinal Nicholas of Cusa, and other thinkers from the thirteenth century on. However, not only were they an uninfluential minority, but they were not at all clear as yet as to what new knowledge and principles would replace the questionable ones. But the return to nature slowly broadened from a trickle to a flood.

Most famous of the first modern scientific investigators of nature was Leonardo da Vinci. Leonardo, among others, distrusted the knowledge that the older scholars professed so dogmatically. He described these men with book learning as strutting about, puffed up and pompous, and adorned not by their own labors but by the labors of others whom they merely repeated. Leonardo almost boasted that he was not a "man of letters." It was a bigger and far more worthy thing, he indicated, to learn from experience, and he conducted extraordinary studies of the flight of birds, the flow of water, the structure of rocks, and the structure of the bones, muscles and vessels of the body.

The new school of biologists, led by Vesalius, ceased quoting Galen and other ancient scholars and looked at man himself. Vesalius asserted that the true Bible was the human body. To convince his students at Padua of the correctness of his facts, he dissected corpses and showed them what he taught. The artists, too, turned to the study of anatomy, and to perspective, light, and mechanics in order to depict man and nature accurately. Paracelsus, the Swiss mystic, chemist, and doctor, proclaimed, "Never did I write a line without experience." William Harvey, in the preface to his book *On the Movement of the Heart and the Blood,* voices the spirit of Vesalius: "I profess to learn and to teach anatomy, not from books, but from dissections; not from the positions of philosophers, but from the fabric of nature."

A second doctrine of importance for the growth of mathematics and science was the renascence of reason at the expense of faith. Though medieval theologians, particularly since the thirteenth century, had likewise attempted to found their subject on reason, faith had to supplement reason where the latter was helpless. But the Renaissance intellectuals, inspired by the Greek emphasis on reason as the path to truth, resolved to apply reason alone to the facts offered by observations and experience.

The revival of interest in nature and reason naturally brought to the fore the subject that had been pre-eminent in Greek thought. Mathematics was readopted as the medium by which reason explores nature. The conviction that nature is mathematical and that every natural process is subject to mathematical law began to take hold in the twelfth century when Europeans first obtained it from the Arabs, who in turn were quoting the Greeks. Roger Bacon, for example, believed that the book of nature is written in the language of geometry. In his day this doctrine sometimes took a rather unusual form. It was believed, for example, that the divine light was the cause of all phenomena and the essence of all bodies. Hence, the mathematical laws of optics were the true laws of nature. Leonardo da Vinci, though his knowledge of mathematics was scanty and his understanding of the nature of mathematical proof erroneous, was nevertheless infected by the new spirit. He wrote that only by holding fast to mathematics can the mind safely penetrate the maze of intangible and insubstantial thought: "No human inquiry can be called true science unless it proceeds through mathematical demonstrations." Kepler, too, affirmed that the reality of the world consisted of its mathematical relations: Mathematical laws are the cause of things being what they are. Mathematical principles are the alphabet in which God wrote the world, declared Galileo; without their help it would be impossible to comprehend a single word; one would wander in vain through dark labyrinths.

Descartes, father of modernism, related a dream that occurred to him on November 10, 1619. In it the truth stood forth clearly that mathematics was the Open Sesame! Thereafter he was convinced that all of nature was a vast geometrical system and he "neither admits nor hopes for any principles in Physics other than those which are in Geometry or in abstract Mathematics, because thus all the phenomena of nature are explained, and some demonstrations of them can be given." To the Renaissance scientist as to the Greek, mathematics was the key to nature's behavior. Thus by 1600 the Greek interest in understanding the mathematical design of nature had taken firm hold in Europe.

The re-establishment of mathematics was advocated not merely because it is a powerful instrument for the study of nature. The undermining of established theological systems and philosophies had resulted in complete uncertainty as to what was true. Keen minds sought to establish new intellectual systems that could be based on certain, cogent knowledge, and they were attracted by the certitude of mathematics. For the truths of mathematics, however much they may have been ignored in medieval times, had really never been challenged or doubted. Said Descartes, "I was especially delighted with the mathematics on account of the certitude and evidence of their reasonings. . . . I was astonished that foundations, so strong and solid, should have had no loftier superstructure reared on them."

Several positive doctrines, then, were re-established in the Renaissance. The study of nature became a major goal, and the reliance upon reason and mathematics to achieve knowledge and certitude was readopted as the most assured means. The pursuit of new goals and the use of new vehicles did lead to radical changes in the social and political order, in religious and philosophical thought, and in artistic and literary activity. Here, we must confine ourselves to those developments that affected the course and significance of mathematics. In this connection we must note some pressing scientific problems that confronted the Europeans.

The geographical explorations raised a number of scientific and mathematical problems. For example, since they involved sailing long distances out of sight of land, mariners needed accurate methods of telling latitude and longitude and of plotting courses. To use the sun, moon, and stars efficiently for these purposes a more accurate astronomy was found necessary. This need hastened the adoption of the simpler heliocentric astronomy, but by 1600 the theory was still not sufficiently developed and exploited. While the determination of latitude could be made relatively simply by noting the direction of the sun when it was highest in the sky, the determination of longitude proved to be a difficult task. One of the accepted methods in the late Renaissance was to note when the moon was on the meridian through the ship's position. Since noon was the instant when the sun was highest in the sky, the time of the moon's passage over the local meridian could be estimated with the aid of the crude portable clocks that ships then carried. This time was compared with the time at which the same occurrence was known to occur at fixed positions on land. The difference in time could be translated into a difference in longitude. Other schemes involving the angular separation of the moon and a planet were also in use. However, all of these schemes yielded no more than an approximate location, and ships not only had to travel longer distances than should have been necessary but often found themselves in dangerous waters or even on rocks. British shipping losses incurred thereby were a national problem during the seventeenth century. At that time the error in determining a position was as much as sixty miles. The rulers of Spain, France, and Portugal offered prizes for advances in navigation. In that same century England established a Commission for the Discovery of Longitude. Not only did the moon's motion have to be known more accurately but a clock was needed that would keep accurate time on a moving ship. For over a century this problem was a major one.

The introduction of gunpowder, we recall, made possible the use of cannon and muskets that could fire projectiles great distances and with high velocities. It was desirable—insofar as the study of cannon fire is desirable—to learn the paths of these projectiles, their range, the heights they could reach, and the effect of muzzle velocity. Even more important was the fact that the very principles of motion had to be reconsidered. Though criticism of Greek doctrines on motion can be found even in the late medieval period, the adoption of the heliocentric theory made it clear that a totally new science of motion was needed. In the sixteenth century we find the mathematicians Jerome Cardan and Tartaglia arguing with each other on the nature of paths taken by moving objects. The latter also undertook a serious study of projectiles, which he incorporated in his *The New Science of Artillery*. A little later Galileo undertook the same study and, as we shall see, with more success. Incidentally, he also taught the new art of fortifications to noblemen.

The importance of these military problems in stimulating mathematical research was, of course, all out of proportion to their general scientific importance. But the princes then, as nations now, were ready to spend great sums on their solution. The needs of war have always aroused nations to put forth money and efforts unimaginable during times of peace. No better example of the readiness of nations to spend money on military problems can be found than the one furnished by our own times.

The study of light had been a continuing one since Greek times, but the new appreciation of the power of lenses to bring the distant heavens and the tiniest particles into the range of the human eye stimulated an expanded interest in light and in the scientific design of lenses. We shall find almost all the major mathematicians of the next few centuries engaged in the study of optics and in the application of mathematics to that branch of physics. In the seventeenth century alone one encounters such great names as Kepler, Descartes, Fermat, Barrow, Huygens, Newton, and Leibniz devoting a large part of their scientific effort to optics.

The Renaissance witnessed the transformation of Europe from a society of nobility and serfs to a society of merchants, industrialists, free laborers, and artisans. These men engaged in a variety of arts, crafts, and technical pursuits, the roster of which seems unbelievably large in view of common impressions as to the state of Renaissance Europe. Mining for coal and metals, metalworking, the making of glass and pottery, weaving, dyeing, brewing, printing, map-making, and the manufacture of gunpowder and armaments were but a few of the industries that proposed technical problems. The artists, architects, builders, jewelers, enamelers, lens designers, navigators, druggists, and doctors raised problems peculiar to their activities. Water power, wind power, and air pressure, exploited more systematically and more fully, brought new questions to men's minds.

Not only was the variety and extent of these activities greater than that found in other civilizations, but the merchants and industrialists sought to improve their products either to meet competition or to increase profits. Independent artisans, in contrast to serfs and slaves, naturally showed greater interest in the quality of their work and in labor-saving devices. As a consequence, all of these groups showed increasing interest in materials and machinery, and in natural phenomena that could be employed to advantage.

The ensuing investigations of nature raised further doubts as to the soundness of medieval explanations of physical happenings. The purposive explanations that had satisfied medieval man—such as that the essence of wood is to provide fuel for fires, which in turn serve to warm man and his food—did not help to make better products out of raw materials or to design and operate machinery. Direct studies of physical phenomena and of causal connections were demanded and pursued.

The vital problems facing the awakened and changing civilization in the dawn of modern times were almost certain to bring about new mathematical and scientific research. But it is questionable whether these problems alone would have produced the highly successful alliance of mathematics and science and the consequent effects on modern civilization. Other factors were operative whose significance becomes clear if we compare our period with the one other period highly active in mathematics and science, the Greek period. The Greek of the classical and Alexandrian periods investigated nature and not only employed mathematics in this endeavor but, as we know, regarded mathematics as the key to the structure of nature. Hence, among the Greeks, too, enormous intellectual activity and close cooperation between mathematics and science took place. Yet this civilization could hardly be called a scientific one or be said to have been dominated by mathematics and science.

There is a story told about Thales, the father of Greek philosophy and mathematics, that points up the essential differences between the Greek and modern European civilizations. While taking a walk one evening with a member of the fair sex Thales became so engrossed in the study of the heavens that he stumbled and fell into a ditch. His companion, no doubt piqued by Thales' preference for some of the beauties of nature at the expense of others, said sharply, "How can you expect to learn anything about the stars if you do not know what is going on at your feet?"

The figure of Thales is symbolic in that he typifies the Greek attitude toward mathematics. Mathematics was the concern of philosophers. They valued the subject because it was the ideal preparation for philosophy, it granted understanding of the structure of the universe, and it was delightful food for minds that enjoyed intellectual challenges. But the intellectual life of the thinkers of Greece had no connection with the daily life and problems of the populace at large. While the philosophers talked about the "great visible gods," as Plato called the planets and stars, the ordinary people accepted the charming but purely fanciful mythology that is familiar to us through their art and literature. The technical pursuits and crafts were practiced by slaves. Slaves even performed medical services. Of course these humble men, like the builders of the Gothic cathedrals in medieval society, were ignorant of science and philosophy. As a matter of fact, Plato made the explicit distinction between the liberal arts, the arts of free men, and the mechanical arts, the arts of servants, and, as we noted elsewhere in connection with arithmetic, distinguished between the mathematics needed for military problems and trade on the one hand and the mathematics that would prepare for the study of philosophy.

The transformation of society in the Renaissance stimulated the common man to participate in learning. He acquired political freedom and an economic independence that supplied incentive for improvement. The printing of books aided his quest for knowledge. Though the books were held by scholars, wealthy princes, and church leaders, the common people sensed the significance of the ideas discussed by the learned and obtained fragments of this knowledge. They heard of the wonderful arts of mathematics and science and caught glimpses of their power and usefulness.

The artisans were generally poor and could not spend years in formal learning. Nevertheless they sought knowledge, often under great hardships, and taught themselves. A few more fortunate ones were supported by their families while acquiring an education. Leonardo da Vinci was entirely self-taught. He, too, heard of the Greek works and managed to glean some ideas from them, as well as from scholars with whom he came into contact at the courts for which he performed services. But it is quite certain that Leonardo hardly mastered these works. Nevertheless, his genius enabled him to divine what might be accomplished with mathematics. Tartaglia, one of the great mathematicians of the sixteenth century, too poor to pay tuition in school, picked up a knowledge of Greek, Latin, and mathematics by himself. Some of these great minds learned enough to teach others of their own social group.

The lowly family origins of many of the great minds of the Renaissance and later times show the entry of a new group into the world of mathematics and science. Leonardo was an illegitimate child born to a peasant woman and to a notary of the small town of Vinci. Tartaglia was born to a poor widow. Copernicus was the son of a merchant, and Cardan and Pascal were the sons of local jurists. Kepler was the son of a common soldier, and Galileo, the son of a cloth merchant. Fermat was the son of a leather merchant; Isaac Barrow, Newton's predecessor as professor of mathematics at Cambridge, was the son of a draper; and Newton was born to a family of farmers. These men, stemming from families of workers, were generally in close contact with the technical needs of artisans. Having glimpsed the potential in scientific knowledge they devoted themselves to the diffusion of that knowledge to the people who could use it profitably.

One of the great barriers to the spread of the newly recovered Greek knowledge was language. The masses spoke the languages of their locales—Italian, French, German, or English—and could not read Greek or Latin. Hence the scholarly knowledge had somehow to be made accessible. Leonardo prepared a plan for a dictionary of the vulgar Italian languages and one for the popularization of the sciences. Albrecht Dürer, the sixteenth-century German artist, wrote in a classic book on mathematics and perspective that he wished to explain ideas that the learned knew and kept to themselves. He therefore not only wrote for the benefit of artists and artisans but wrote in German. To reach the masses of people Galileo wrote in Italian, and Descartes, who feared his new ideas would not be received by the scholars who held to ancient or medieval learning but would be appreciated by the masses, wrote in French. During the second half of the sixteenth century all of Plato and Aristotle was translated into the popular Italian languages. Florence set up a school of mathematical and technical studies that invited learned scientists to lecture in the popular languages.

A number of academies, such as the Florentine Academy of Design, in which Vasari, the famous sixteenth-century artist and biographer, was a key figure, were formed to bridge the gap between the ancient works possessed by the learned and the needs of the people. These academies popularized mathematics, linked theoretical knowledge and the knowledge derived from experience, and sought to relate the existing knowledge to the problems faced by artists, engineers, and artisans generally. Individuals also helped to bring the knowledge to bear on the problems of the times. Tartaglia singled out the useful problems arising in practical work and the knowledge gained empirically and tied them to the newly acquired mathematics and science. In particular he took up the dynamic problems raised by the use of cannon. During the seventeenth century the activity of relating mathematics and the knowledge and problems gained by experience became extensive. In the opening sentence of his *Dialogues Concerning Two New Sciences* Galileo acknowledged the inspiration for some of his investigations:

> *The constant activity which you Venetians display in your famous arsenal suggests to the studious mind a large field for investigation, especially that part of the work which involves mechanics; for in this department all types of instruments and machines are constantly being constructed by many artisans, among whom there must be some who, partly by inherited experiences and partly by their own observations, have become highly expert and clever in explanation.*

The growing contacts between scholars and artisans broke down the previously existing social distinction between those who reasoned and those who learned from experience or applied the arts to practical problems. The laborer felt free to investigate scientific problems while the thinker grappled with the technical problems of the artisan. Science and pure thought became the concern of businessmen, technicians, mechanics, engineers, and artists. In turn the engagement of free men in labor and the approbation of labor by the Church so dignified work that even noblemen turned to experimentation and invention without fear of social stigma.

Thus the separation and even opposition between reasoning and learning on the one side and practical needs on the other were turned into an alliance of speculative thought and application. The Greeks had shown mankind some of the vast realms and power of mathematics but had kept the subject hitched to the stars. The Renaissance retained the Greek concept of mathematics; it took over the Greek enthusiasm for mathematical study of nature. But the study of nature became not merely the explanation of the heavenly motions and the abstract laws of space but the study of mundane phenomena. By the transformation of society that took place during the Renaissance one might say that science and mathematics were brought down to earth. By the end of the sixteenth century mathematics covered not only arithmetic, geometry, music, and astronomy—all traditionally mathematical subjects since Greek times—but also astrology, goniometry (the measurement of volumes), meteorology, optics, geography, hydrography (mapping the seas), mechanics, magnetism, architecture, perspective, military architecture, and map-making. Science embraced applied science, technology, industry, the arts and crafts.

The Renaissance is often described as the discovery of man and nature. It was a rediscovery of nature through the guidance of the Greeks, but it was the discovery of man in that the common man first began to take his place among the learned and to express his interests. The significance of this movement is clear even today. Every country is now concerned with finding and educating its talented youth.

The projection of the problems of industry, the crafts, navigation, and the military into the pursuit of science and mathematics proved to be a boon to the study of science and a benefit to civilization. But one must not infer that mathematics and science thereafter concerned themselves entirely with practical problems and became handmaids to the engineers. If one reads the great works of Descartes, Galileo, or Newton he will find little that shows a relationship between the problems of technology and the scientific and mathematical subjects these men pursued. Descartes's writings on mathematics, optics, meteorology, and cosmology do not treat with practical problems. Newton's *Principia* hardly mentions the problems of navigation, and his treatment of the motion of projectiles in that book is purely academic. However, the larger directions these men pursued were determined or at least influenced by the prevailing problems of the times. Descartes's interest in optics was certainly inspired by the new and exciting uses of lenses in telescopes and microscopes, and in other writings Newton said that he was concerned with studying the moon's motion to improve the methods of determining location at sea. Moreover, these same men or their followers applied their work to the problems facing the applied arts and sciences. If one studies the broad context or framework in which these men and modern mathematicians and scientists have worked, he will see the interplay of theory and practice.

One must also keep in mind that great thinkers seek far more than the solution of practical problems. They seek to understand the workings of nature, and they use the concrete problems and experiential facts discovered by technicians and craftsmen as clues to fundamental physical principles and as suggestions for directions of research. These same experiential facts, and in more recent times elaborately planned experiments, are also used to check deductions and inferences from the theoretical work.

There is one other characteristic of the work of the great mathematicians and scientists that often baffles the student who seeks to fathom the workings of great minds. Any kind of problem, practical or otherwise, may suggest a line of inquiry, and the mathematician or scientist will pursue the idea simply because it is intrinsically interesting. Yet his results may prove to be immensely valuable to his contemporaries or to future generations. Modern mathematicians have learned the lesson taught by the Greeks. These scientists did not hesitate to pursue any ideas that seemed exciting and attractive without regard for utility. Fifteen hundred and two thousand years later their discoveries proved to be just the tools man needed for the successful resolution of important scientific problems. The classic example in elementary mathematics is the Greek work on the curves called the conic sections: the parabola, ellipse, and hyperbola. The primary Greek interest in these curves was to solve the famous but totally impractical construction problems, squaring the circle, doubling the cube, and trisecting any angle. As a consequence the conic sections were familiar to Kepler, and the ellipse supplied the answer to the paths of the planets. All three curves are now used in a host of applications. The value of purely speculative thought, of mathematical and scientific theories pursued for their intrinsic interest and with no concern about use, is now clearly established. The freedom to think along lines one chooses is not just a political right but one of the recognized values in scientific thought.

The alliance and interaction of theory and practice, then, was one of the new factors in the development of modern mathematics. There was another. The Greeks sought understanding of nature, and modern mathematicians and scientists have continued to pursue this objective. But in the seventeenth century another goal was envisaged and proclaimed, mastery of nature. The Renaissance artisan may have seen how he could earn his daily bread more readily or increase the size of the loaf by utilizing some principle of mathematics. But during that period, and even before, a few great men had already foreseen that mathematics and science were powerful enough to make nature serve man. Instead of being content to sail with the wind when it was favorable, Francis Bacon and Descartes proposed to make the winds blow in the directions man chose.

Bacon affirmed that the interrogation of nature was not to delight scholars but to serve man. It was to relieve suffering, to better the mode of life, and to increase happiness. Bacon criticized the speculative thought of the Greeks that resolved nature into abstractions. This, he said, is less to our purpose than to dissect her into parts and to put the parts to use. It is interesting to compare two ideal states, the one described by Plato in his *Republic* and the other by Bacon in his *New Atlantis*. Whereas Plato's Republic was to be ruled by philosophers, the most important institution in Bacon's ideal state was to be a college of scientific investigators who were to discover new truths applicable to the improvement of conditions of life. By establishing the reign of man over nature, Bacon foresaw that science would provide man with "infinite commodities," endow "human life with new inventions and riches," and minister to the material conveniences and comfort of man.

Though Descartes placed less emphasis on experimentation and far more on mathematics he conceived the same goal for science. The title Descartes first planned to use for his great work, the *Discourse on Method,* was "The Project of a Universal Science which can elevate our Nature to its highest degree of Perfection." Moral and material improvement of man would result from the application of science and philosophy. Descartes wrote:

> *It is possible to attain knowledge which is very useful in life, and instead of that speculative philosophy which is taught in the Schools, we may find a practical philosophy by means of which, knowing the*

force and the action of fire, water, air, the stars, heavens and all other bodies that environ us, as distinctly as we know the different crafts of our artisans, we can in the same way employ them in all those uses to which they are adapted, and thus render ourselves the masters and possessors of nature.

These men had seen one or two gold nuggets in their time and realized that there must be a gold mine from which they came. They called attention to this source of wealth and started a gold rush that not only located some lodes but drew more and more people into the search for new strikes. Scientific societies were organized in England, France, and Italy, of which the Royal Society of London and the Academy of Sciences at Paris are perhaps best known, to press for new knowledge, disseminate the information, and explore it for the good of man. As we shall see, the vision of the power of mathematics and science to master nature in man's behalf proved to be more than the outpouring of facile, imaginative minds.

10: THE WEDDING OF CURVE AND EQUATION

I have resolved to quit only abstract geometry, that is to say, the consideration of questions which serve only to exercise the mind, and this, in order to study another kind of geometry, which has for its object the explanation of the phenomena of nature.

—DESCARTES

BY 1600 the stage was set in Europe for new developments in mathematics and science, and the world awaited only great writers who would rise to the occasion and deliver the plays. It was not long before minds that matched in quantity and quality the best of the prolific Greek period studied the new scientific themes and gave fresh interpretations of the omnipresent drama of nature. The century opened auspiciously with the work of Descartes and Fermat.

It seemed clear to these men that mathematics was badly in need of new methods for working with curves. In the light of the problems besetting the seventeenth-century scientists this judgment is readily understandable. The astronomers were already making extensive use of ellipses and were investigating the more complicated curves that planetary satellites were apparently following.

The increased interest in light and in the use of lenses raised numerous problems involving curves. When light travels long distances through the earth's atmosphere it is gradually bent or refracted and hence follows a curved path. For example, when the sun is near the horizon light from the sun travels through a good deal of the earth's atmosphere and is refracted appreciably. When a person at P (fig. 43, p. 148) sees the light from the sun he judges by the direction from which the light seems to come that the sun is at S', whereas actually the sun is at S. In fact the sun may be seen even when it has passed below the horizon. Both Tycho Brahe and Kepler had noted this fact. They saw the sun even when the moon was already partially eclipsed by the earth, though the very fact that the earth blocked the sun's light meant the sun was below the horizon. Hence, observations of the sun's position and of the posi-

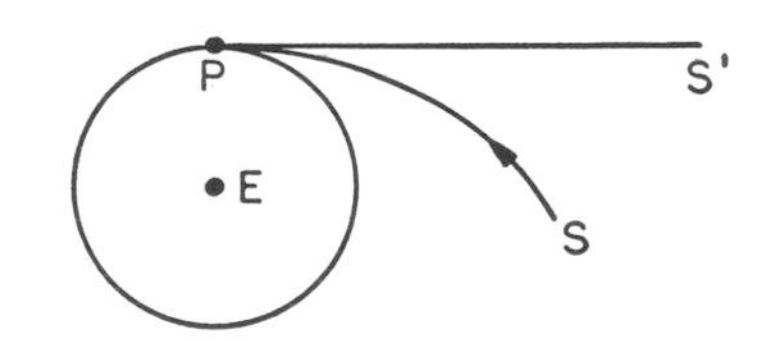

Fig. 43

tions of the moon and planets had to be corrected by knowledge of the actual path of light. The use of lenses for spectacles, telescopes, and microscopes suggested a number of problems involving curves, notably the shapes of the lenses that would focus light from objects at various distances. Both Fermat and Descartes made notable contributions to the science of light.

The use of cannon raised numerous questions concerning the paths of projectiles. What was the precise path? How should projectiles be aimed to hit a given target or clear a wall hundreds of yards away? The paths of projectiles were, incidentally, but one of the problems involving bodies in motion; the adoption of the heliocentric theory had forced the question of why all bodies rising and falling near the surface of the earth stayed with the earth. The problem of devising a reliable clock that could be taken on shipboard and used to tell longitude at sea also led to the study of curves. In the seventeenth century interest centered on the pendulum clock. The motion of the bob was therefore studied to determine its relation to the time recorded by the clock.

It may be apparent, then, that a number of urgent scientific problems of the early seventeenth century demanded a knowledge of the properties of curves. But mathematics already had a considerable body of information on this subject. Euclidean geometry proper contained hundreds of results about the straight line and circle. In addition, the Greeks had also studied exhaustively the class of curves that they called the conic sections because they were originally arrived at by slicing cones. We are all familiar with a cone-shaped figure such as an ice cream cone. If we place two such very long cones as shown in figure 44 we get what mathematicians call a conical surface, or sometimes, just a cone. When the plane cuts entirely through one nappe of the cone the curve of intersection is an ellipse (*DEF* in fig. 44). If the cutting plane is inclined so as to cut both nappes of the cone, the curve of intersection consists of two parts and is called a hyperbola (*RST* and *R'S'T'*). If, finally, the cutting plane is parallel to one of the generators of the

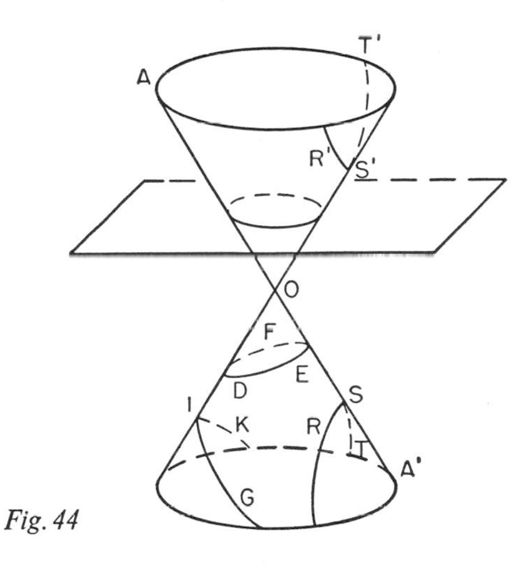

Fig. 44

cone, such as *AOA'*, the intersection is called a parabola (*GIK*). The basic facts about the conic sections were collected and organized by Euclid in a book that has not come down to us. A little after Euclid's time another famous mathematician, Apollonius, wrote a treatise on the subject, most of which is extant. Apollonius is almost as famous for this treatise as Euclid is for his *Elements*. Indeed, as we have already noted, Kepler was able to establish his three laws of planetary motion only because the Greeks had supplied so much information about the conic sections.

In view of the immensity of the Greek knowledge available to seventeenth-century Europe why did Fermat and Descartes decide that mathematics needed new methods of working with curves? The reasons were stated by Descartes. He complained that the geometry of the Greeks was so much tied to figures "that it can exercise the understanding only on condition of greatly fatiguing the imagination." Descartes also appreciated that the methods of Euclidean geometry are exceedingly varied and specialized. Each theorem requires a new type of proof, and much effort and ingenuity were applied to find such proofs. Moreover, the proofs of the important theorems about circles are very complex. To find the area of a circle Euclidean geometry uses inscribed and circumscribed polygons of larger and larger numbers of sides, and the entire argument leading to the familiar conclusion about the circle is quite involved. However, this argument is simple compared to what is needed to treat the conic sections and the few other curves explored in Greek geometry.

There is another limitation inherent in all geometry. One might indeed determine by some geometrical argument what type of curve a projectile follows when shot from a cannon, and one might prove some geometrical facts about this curve, but geometry could never answer such questions as how high the projectile would go or how far from the starting point it would land. The seventeenth century sought the quantitative or numerical information needed for practical applications, and such information is provided by algebra.

Fortunately for the world, a great deal of progress had been made in algebra during the latter half of the sixteenth century and the early part of the seventeenth. Cardan, Tartaglia, Viète, and Descartes and Fermat themselves had extended the theory of the solution of equations (see chapter 5), had introduced symbolism, and had established a number of algebraic theorems and methods. And both Descartes and Fermat, working independently of each other, saw clearly the potentialities in algebra for the representation and study of curves.

Descartes, born in 1596, distinguished himself in philosophy, science, and mathematics. His search for method in geometry was but part of his search for reliable and effective methods of obtaining knowledge in all fields. During his youth he had received a typical school education at the Jesuit College of La Flèche but he was entirely dissatisfied with what he had been taught of logic, philosophy, history, theology, and the other standard school subjects. What bothered him was that no one doctrine in all these fields seemed beyond dispute. Mathematics, however, had established conclusions that were universally accepted and had withstood the test of time. Hence mathematical method attracted him as the sure road to truth in all fields. He began his serious work in philosophy by rejecting all the knowledge that he had been taught and that was accepted by educated people of his time. He said that in order to seek truth it was necessary once in the course of one's life to doubt as far as possible all things. He then began a new construction of philosophy by the deductive method, based on axioms that to him seemed clear and self-evident.

While Descartes's reconstruction of philosophy is one of the landmarks of modern thought, what is relevant here is his general concern for method and his success in introducing method in geometry by means of algebra. As an appendix to his *Discourse on Method* he published his *Geometry*.

Pierre de Fermat, the other celebrated founder of the subject now called analytical geometry, but better termed coordinate geometry, was equally interested in instituting method. But his life and his approach to mathematics were quite different from those of Descartes. As a youth Descartes had pursued an adventurous life as a soldier and man of the world in that most worldly city of Paris; he then secluded himself in Holland for twenty years to study and write. Fermat's life was more conventional, at least in the sense that he prepared for the law at the University of Toulouse and served as a lawyer throughout his life. He married at the age of thirty and fathered five children. He ignored the larger problems of philosophy and relaxed with mathematics. Whereas Descartes's contributions were as great but spread over many fields, Fermat devoted himself entirely to mathematics and science. In addition to being one of the two founders of coordinate geometry and to sharing honors with Pascal in founding the theory of probability, he founded the modern theory of numbers, made contributions to algebra and to the calculus, and created a fundamental principle of optics that still bears his name. All of this original work he accomplished in his spare time.

Since Fermat had also participated in the advancement of algebra, he, too, became aware of the potentialities in algebra for the investigation of geometry. Though there are some differences in what the two men created we shall ignore these and examine their common central idea. Our version will be a somewhat modernized account of both men's work.

Their basic thought was that algebra should be used to characterize any curve and as the means of deducing facts about the curve. Now algebra deals with numbers and operations with numbers. Hence, in some way, numbers had to be brought into the picture. This step proved to be simple. The idea of locating points by numbers was old. For example, Hipparchus had introduced latitude and longitude to locate points on the surface of the earth, and today latitude and longitude are essentially numbers that express position in relation to the equator and the meridian of longitude through Greenwich, England. In place of this system of associating numbers with points on the surface of a sphere Descartes and Fermat, who were interested in curves lying in a plane, introduced two perpendicular lines or axes, and agreed to represent any point

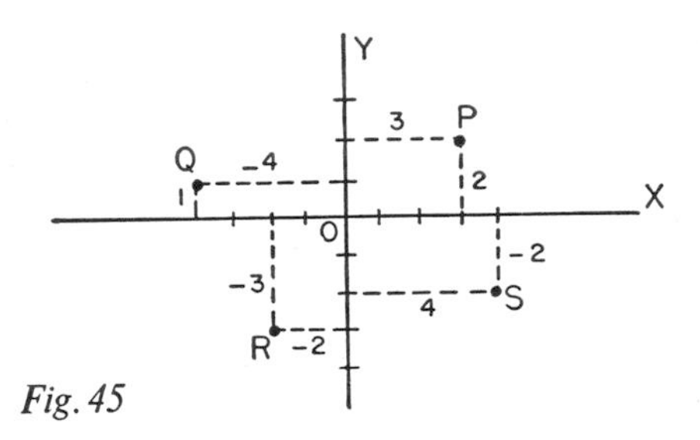

Fig. 45

in the plane by its distances from the two axes. Thus the point P in figure 45 is 3 units from the Y-axis and 2 units from the X-axis. Alternatively we may say that one reaches the point P by starting from O, called the origin, and traveling 3 units to the right in the direction of the X-axis and 2 units upward in the direction of the Y-axis. The location of the point P is then uniquely specified by the numbers 3 and 2. These numbers or coordinates of P are generally written (3,2). The first coordinate is called the abscissa of P and the second, the ordinate.

To distinguish points to the right of the Y-axis from those located to the left, distances to the left are represented by negative numbers. Thus to reach the point Q of figure 45 one proceeds 4 units to the left in the direction of the X-axis and 1 unit upward in the direction of the Y-axis. The coordinates of Q are therefore —4 and 1, and are represented as (—4,1).

The distinction between upward and downward is likewise made by using positive and negative numbers; as a consequence, the coordinates of R in figure 45 are —2, and —3, and the coordinates of S are 4 and —2.

Thus far Descartes and Fermat had a simple scheme for representing the position of any point in the plane by means of numbers, these numbers being distances from two arbitrarily chosen but fixed axes. To each point there corresponds a unique pair of numbers and to each pair of numbers a unique point. But of course their objective was to study curves algebraically. Let us therefore consider a curve, say a circle, and let us place this circle on a set of axes so that we may associate coordinates with each point. For simplicity let us suppose that the center of the circle is at O (fig. 46).

Now the circle is no more than a collection of points, namely those points that are at the same distance, say 5 units, from the center. The quantity 5 is, of course, the length of the radius. Since each point on the circle has a pair of coordinates attached to it, the problem of characterizing the circle algebraically seems to amount to the question, What property or relationship do the coordinates of points on the circle possess that distinguishes them from other points in the plane? If we consider a typical point P on the circle, and let x and y represent its two coordinates, then we see that the lengths x, y, and 5 form a right triangle. According to the Pythagorean theorem of Euclidean geometry

$$x^2 + y^2 = 5^2. \tag{1}$$

This same statement holds also for points on the circle such as Q.

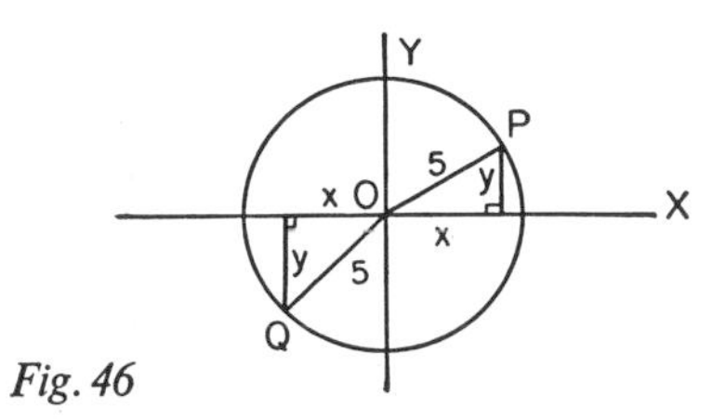

Fig. 46

Even though the coordinates of Q are negative, their squares are positive and so obey equation (1). Hence we have in equation (1) an algebraic statement that holds for each point on the circle and for no other points. This equation says that the square of the first coordinate of any point of the circle plus the square of the second coordinate equals the square of the radius.

It is important to observe the special use of the symbols x and y in equation (1). The letter x is often used in elementary algebra to represent a fixed but unknown quantity whose value one may be trying to determine. Sometimes x and y are used to represent two different but fixed unknowns whose values are then sought. But in equation (1), x and y are not unknowns. They are symbols that can represent the coordinates of *any* point on the curve. Hence x and y can take on many, in fact an infinite number of values. Thus x can be 3 and y can be 4, because $3^2 + 4^2 = 5^2$. Accordingly, the point whose coordinates are (3,4) is a point on the circle. Again x can be 0 and y can be 5, because $0^2 + 5^2 = 5^2$ and the point whose coordinates are (0,5) is a point on the circle. As another example, x can be $-3/2$ and y can be $\sqrt{91/4}$, because

$$\left(-\frac{3}{2}\right)^2 + \left(\sqrt{\frac{91}{4}}\right)^2 = 5^2,$$

that is,

$$\frac{9}{4} + \frac{91}{4} = 25.$$

Equation (1), the algebraic representation of the circle, illustrates what Descartes and Fermat were seeking. This equation singles out just those points that belong to the circle and expresses this locus in algebraic form.

To become more familiar with their idea let us examine the equations of one or two more curves. We shall consider a straight line, which can be considered as a special type of curve, making an angle of 45° with the horizontal. To obtain an equation for this straight line we again place this line on a set of coordinate axes, and for simplicity we may as well have the line pass through O (fig. 47). Euclidean geometry tells us that OQP is an isosceles right triangle and that OQ must therefore equal QP. Hence

(2)
$$y = x$$

is the relation or equation that characterizes the points of the straight line concerned.

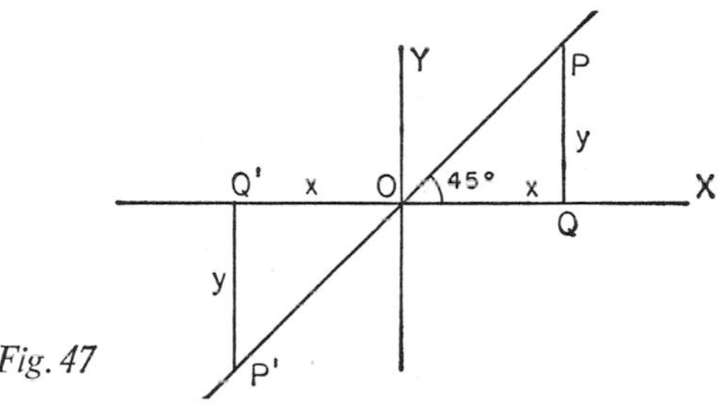

Fig. 47

It will be noted that this equation also represents points on the straight line to the left of O. The x and y values of P', for example, are both negative; nevertheless, for this and other points on the straight line it is still true that $y = x$.

Had we considered a straight line inclined more steeply to the horizontal, for example, one that rises 2 units for each horizontal

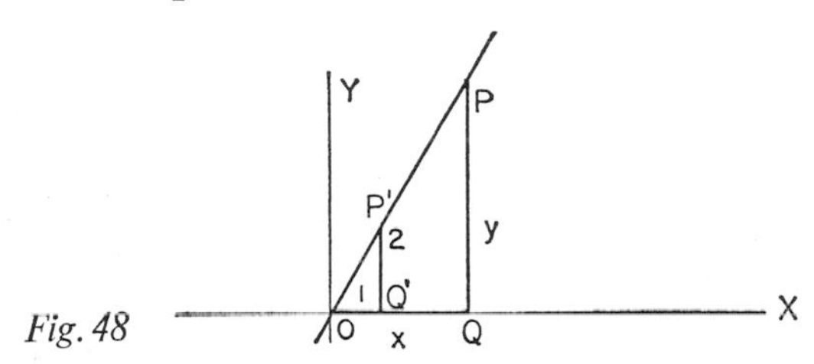

Fig. 48

difference of 1 unit (fig. 48), then from the similar triangles $OQ'P'$ and OQP we might have argued that

$$\frac{y}{x} = \frac{2}{1}$$

or that

(3)
$$y = 2x$$

is the equation of the line.

In deriving the equations of the circle and straight lines above we chose simple positions for these curves on a set of axes. But let us deliberately consider the straight line $O'P'$ (fig. 49, p. 156) parallel to the line whose equation is $y = 2x$ and lying above it. The question we raise is, What is the equation of the line $O'P'$? To answer this question we should again seek the relation between x and y which holds for any point on this line. Let P be a point on the line $y = 2x$ and let P' be a point directly above P and on the line $O'P'$. Now P and P' have the same x-value or abscissa, namely OQ. But the y-value or ordinate of P' is larger than that of P by the amount PP'. Suppose the distance PP' is 3. Since the two lines are parallel, the vertical distance between a point on the lower line

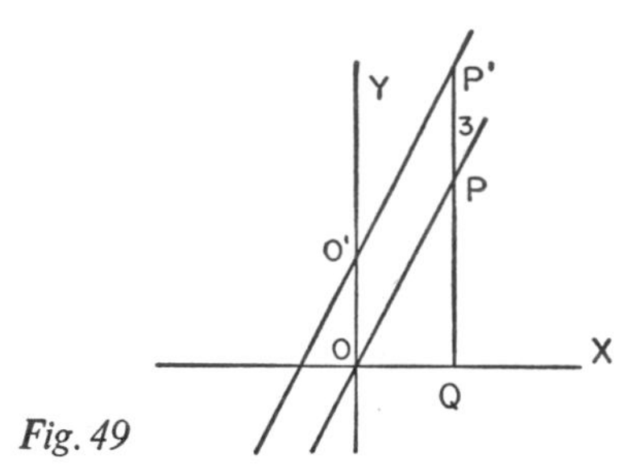

Fig. 49

and a point directly above on the upper line will always be 3. Hence, whereas the y-value of each point on the line OP is always twice the x-value, because the equation of OP is $y = 2x$, the y-value of each point on $O'P'$, for the same x-value, will be $2x + 3$. That is, the equation of $O'P'$ is

$$y = 2x + 3. \tag{4}$$

There are several significant facts to be noted about this result. The first is that though the straight line $O'P'$ is identical except for position with the straight line OP, its equation is different. The difference results from the fact that OP passes through the origin O of the system of coordinates whereas $O'P'$ does not. Hence the very same curve may have a different equation representing it if its position with respect to the coordinate axes is different. If, therefore, we should seek to identify a curve by its equation we shall have to take into account these possible differences in the equations due to differences in position.

The second fact we should note about equation (4) is the following. If a straight line having a slope of 2 can always be placed on a coordinate system so that its equation is $y = 2x$, why do we bother with a more complicated form such as $y = 2x + 3$? The answer is that if one wished to deal with two such straight lines or two curves simultaneously, and to keep these curves in the relative positions in which they may happen to be in some physical application, we could not place both of them in the identical position on the set of axes. Moreover, even when dealing with a single curve we shall find that we may wish to place it in a position with respect to the axes that suggests the physical situation more readily.

Let us summarize the preceding discussion. Any curve may have different equations depending upon its position on the set of coordinate axes. The position chosen will depend upon convenience for application or may depend upon the relative positions of the curves if more than one curve is being studied.

The curves we have considered thus far, namely, the straight line and circle, are rather readily represented by equations. How easily can we obtain equations for more complicated curves? Let us consider the parabola as an example of how the process of obtaining the equation of a curve works in the case of more complicated curves. To know a curve means mathematically to know some property that characterizes all the points of that curve. The parabola can be defined as the set of all points that are equidistant from a given point and a given line. The given point, incidentally, is called the focus, and the given line is called the directrix. Thus to treat a parabola we must start with a given point and a given line, F and d, say, in figure 50, p. 158. Then any point such as P for which PF equals PD, PD being the (perpendicular) distance from P to d, will be a point on this parabola.

To obtain the equation of this parabola we must introduce a set of axes. In view of what we learned about the straight line, the choice of the position of these axes will determine the form of the resulting equation. Experience has shown that a particularly simple equation results if we choose the axes in the following way. Let the X-axis be the line through F and perpendicular to d; let the Y-axis be the line perpendicular to the X-axis, of course, and halfway between F and d. Since the given point F and line d are fixed, there is a fixed distance from F to d that we shall suppose is 10 units. Because the Y-axis is halfway between F and the line d, the distance FO is then 5 units and the perpendicular distance between d and the Y-axis is also 5 units.

We are now prepared to express algebraically the property that defines the parabola. Let P be any point on the curve and let its coordinates be (x, y). We must now express the fact that PF equals

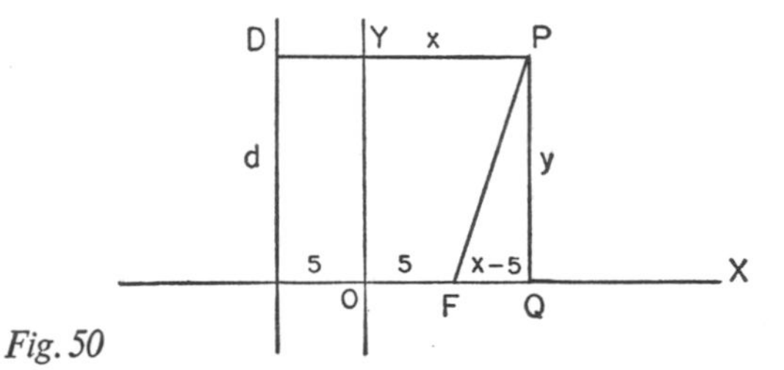

Fig. 50

the distance from P to d. A glance at figure 50 shows that PF is the hypotenuse of a right triangle whose arms are $x - 5$ and y. Hence

$$PF = \sqrt{(x-5)^2 + y^2}.$$

The perpendicular distance from P to d is clearly $x + 5$. Hence the algebraic statement of the condition P satisfies is

$$\sqrt{(x-5)^2 + y^2} = x + 5. \tag{7}$$

We could of course stop at this point and claim that equation (7) is the equation of the parabola. But one of the advantages of algebra is that one can use it to simplify algebraic expressions and equations. In this case, if we square both sides of equation (7), which is justified by the axiom that one may multiply equals by equals, we obtain

$$(x-5)^2 + y^2 = (x+5)^2.$$

By multiplying $x - 5$ by itself and $x + 5$ by itself, we have

$$x^2 - 10x + 25 + y^2 = x^2 + 10x + 25.$$

If we now subtract x^2 and 25 from both sides and then add $10x$ to each side of the resulting equation we obtain

$$y^2 = 20x. \tag{8}$$

Equation (8) is obviously a much simpler form of the equation of the parabola than equation (7). The number 20 in the final equation is, incidentally, twice the distance from F to d. Had we wished to be general we might have called this distance a and the resulting equation would have been $y^2 = 2ax$.

We have now obtained the equation of the parabola, and of course we know what a parabola looks like, but we do not know how the parabola lies in relation to the fixed line d and the fixed point F, or alternatively stated, how the parabola lies in relation to the coordinate axes. We could seek those pairs of coordinates that satisfy equation (8) and plot the corresponding points. For example, when $x = 1$, then, according to (8), $y^2 = 20$ and $y = +\sqrt{20}$ or $y = -\sqrt{20}$. What this calculation shows is that $(1, \sqrt{20})$ and $(1, -\sqrt{20})$ are two pairs of coordinates that belong to points on the curve. We can plot these points (fig. 51) and by obtaining many more such points ultimately get some indication of how the curve lies in relation to the axes.

But often a little reasoning supplants a lot of hard labor. Let us analyze equation (8). By taking the square root of both sides of equation (8) but remembering that the square root of a number can be positive or negative we have

$$y = \pm \sqrt{20x}. \tag{9}$$

This equation shows that to each positive value of x that one may choose there are two corresponding values of y. These two values

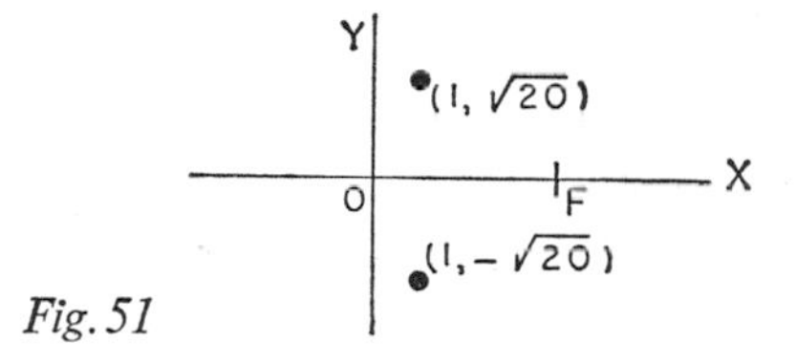

Fig. 51

of y are numerically the same but differ in sign. Geometrically interpreted this result means that to each positive value of x there are two points on the curve, one with that x-value as abscissa and the corresponding positive y-value as ordinate and the other with the same x-value as abscissa but with the negative y-value as ordinate. Moreover, since the numerical values of the two ordinates are the same, one point will be as far below the X-axis as the other is above.

Since the reasoning we have just pursued applies to any positive x-value we might choose, we may draw the general conclusion that along with whatever part of the curve lies above the X-axis there will be another part below that is the mirror image of what lies above. Hence, if we can figure out what the part of the curve above the X-axis looks like, we can draw the part below. We shall therefore concentrate on that part of the curve above.

Equation (9) says that the ordinate of any point is the square root of 20 times its abscissa. Hence, if we consider a range of abscissas starting with a small value and continually increasing, the corresponding ordinates will at first likewise be small and con-

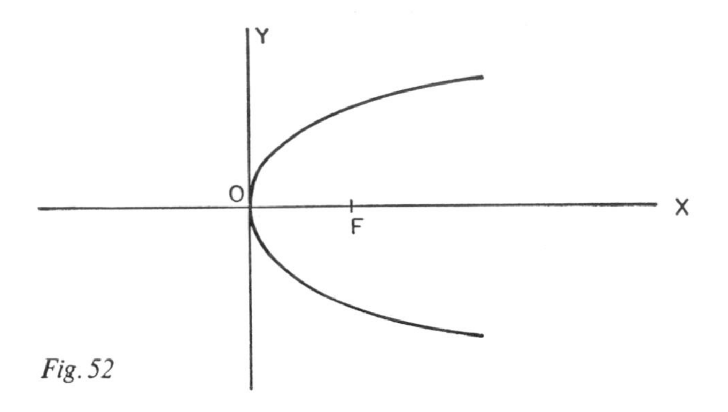

Fig. 52

tinually increase. We see this conclusion from the fact that the square root of a larger number is larger than the square root of a smaller one. Thus far, then, we can say that as the abscissas increase the ordinates must increase. But now we must observe one more fact about square roots. If we take a sequence of numbers such as 4, 9, 16, 25, 36 . . . , the corresponding square roots are 2, 3, 4, 5, 6. . . . We note that the square roots increase but not so rapidly as the original members of the sequence. For example, as we pass from 16 to 25 in the original sequence, the square root changes from 5 to 6. What all this means as far as the curve is concerned is that as the abscissas of the points on it increase, the ordinates increase but not so rapidly. The picture of the curve that results is shown in figure 52. Of course this picture shows just the general shape of the curve. If one wishes to know precisely how large the ordinate is for any given abscissa he would use equation (9) above to calculate it. We should note finally that we did not consider negative values of x. The reason is simply that there is no

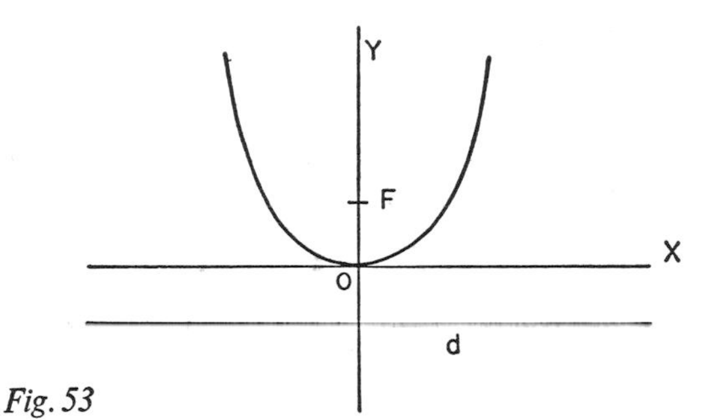

Fig. 53

square root of a negative value of x that is an ordinary number.*

For subsequent applications we must remember that the equation of a parabola need not take the form (8). If the given line d and the given point F should happen to lie as shown in figure 53, it would be desirable to take the Y-axis through F and perpendicular to d, and to take the X-axis halfway between F and d. The process we went through to obtain equation (8) could still be performed but the result would be different. We could go through the process but a little reasoning will tell us the answer at once. In figure 53 the Y-axis plays precisely the role the X-axis played in figure 52, and the X-axis in figure 53 plays the role the Y-axis previously played. All this means is that abscissa and ordinate are interchanged and the equation of the parabola is

$$x^2 = 20y. \tag{10}$$

There is another readily derived variation on the form that the equation of the parabola may take. If the position of F and d should be as in figure 54 (p. 162), and if we choose the axes as shown, then we have but to compare figure 54 with figure 53 to see that the roles of positive and negative ordinates are interchanged. Whatever holds in (10) for positive y should now hold for negative y. The equation of the parabola placed as in figure 54 is therefore

$$x^2 = -20y. \tag{11}$$

We have seen thus far how Descartes and Fermat associated an equation with any given curve. But these men also pointed out

* There are numbers in mathematics that are roots of negative numbers. These numbers are called complex numbers, but they play no role in the applications we shall consider.

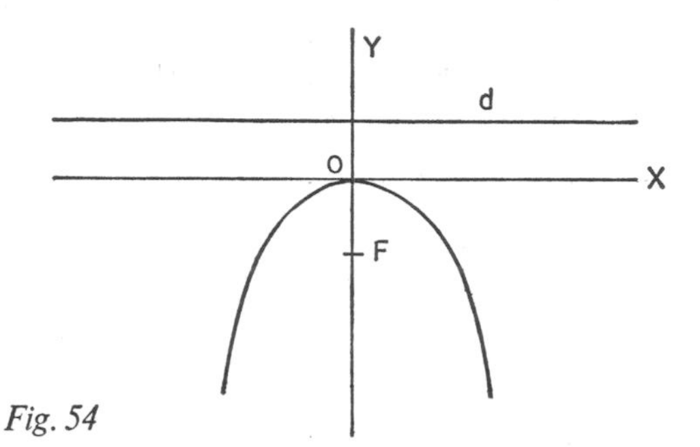

Fig. 54

that their idea could be applied in reverse. Given an equation involving x and y, it is possible to associate a curve with that equation. Let us try to determine what curve corresponds to the equation

$$y = x^2 - 6x + 9. \tag{12}$$

To determine this curve we could of course choose a value of x, say 2, calculate the corresponding y-value, which in this case would be 1, and thus obtain one point on the curve, namely, the point whose x-value is 2 and y-value is 1. By locating this point and a number of others with respect to an X-axis and a Y-axis we might begin to perceive the shape of the curve.

But there is a more sophisticated approach that tells us much more and tells it more readily. A person who has familiarized himself with simple processes of algebra will recognize that the right-hand side of equation (12) is just $(x - 3)^2$. Hence we have thus far that

$$y = (x - 3)^2. \tag{13}$$

We have already considered equations such as $20y = x^2$ [see equation (10) above]. As we remarked earlier in connection with equation (8) the number 20 is incidental and is twice the distance between focus and directrix. No matter what this number is, the corresponding curve is a parabola. Hence the equation

$$y = x^2 \tag{14}$$

also represents a parabola. The similarity between the equation (13) and equation (14) is striking. If in some way $x - 3$ could be replaced by x, the two equations would be the same. Let us therefore try the following idea. We shall introduce a new letter x', such that

$$x' = x - 3 \tag{15}$$

and, for symmetry, a new letter y', which equals y. Then equation number (13) reads

$$y' = (x')^2. \tag{16}$$

We know that the curve corresponding to equation (16) is the parabola shown in figure 55 with the origin O' at the intersection of the X'-axis and the Y'-axis.

Now let us compare equations (13) and (16). Equation (15) says that x is always 3 units more than x'. In other words, to each point on the curve of equation (16) there is a point on the curve of equation (13) whose y value is the same but whose x-value is 3 units more. Where are the latter points in relation to the former? Just 3 units to the right. If, therefore, we were to take the curve shown in figure 55 and slide it three units to the right we would have the curve corresponding to equation (13). But sliding the curve to the right means in effect leaving the curve where it is and choosing a new origin O which is 3 units to the left of O' and a new Y-axis through O to replace the Y'-axis through O'. If the desired curve of equation (13) can be pictured merely by choosing a new origin, then of course the shape of the curve is unaltered and *equation (13) still represents a parabola* but in a different position, with relation to the origin and Y-axis, than the curve of equation (16) is with respect to the X'-axis and Y'-axis.

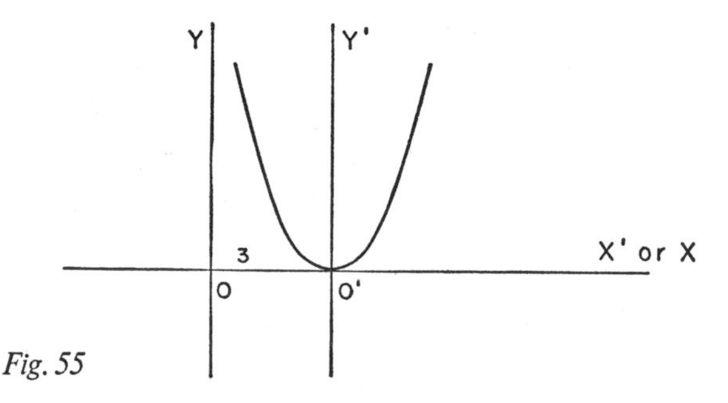

Fig. 55

Thus there is a curve that corresponds to the equation $y = x^2 - 6x + 9$ and in this particular case the curve happens to be a familiar one, the parabola. The curve determined by any given equation in x and y would not necessarily be a familiar one. In fact, the broad significance of the fact that an equation in x and y determines a curve is that since the number and variety of equations that can be written down is unlimited, so is the variety of curves. The study of the curves corresponding to various equations in x and y introduced a large number of new curves into mathematics, many of which proved to be useful and some of which have not even yet been fully explored. Through the notion of equation and curve Descartes and Fermat disclosed a new universe of curves.

The heart of Descartes's and Fermat's idea is now before us. To each curve there belongs an equation that describes the position of any point on the curve. Moreover, each equation relating x and y can be pictured as a curve. The association of equation and curve, then, is the brand-new thought. By combining algebra and geometry Descartes and Fermat introduced an entirely new approach to the study of curves.

The reader may be willing to grant that coordinate geometry permits an entirely new representation of curves, but he might very well ask, What good does the equation of a curve do beyond, perhaps, what has just been pointed out, namely, that equations lead to new curves? One value of the new representation, indeed a value that motivated Descartes and Fermat, is that it is now possible to use algebra to solve problems involving curves. The material we shall examine throughout the rest of this book will provide ample illustration of such uses of coordinate geometry. We might consider one example at this point.

Suppose one wished to determine the shape of a reflector that will concentrate the light rays emanating from a point F into a set of rays all of which will travel outward in the same direction. A typical ray such as FP (fig. 56) issuing from F should be reflected at P on the reflector and proceed to the right in some direction that is to be the same for all rays. If we can achieve the desired shape for rays that travel upward from F, then the same shape should work for rays that travel downward. Hence we may assume that the reflector shape is symmetrical about some line OF, say, and we shall require that the direction of the emerging rays be parallel to OF. We shall impose an additional condition. Since after reflection all the light rays will go forward, we shall require that the light traveling along the various rays all reach the vertical line FE

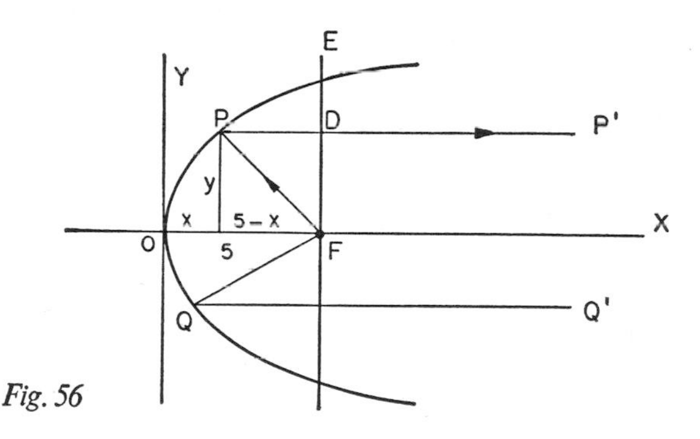

Fig. 56

at the same time. Thus all the light that emanates from F at any instant will proceed to the right simultaneously in a beam having the direction OF.

With these physical requirements set forth we may now proceed to determine the shape of the reflector. Let P be a typical point on the curve. We introduce a set of axes as shown in the figure, the Y-axis being constructed at O, the farthest point to the left on the desired curve. To be concrete we suppose that the distance OF is 5 units. Now light from F will travel the distance FP and then PD. But FP is the hypotenuse of a right triangle whose arms are y and $5 - x$. Hence

$$FP = \sqrt{(5 - x)^2 + y^2}.$$

The distance PD from P to the line FE is $5 - x$. Thus the light along any ray will travel the total distance

$$\sqrt{(5 - x)^2 + y^2} + 5 - x \tag{17}$$

to reach the "front" or line FE.

In order to set up an equation that characterizes the curve of the reflector we now try to find a quantity that equals the quantity in (17). We do this by considering the ray of light that leaves F and travels along FO. Since the Y-axis was constructed at the extreme left point of the curve, this axis just touches but does not cut the curve. Hence the Y-axis is tangent to the curve at O. Now a tangent to a curve and the curve have the same direction at the point of tangency. Since FO is perpendicular to the Y-axis it must also be perpendicular to the curve at O. Thus the light that travels from F to O will be reflected along OF because the angle of incidence of the light ray is O° and so the angle of reflection must be O°. This light must travel the distance $FO + OF$ to reach the line FE, and this distance is 10 units. Hence, if each of the other light rays emanating from F is to reach FE at the same time, each must likewise travel a distance of 10 units.

We know then that the expression (17) must equal 10; that is,

$$\sqrt{(5 - x)^2 + y^2} + 5 - x = 10.$$

We now subtract 5 from both sides and add x. The result is

$$\sqrt{(5 - x)^2 + y^2} = 5 + x.$$

By squaring both sides of this equation we obtain

$$(5 - x)^2 + y^2 = (5 + x)^2.$$

Multiplication of $(5 - x)$ by itself and of $(5 + x)$ by itself yields

$$25 - 10x + x^2 + y^2 = 25 + 10x + x^2.$$

By subtracting 25 and x^2 from both sides and by adding $10x$ to both sides we obtain

$$y^2 = 20x \tag{18}$$

as the equation connecting the coordinates of any point on the curve.

What curve does equation (18) represent? Our knowledge of coordinate geometry tells us that the curve is a parabola. The actual reflector will of course be just an arc of the parabola. Thus we have gained the immensely useful fact that a parabolic arc will concentrate light issuing in various directions from one point into a beam of parallel rays.

Application of this fact can be seen, of course, in automobile headlights. The reflector used there is not a curve but a surface. However, every section of this surface (fig. 57) cut out by a plane through the axis FO is a parabola and the light that issues from

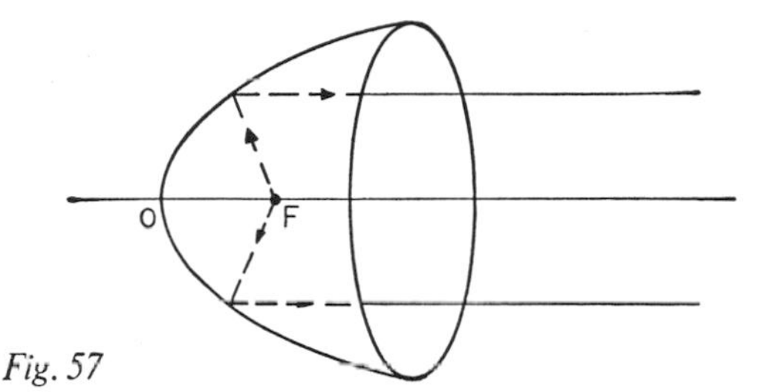

Fig. 57

F in various directions is reflected in a beam of parallel rays that strongly illuminates the road ahead of the automobile.*

Descartes and Fermat accomplished far more than they realized. The forms of all physical objects that are studied for any reason whatsoever are, at least when idealized, curves and surfaces. The fusilage of an airplane, its wings, the hull of a boat, the cable of a bridge, the trajectory of a projectile, the shape of the projectile itself, and the paths of a ball thrown by a child, an electron expelled from an atom, the planets, and light rays are curves or surfaces. These can be represented by equations and the shapes or motions studied by applying algebra to these equations. Hence it was not just the classic curves of geometry that Descartes and Fermat algebraicized, but all paths, curves, and surfaces that occur in the physical world. In other words, they made possible the algebraic representation and the study by algebraic means of the various objects and motions of interest to scientists. Their method is so basic in science that they may very well be called the founders of applied mathematics. The story of coordinate geometry illustrates how an interest in geometric methods proved to be immensely rewarding for science and engineering.

* See chapter 17 for other applications.

11: EXPLANATION VERSUS DESCRIPTION

Nor should it be considered rash not to be satisfied with those opinions which have become common. No one should be scorned in physical disputes for not holding to the opinions which happen to please other people best.

—GALILEO

DESCARTES and Fermat created one of the key mathematical tools for the development of modern science. Galileo Galilei created modern science. In one respect Galileo responded to the needs of his time. He took up the study of motion, the most prominent and aggravating scientific problem of the seventeenth century. But Galileo was an innovator of highest rank in creating the methodology of modern science, the methodology that has made possible its extraordinary fecundity. Though a few great minds that preceded Galileo had caught visions of how science should proceed, no one saw it so clearly, rid himself so thoroughly of antiquated and ineffectual encumbrances, and demonstrated by concrete investigations the new approach to nature. With Galileo science stepped over the threshold into modern times.

Galileo was born in Pisa in 1564. After some preliminary schooling he entered the University of Pisa to study medicine; however, curious about mathematics, he took lessons privately in the subject from a practical mathematician. His interest became so strong that he turned away from medicine. By the age of twenty-five, he had produced significant mathematical papers and was appointed lecturer at the university. Convinced that Aristotelian physics was wrong, Galileo did not hesitate to say so. These attacks on the accepted science of the time did not endear him to his colleagues. When, in 1592, the Republic of Venice offered him a professorship of mathematics at the University of Padua, Galileo left his native city to accept it. After eighteen years at Padua he left to live in Florence under the patronage of the Duke of Tuscany.

Galileo was great in many fields. He was an inventor of first magnitude. Having heard of the existence of the telescope, he immediately set about constructing one and soon made many improvements. He invented the microscope, though others did so independently. While watching a chandelier swing in a church, he noted that the fixture took the same time to make one complete swing whether of small or large amplitude (see chapter 18); he is supposed to have used his pulse beat to time these swings. Recognizing that he had discovered a useful principle, Galileo proceeded to construct a pendulum clock. Then he used the principle of the pendulum to design the pulsimeter, a device for mechanically recording pulse rates. He constructed the first thermometer. He also designed a special type of compass with scales on it that yielded automatically the numerical results of extensive mathematical computations. This device was so successful that he made many for sale.

Galileo was one of the famous astronomical observers. We have already had occasion, in connection with the work of Copernicus and Kepler, to mention the discoveries that startled his contemporaries. These discoveries were made with a telescope which by modern standards was extremely limited. It has been said that it was remarkable that he could find Jupiter in the sky, let alone discover the satellites of Jupiter.

Along with a few other mathematicians, such as Descartes and Pascal, Galileo wrote works that rank among the world's literary masterpieces. He wrote the best Italian prose of the seventeenth century. His first outstanding work was the *Sidereal Messenger* (1610), which contains an account of his observations of the heavens. His most famous works are his *Dialogue on the Great World Systems* (1632), in which he attacked the astronomical views of Aristotle and defended the heliocentric theory of Copernicus, and his *Dialogues Concerning Two New Sciences* (1638), the two new sciences being the strength of materials and motion. Galileo's courage to espouse new ideas is evident throughout as, tenaciously and with clever dialectic, he demolishes his opponents' arguments. His writing in these dialogues is polemic but cool; the prose is clear, direct, witty, simple, and yet profound. Galileo deliberately addresed the public, for which reason he wrote in Italian, and succeeded in reaching it through the spirit and force of his arguments. His preference for a direct style caused him to dislike the more mystic, philosophical, and fervent writings of his friend Kepler. Galileo wished to separate science and mathematics from theology and mysticism. He said in one of his letters, "Yet for my part any discussion of the sacred Scriptures might have lain dormant forever; no astronomer or scientist who remained within proper bounds has ever got into such things."

Galileo's finest contribution to thought is the invention of the methodology of modern science. It is his approach to nature which not only engaged mathematics more heavily but which led to some of the finest mathematical creations since the year 1600. To appreciate the changes Galileo introduced we must go back a bit.

The established system of scientific thought and the nature of scientific activity before and in Galileo's time derived from Aristotle, and the chief characteristic of the Aristotelian approach to nature was the search for material or qualitative explanations. The Aristotelians tried to explain phenomena on the earth in terms of qualities or substances they had come to believe were basic, for example, hot and cold, wet and dry. Combinations of these qualities were thought to give rise to the four elements earth, air, fire, and water. Thus, hot and dry qualities produced fire, hot and wet produced air, and so on. Each of the four elements had a characteristic motion. Fire, the lightest, naturally sought the heavens, while earthy matter sought the center of the earth. Solids, fluids, and gases were regarded as three different kinds of substances possessing different fundamental qualities (rather than, as we say, different states of the same substance). The transition from fluid to gas meant to the Greeks and medievalists the loss of one quality and the acquisition of another. Different objects differed in fundamental qualities. For example, it was thought that in changing mercury into gold one took from mercury the quality that contributed fluidity and substituted the quality that possessed rigidity. This idea of basic qualities was still pursued during the early stages of modern chemistry. Sulphur possessed the substance of combustibility, called phlogiston; salt, the substance of solubility; and metals, the basic substance of mercury. Heat, until the nineteenth century, was a substance, called caloric, which could be gained or lost by bodies as they acquired or lost heat.

The Aristotelians sought to classify objects according to the qualities or basic substances they contained; hence, one of their major goals was classification, a method still basic in biology. To explain how one event brought about another the Aristotelians built an elaborate scheme. All phenomena came about because there were four types of causation, the material cause, the formal cause, the effective cause, and the final cause. Just to differentiate these causes let us consider an artist making a statue. The material cause would be the stone and the artist's tools; the formal cause would be the design the artist had in mind; the effective cause would be the artist actually chipping away at the stone; and the final cause would be the purpose the statue would serve in beautifying some room or building. The final cause was the most important because it gave meaning to the entire activity.

Where was mathematics in this scheme? Since mathematics to the Greeks and medievalists was largely geometry, and geometry dealt with figures, mathematics was of use mainly in describing the formal cause—a rather limited role.

For several reasons the Aristotelian approach to nature dominated in the medieval period and the Renaissance. Aristotle's writings were comprehensive and they were more widely disseminated than those of the other Greeks. Further, Aristotle's theory of final cause had been adopted and espoused by Catholic theology. The explanation of man's life on earth was that it prepared him for Heaven, and quite generally the Church explained earthly phenomena as serving purposes intended by God.

Though explanations of natural phenomena in terms of purpose or in terms material or effective cause may satisfy most minds, it was Galileo's great insight that such explanations did not greatly advance the cause of science. Galileo decided that he would seek descriptions of *how* things worked rather than explanations of *why* they worked or what purpose they served. He declared in his *Two New Sciences* that he would investigate and demonstrate some of the properties of accelerated motion without regard to what the cause of the acceleration might be.

To become a little surer of the distinction between the Aristotelian and Galilean approaches, let us consider a simple example. Suppose a ball is released from the hand and allowed to fall. The Aristotelians attempted to explain why the ball fell, and their answer was that it fell because it had weight. Why should the ball fall toward the earth? The answer to this question was that each object in the universe had a natural place and unless hindered would seek that place. The natural place of heavy objects being the center of the universe—in Aristotle's astronomy the center of the earth—the ball fell to earth. Also, according to the Aristotelian explanation of motion the only resistance to the fall of the ball was the resistance of the air. Since this resistance was constant the ball fell with constant velocity. Actually, balls were observed to fall with constantly increasing velocity, and the Aristotelians had to introduce further explanations as to how the action of air increased the velocity.

In place of such explanations Galileo proposed to describe the motion of the ball. Now description could have been confined to describing the path taken by the ball, in the present case, a straight line. This form of description was utilized by Kepler in his first law of motion, which merely described the shape of the paths taken by the planets. But Galileo sought *quantitative* description. He proposed to find out how far the ball fell in a given time, what velocity the ball acquired in that time, how much the velocity increased in successive units of time, the effect of any applied force on the distance and velocity, and the like. Moreover, he proposed to ignore completely such questions as why the motion took place or what mechanism created the motion.

It may appear that Galileo was asking for rather little compared to what the Aristotelians had sought. A mere description of what takes place hardly seems to be more than the first step in a scientific inquiry, and one would expect it to be followed by the really important step of examining the phenomenon for the inner mechanism or the causal forces at work. We have therefore yet to understand the wisdom of Galileo's preference for quantitative description. But before we see just what it accomplished, let us note other basic doctrines in the Galilean approach.

His next thought was that any branch of science should be patterned on the model of mathematics; that is, one should obtain a few basic principles applicable to a class of phenomena and then deduce new truths by reasoning. The basic principles would serve as do the axioms of mathematics, and the deductions from them would be analogous to the theorems of mathematics. This particular idea was not especially novel. Even Aristotle had aimed at a deductive system for the various branches of science with the mathematical model also in mind. Nevertheless, one is a little surprised on opening Galileo's *Two New Sciences* to find axioms, theorems, and proofs arranged as in any mathematics textbook.

But how does one go about finding fundamental principles? They are not available merely for the asking. In this search Galileo once again departed radically from the Greeks and the medievalists, who believed that reason alone could determine fundamental principles. Since they believed that the axioms of mathematics were self-evident truths they could and did point to mathematics to demonstrate this power of reason. Reason seemed equally effective in other domains. Since the heavenly bodies were perfect and repeated their motions in definite periods of time, and since the circle was the perfect curve that allowed repeated behavior, it was clear that the heavenly bodies must move in circles, or, at worst, combinations of circles. That all objects in the universe had a natural place was a compelling principle. That force should be required to keep a body in motion was beyond dispute. Some minor observations may have assisted these predecessors of Galileo in obtaining numerical data such as astronomical distances, but on the whole, observation of nature played little part.

Nor had it been deemed necessary to check the deductions from these principles against the behavior of nature. Any deviation observed in practice could be explained by special circumstances. Moreover, since there were no large and complex engineering projects whose design was based on scientific principles, one might almost say that there was no occasion to check the principles. Some of Galileo's contemporaries, notably Kepler, were more concerned to make their theories and deductions fit observations. Descartes, too, experimented and insisted on having his theories accord with facts. But one might say of the pre-Galileans that they first decided how the world ought to be and then fitted what they saw into their preconceived schemes. Their eyes saw what their minds permitted them to see.

Galileo, very much aware of this proclivity of his predecessors, criticized these scientists and philosophers. He wrote that nature did not first make men's intellects and then arrange the world to be acceptable to them. Rather, nature made things as she liked and let man's reason struggle as best it could to understand her ways. "Nature nothing careth whether her abstruse reasons and methods of operating be or be not exposed to the capacity of men." The way to obtain correct and basic principles, Galileo indicated, was to pay attention to what nature says rather than to what the mind believes.

Such criticisms had been voiced by many predecessors of Galileo. Leonardo da Vinci had said that sciences that arise and end in thought do not give truths because in these mental considerations no experience enters, and without this nothing is sure. "If you do not rest on the good foundation of nature you will labor with little honor and less profit." Galileo's contemporary, Francis Bacon, had spoken sharply for banishing the various idols that preoccupied men's minds and prevented them from seeing the truths. But prior to Galileo the use of experience to build scientific doctrine was fumbling and without direction.

Even though Galileo experimented purposefully and with telling effect, we must not conclude that experimentation was undertaken on a large scale and that it became the new and decisive force in science. This was not to come until the nineteenth century. There were of course some famous experimenters in the seventeenth century: Robert Hooke, the physicist, Robert Boyle, the chemist, and Christian Huygens, the mathematician and physicist, to say nothing of Galileo himself and Isaac Newton. Galileo, however, was a peculiar mentality and a transitional figure insofar as experimentation is concerned. He was not the experimenter he is often reputed to be. He, and even Newton, believed that a few critical experiments and keen observation would readily yield correct fundamental principles. Newton emphasized his reliance upon mathematics and said he used experiments largely to make his *results* physically intelligible and to convince the "vulgar." Many of Galileo's so-called experiments were really "thought-experiments"; that is, he used his experience to imagine what would happen if an experiment were performed and then drew an inference as confidently as if he had actually carried out the experiment. In his writings he describes experiments he never made. He advocated the heliocentric theory even though, in the stage in which Copernicus left it, the theory did not yet accord well with observations. In describing some experiments on motion along an inclined plane Galileo did not give actual data but said that the results agreed with theory to a degree of accuracy that was incredible in view of the poor clocks available in his time. A few fundamental principles derived from nature and much mathematical reasoning constituted Galileo's and Newton's scientific method.

Galileo did have some preconceptions about nature that made him confident a few experiments would suffice. He believed that nature was simple, orderly, and mathematically designed and that nature follows immutable laws. God, he believed, put into the world that rigorous mathematical necessity which men discover only laboriously. "Nor does God less admirably discover himself to us in Nature's actions, than in the Scripture's sacred dictions." When Galileo apprehended a clear mathematical principle, then he was sure that it was the correct one and proceeded to use it. For example, when he undertook to study accelerated motion, that is, motion with changing velocity, he assumed as the simplest principle that the increases in velocity in equal intervals of time were equal. This he called uniformly accelerated motion. Of course he agreed that the results of reasoning on this basis must be checked by experiments.

In minimizing the resort to experimentation by Galileo and others of his century we must, however, differentiate between experiment and experience. Galileo had available and used the practical knowledge that artisans, engineers, manufacturers, navigators, and others had amassed. To an extent this experience substituted for experimentation since it supplied reasonably reliable knowledge of natural phenomena. Nevertheless, this sort of work must be distinguished from the vast, directed, and pure scientific experimentation that is so prevalent in our time. Galileo's brilliant success in launching modern science stems more from other features of his mode of investigation.

We have been discussing Galileo's decision to search for axiomatic scientific principles, and we have seen thus far that he resolved to derive these from nature herself. However, if one seeks to study motion on and near the earth, say, and observes the fall of many different sizes, shapes, and weights of bodies, or the flight of a variety of projectiles, he will end up with a mass of observations from which no clear pattern or principle will emerge. Galileo did otherwise. He attempted, insofar as thought and experiment would permit, to strip away all the incidental and minor effects. To eliminate the effects of friction he considered smooth balls rolling down a smooth slope. Having discovered that pendulum motion is little affected by air resistance, he studied pendulum motion to obtain fundamental principles of motion. This approach is precisely the one the mathematician makes when, for example, he studies geometrical figures. He strips away molecular structure, color, and thickness to get at the basic properties of these figures. Thus, insofar as he observed and experimented, Galileo idealized just as the mathematician does and thereby penetrated the phenomena to obtain the basic physical principles.

Galileo undertook to study motion for reasons already mentioned. Since he had decided to be quantitative, that is, to seek numerical relationships, he had to select those characteristics of motion that were basic and yet accessible to measurement. The principle he enunciated was, Measure what is measurable and render measurable what is not yet so. He began by adopting an older philosophical doctrine that distinguished between primary and secondary qualities of matter. He said: "White or red, bitter or sweet, sounding or silent, sweet-smelling or evil-smelling are names for certain effects upon the sense-organs; they can no more be ascribed to the external objects than can the tickling or the pain caused sometimes by touching these objects." Thus, colors, tastes, and the differences among sounds and smells he regarded as the effects of real properties on the sense organs but as nonexistent in nature. He said further: "If ears, tongues, and noses are removed, I am of the opinion that shape, quantity, and motion would remain, but there would be an end of smells, tastes, and sounds, which abstractedly from the living creature, I take to be mere words." Shape, quantity (size), and motion, then, are the primary or physically basic properties of matter. These are real, external to human perception, and permanent.

But what properties of shape, quantity, and motion are measurable or potentially measurable? After further analysis and reflection Galileo decided to concentrate on such concepts as space, time, weight, velocity, acceleration, inertia, and force. These characteristics of matter in motion are measurable and they indeed proved to be most significant in the rationalization and conquest of nature. The emphasis on measurable properties in Galileo's scientific method was so strong that his philosophy is often summed up with the statement that science is measurement. However, as is already partly evident, this principle was but one of many that Galileo introduced.

In the selection of measurable characteristics Galileo again showed his genius, for the ones he selected are not clearly the most important. In fact, choices such as space, time, and weight do not strike us as remarkable or as evidence of keen insight. But we are already accustomed to these concepts. Medievalists spoke of bodies possessing the potentiality of motion and of motion itself as the actuality. They spoke of efficient causes, essences, action, ends, natural places, natural motion, and violent motion. We would have to dispossess our minds of modern ways of thought and immerse ourselves in such occult medieval concepts to appreciate what Galileo banished and how revolutionary his choices were.

It is perhaps evident that Galileo envisioned and undertook an entirely new program for science. But just how was mathematics involved, and why was Galileo's program so instrumental in strengthening the alliance of mathematics and science? To be sure, his approach was mathematical in character in that he idealized and he proposed to use deductive reasoning, but these characteristics in themselves do not involve mathematics proper. It is also true that he wished to be quantitative, and this would mean the use of numbers; however, any bookkeeper uses numbers but does not necessarily use mathematics. Galileo's program contemplated a more basic role for mathematics.

We have mentioned that he sought fundamental physical principles from which he hoped to deduce new scientific truths. The principles he sought were to be almost as thoroughly mathematical as the axiom that equal numbers added to equal numbers give equals. Of course, since Galileo was to deal with physical phenomena his physical principles would also contain physical concepts tied together in some mathematical statement. Let us see just what all this means.

Consider the simple case of a ball released by the hand and allowed to fall, or, as it is often described, a ball that is dropped. We have already noted that Galileo proposed to find a quantitative description of this motion. He suggested that we concentrate on the time that the ball has been falling and the distance that the ball has fallen in that amount of time. Now as the ball falls, the time that it has been falling continually increases; likewise, the distance that it has fallen also increases as the time does. These quantities, time and distance, are then *variable* quantities. The time that the ball has been falling can be any number from 0 to the final value, namely, the time at which the ball strikes the ground. Likewise, if we measure distance from the position of the hand downward, the distance the ball has fallen at any time can be any number from 0 to the distance from the hand to the ground.

The principles Galileo sought were to be significant relationships between variables. In the present example he sought to relate the distance the ball falls in any given time to the time. He found that *the distance is always 16 times the square of the time* provided that the distance is measured in feet and time in seconds. Suppose the ball has fallen for 3 seconds; the square of 3 is 9; 16 times 9 is 144; and this is the distance the ball falls in 3 seconds. The relationship Galileo expressed in words we now write in algebraic symbolism as follows. Let t be a symbol for the variable time that the ball falls; let d be the symbol for the variable distance that the ball falls. Then the italicized statement above can be written as

$$d = 16t^2. \tag{1}$$

Thus we see that Galileo sought mathematical statements about the physical concepts he wished to investigate.

The relationship (1) between distance and time is nowadays called a *function,* or a functional relationship between the variables distance and time. The algebraic representation of the functional relationship, as opposed to the italicized verbal statement, is called a *formula*. Several facts about formula (1) are important for the proper understanding and use of formulas. We must be careful to note that the letters d and t represent not just one value of distance and one value of time but any value within their respective ranges. Thus, if the ball reaches the ground in 5 seconds, say, then the variable t represents all values from 0 to 5; and if the distance from the hand to the ground is 400, then the variable d represents all values from 0 to 400. Moreover, if we substitute any value from 0 to 5, say $5/2$, for t in (1), and calculate accordingly $16(5/2)^2$, we obtain 100 as the value of d, which corresponds to the value of $5/2$ for t. Thus the formula says that when the ball has fallen $2\frac{1}{2}$ seconds, the distance it has fallen in that time is 100 feet. For each value of one variable, *time* in this case, one may calculate exactly the corresponding value of the other, *distance*. This calculation can be performed for millions of values of the time variable, or strictly, an infinite number of values. The simple formula $d = 16t^2$ contains, then, an infinite amount of information. Thus the formula is compact, precise, and replete with information.

It is most important to note that the mathematical formula is a description of what occurs and not an investigation into causal relationships. The formula $d = 16t^2$ says nothing about *why* a ball falls nor whether balls have fallen in the past or will continue to fall in the future. It gives merely quantitative information on *how* a ball falls. And even though such formulas are used to describe a phenomenon in which some cause or physical mechanism is at work, the scientist does not have to investigate, let alone understand, the causal connection in order to apply the formula to practical problems. This was the truth Galileo saw clearly when he emphasized mathematical description against his predecessors' less successful qualitative and causal inquiries into nature.

After finding physical axioms in the mathematical form that (1) exemplifies, Galileo planned to use the machinery of mathematics to deduce new truths. Thus one might apply algebraic techniques to formula (1) or one might combine several such formulas by proper algebraic steps to obtain a totally new fact, just as in geometry one deduces new theorems from the axioms and previously established theorems. We see, therefore, how Galileo's program for science was to involve mathematics not only as a language in which the basic laws were to be framed but as the vehicle for carrying the burden of reasoning toward the demonstration of new truths. The scientific knowledge that Galileo envisioned was to be a series of mathematical formulas relating physical variables.

Galileo's program for science may seem thus far to be unimpressive. Of what use can a lot of barren formulas be in studying physical phenomena? They do not explain what occurs, nor do they indicate how one may apply scientific knowledge to the construction of useful devices. The full significance of Galileo's ideas has yet to be revealed. We shall see that such formulas have proved to be the most valuable knowledge man has ever acquired about nature. We shall find that the amazing practical as well as theoretical successes of modern science have been achieved mainly through the quantitative descriptive knowledge it has amassed and manipulated rather than through metaphysical, theological, fanciful, and even mechanical explanations of the causes of phenomena. The history of science is the history of the gradual elimination of gods and demons and the reduction of vague notions about light, sound, force, chemical processes, and other phenomena to number and quantitative relationships. As we pursue the furtherance of Galileo's program by Galileo himself and his successors, notably Isaac Newton, we shall see that they revived the Pythagorean-Platonic idea of substituting mathematics for nature. Newton was later to admit that if he saw further than others it was because he stood on the shoulders of giants. Of these the tallest was Galileo.

12: VERTICAL MOTION

To give us the science of motion God and Nature have joined hands and created the intellect of Galileo.

—Fra Paolo Sarpi

Galileo had formulated his program and he proceeded to put it into effect. Whereas scientists and philosophers of Greek and medieval times had tried to embrace the whole of man and nature, Galileo, "with the restraint that shows the master," decided to select a few relatively limited phenomena and study these intensively. His principal choice, for reasons already cited, was the subject of motion.

His program called first of all for fundamental principles that could serve as the basis for reasoning. To appreciate what he sought we should be clear about what is fundamental. Everyone knows that a triangle with sides of relative length 3, 4, and 5 is a right triangle. This fact is certainly true but it is not fundamental. Very little can be done with it toward proving new conclusions. On the other hand, Euclid's axioms, perhaps unimpressive because they say rather little in themselves and seem to be mere truisms, are fundamental because with them we may prove the Pythagorean theorem for all right triangles and prove hundreds of other significant theorems.

The task of finding correct fundamental principles of motion was immense because Galileo had to break sharply with the two-thousand-year-old mechanics of Aristotle. All thinkers prior to Galileo's time believed that a force must be applied not only to set an object into motion but to keep it in motion. Now, in a sense, this is correct. To keep an automobile moving along a road or a ball rolling along even a flat surface at a constant speed some force has to act. The automobile and the ball are hindered by the resistance of the air and some friction between the ground and the object; hence some force must be applied to overcome the air resistance and friction. But Galileo decided to view these same phenomena differently. Like a mathematician, he idealized and asked, What if air resistance, friction, and any other hindering forces were not present and an object were set into motion? Would a force be needed to keep the object in motion?

The Aristotelians would not have entertained such a question. They treated the phenomena they experienced, and in experience air resistance is always present. By generalizing on their experience the Aristotelians concluded that the natural state of an object, that is, the state when no force acts, is one of rest. For an object to remain in motion, they believed, a force must constantly act on it. Thus, to account for the continued motion of an arrow shot from a bow, the Aristotelians built up an elaborate explanation of how the action of the air on the arrow supplied a constant force.

Galileo's decision to consider motion in a vacuum was clearly a radical one, and his answer to the question of whether a force is needed to keep an object in motion under idealized conditions was equally radical. He concluded that if no force acts on an object and the object is at rest, it will continue at rest; if no force acts and the object is in motion *it will continue indefinitely to move at a constant speed in a straight line*. This fundamental law of motion is now known as Newton's first law of motion. The law says that an object in motion will *not* change its speed or direction unless acted upon by a force. Hence, objects persist in maintaining their speed and direction. This property of matter, namely, resistance to change in speed and direction, is called its *inertial mass* or simply its mass. To put the matter somewhat more crudely, even inanimate objects object to being pushed around.

Just to be clear about the meaning of this first law of motion let us imagine a very smooth object sliding along very smooth ice in an atmosphere so rare that the air resistance is negligible. Then if the object happens to have a velocity of 30 miles per hour, say, and no force acts on it, it will continue to slide along indefinitely at 30 miles per hour.

What can one say about the motion of an object if some force is applied to it? Here Galileo made a second fundamental discovery. The application of a force causes an object to gain or lose velocity or to change the direction of its motion. For example, force applied to an automobile at rest sets it into motion, and air resistance, which is a force, causes an object sliding along smooth ice to lose velocity.

Let us call gain or loss in velocity per unit of time acceleration. Thus, if an object or body gains velocity at the rate of 30 feet per second each second, its acceleration is 30 feet per second each second, or in abbreviated form 30 $\mathrm{ft/sec^2}$. An automobile that starts from rest and acquires a velocity of 30 miles per hour in 10 seconds has accelerated its motion at the average acceleration of 3 miles per hour each second or 4.4 feet per second each second.

Thus far, Galileo's second discovery is a qualitative statement to the effect that force causes acceleration. The full and precise quantitative statement of this relationship, now known as Newton's second law of motion, is that the force equals the product of the mass and the acceleration. Expressed as a formula this law reads

$$\mathrm{F} - ma, \tag{1}$$

where F is the amount of force applied, m is the mass of the object being accelerated, and a is the acceleration imparted by the force. Of course, formula (1) presupposes that the three quantities involved are measured in suitable units, just as the statement that the area of a rectangle equals the length times the width presupposes that if the sides are measured in feet the unit of area is the number of square feet. In the system in which acceleration is measured in feet and seconds and the mass is measured in pounds, the force is measured in poundals or sometimes also in pounds, a pound of force equaling 32 poundals.

Let us examine the reasonableness of formula (1). It implies, for example, that if a larger mass is to be given the same acceleration as a smaller one, the applied force must be larger, that is, if a is constant, then when m increases, F must increase. This conclusion agrees with experience. Although, as we shall see more clearly in a moment, weight and mass are not the same, nevertheless a heavier object has more mass, that is, more resistance to change in velocity, and therefore more force must be applied to produce a given acceleration. For example, more force is required to set a heavier automobile in motion than a light one.

Formula (1) also implies that a constant force produces a constant acceleration on a constant mass for if F and m are fixed,

a must be also. For example, a constant air resistance causes a constant loss in velocity. This accounts for the fact that a ball rolling on a smooth floor will lose velocity continually until it has zero velocity. If no force is applied to a body, that is, if $F = 0$, then the product ma must be equal to zero. Since the mass m of any object is not zero the acceleration must be zero. Hence the body will neither gain nor lose velocity and will move with constant speed.

The most familiar simple example of motion is that of an object which is dropped and falls to earth. By Galileo's time the concept that the earth attracts all objects to it, that is, the concept of gravity, was known and accepted. The earth exerts a force on any object, and this force, if not offset by any other force, gives the object an acceleration. The surprising fact which Galileo discovered is that if one neglects air resistance, then any object falling to earth has a constant acceleration and, moreover, this constant is the *same for all objects,* namely 32 ft/sec^2. Thus another fundamental law which applies to objects falling to earth is

$$a = 32.* \tag{2}$$

The acceleration of 32 ft/sec^2, which all falling objects possess, is caused by the force of gravity. When we speak of this force as applied to objects near the surface of the earth we call it weight. Since the a of formula (1) is in this special case 32 ft/sec^2, and since the force is now weight, the formula says that the weight of any object is always 32 times its mass. In symbols,

$$W = 32m. \tag{3}$$

Thus two distinct properties of objects, weight and mass, are simply related by the fact that one is always 32 times the other. Because of this constant relationship we tend to confuse the two properties but we should be clear about the distinction. Mass is the property of resistance to change in speed or direction; weight is the force with which the earth attracts the object. If an object rests on a horizontal surface the surface opposes the force of gravity. Hence, insofar as motion along the surface is concerned, the weight of the object plays no role. Nevertheless, the mass of the object is

* See, however, chapter 15.

still present; that is, the object resists change in speed, and a force is required to set the object in motion or to change its speed or direction. We shall see later how important it is to distinguish between mass and weight.

We now have a few fundamental principles about motion and perhaps should see next whether, in accordance with Galileo's plan, mathematical reasoning can lead to new information. Let us consider the motion of an object which is dropped. At the outset this object will have zero speed or velocity. However, it gains velocity at the rate of 32 ft/sec^2. Hence at the end of one second its velocity is 32 ft/sec. At the end of two seconds its velocity is 64 or $32 \cdot 2$ ft/sec. Then at the end of t seconds its velocity is $32t$ ft/sec. In symbols,

$$v = 32t. \tag{4}$$

This formula tells us that an object does not fall at a constant velocity but rather at a velocity which increases as t increases. This fact is familiar, for everyone has observed that objects dropped from high altitudes hit the ground at higher speeds than do objects dropped from low altitudes.

Let us seek next the distance a dropped object falls in some given number of seconds, say 6. Ordinarily one obtains distance traveled by multiplying the velocity by the time. However this procedure is correct only if the velocity is constant. In our case the object starts with 0 velocity and, according to formula (4), after 6 seconds has a velocity of 192. Which velocity should we use to compute the distance traveled? A natural suggestion would be to use the average velocity, and presumably the average velocity is half the sum of the initial and final velocities, that is,

$$\frac{0 + 192}{2} \text{ or } 96.$$

Now it is not always correct to obtain the average velocity by taking one-half of the sum of the initial and final velocities. If an automobile travels at 10 miles per hour for a period of 59 minutes and then travels at 50 miles per hour for one minute, the average velocity is not one-half of $10 + 50$ or 30 miles per hour. The automobile does not cover 30 miles in that one hour. However, the average velocity or speed of 96 conjectured above is correct in our case because the object falls with constant acceleration. Let us see why the average of initial and final speeds is correct in the present instance. The speed of 96 feet per second is attained by the falling object at the end of 3 seconds. We learn this fact by means of formula (4), for when v is 96 then t is 3. Moreover, if h is any interval of time, the speed of the falling body at $3 - h$ seconds is, by formula (4), $32\ (3 - h)$ or $96 - 32h$. On the other hand, the speed of the falling body at $3 + h$ seconds is $32\ (3 + h)$ or $96 + 32h$. Thus for every instant of time before 3 seconds the speed is as much below 96 as it is above 96 at the corresponding instant after 3 seconds, that is, h seconds later. In other words, for every instant that the object falls at a speed less than 96 feet per second, there is another instant at which it falls at a speed as much greater than 96 feet per second as it was previously less. Hence 96 is the average speed.

This argument can readily be generalized. If the object falls for exactly T seconds, its final speed is $32T$ and its initial speed is, of course, 0. The average of the initial and final speeds is

$$\frac{0 + 32T}{2} \text{ or } 16T.$$

Moreover, this average speed is attained at the time $T/2$ because if we substitute $T/2$ for t in formula (4) we obtain $16T$. If we now review the above argument with 3 replaced by $T/2$ and the 96 replaced by $16T$ we see that it still holds. Hence we have the conclusion that if an object travels with constant acceleration, its average speed is one-half the sum of the initial and final speeds.

Then if an object falls for any number t of seconds, its average speed is $16t$ and the distance fallen is the average speed times the time t it has traveled. In symbols, if d is the distance fallen then

$$d = 16t^2. \tag{5}$$

Formula (5) says, for example, that in 3 seconds the object falls $16 \cdot 3^2$ or 144 feet. With a little mathematics we have derived an important law of falling bodies.

It is important to note that formulas (4) and (5) are of physical interest and physically significant only for zero and positive values of t. For negative values of t, the object does not fall and hence the formulas do not represent the motion. However, the mathematical formulas as formulas do have significance for negative values of t. All of this means that we must constantly be careful to distinguish the mathematical representation of a physical phenomenon from the physical phenomenon itself.

With the formulas we have derived we can now deduce a number of valuable conclusions about falling bodies. By dividing both sides of formula (5) by 16 and taking the square root of both sides we obtain for positive t (which is the one of physical interest)

$$t = \sqrt{\frac{d}{16}} \cdot \tag{6}$$

Hence, given the distance a body falls, we can calculate the time required to fall that distance.

Formula (6) is equally significant for what it omits. The mass of the object falling does not appear in the formula. Hence, if air resistance is neglected, *all* bodies take the *same* time to fall a given distance. This is the lesson Galileo is supposed to have learned by dropping objects from the leaning tower of Pisa. Accordingly, a feather and a piece of lead will take the same time to fall a given distance in a vacuum, a fact that many people still find hard to believe because they observe falls through air, and the resistance of air to a moving feather is considerable.

From formulas (4) and (6) we can derive another useful formula. Formula (6) tells us the time t required by a dropped object to fall the distance d. Formula (4) tells us the velocity a dropped object acquires in t seconds. Hence, if we substitute the value of t given by (6) in (4), we obtain the velocity acquired by a dropped object in falling d feet. Let us substitute the value of t in (6) for t in (4). The result is

$$v = 32\sqrt{\frac{d}{16}} \cdot$$

Now

$$\sqrt{\frac{d}{16}} = \frac{\sqrt{d}}{\sqrt{16}} = \frac{\sqrt{d}}{4} \cdot$$

Hence

$$v = 32\frac{\sqrt{d}}{4} = 8\sqrt{d}\,. \tag{7}$$

Thus, an object that falls from the top of the Empire State Building, which is about 1000 feet high, will have a velocity of $8\sqrt{1000}$ or about 256 feet per second, which is about 174 miles per hour. Of course the above formulas neglect air resistance.

The derivation of formulas such as (5), (6), and (7) from (2) and (4) illustrates in a small way how Galileo hoped to carry out his program of deriving important laws of nature from a few basic ones. It will be noted that mathematical reasoning based on the physical axioms and the axioms of mathematics permits the deductive derivation of the laws. These examples, as well as others we shall examine shortly, illustrate how the mathematician can sit back in his armchair and obtain dozens of significant laws of nature. His tools, aside from paper and pencil, are the axioms and theorems of mathematics and the axioms of physics, such as the laws of motion. Mathematical deduction is the essence of his work.

Galileo's results thus far tell us what happens when a body which is dropped, or just released from the hand, falls straight down. But suppose an object is thrown downward. This means that the hand exerts a force and imparts a velocity to the object over and above the velocity it gains from the attraction of the earth. Let us determine the velocity and distance fallen by the object in time t. If there were no gravitational force and if the hand gave the object a velocity of, say, 128 ft/sec, then, according to the first law of motion the object would continue at that constant velocity indefinitely. But of course the force of attraction gives the object an additional velocity of $32t$ feet per second in t seconds. These two velocities operate simultaneously to cause downward motion and produce a resultant velocity which is the sum of the two separate ones. Hence the total velocity v after t seconds of fall is

$$v = 128 + 32t. \tag{8}$$

The distance fallen in t seconds is also readily obtained. The distance the object would fall because of the constant velocity given to it by the hand is $128t$. The distance it falls during the same t seconds because of the velocity acquired by the action of gravity is, as we saw above in (5), $16t^2$. The object will therefore fall a total distance of

$$d = 128t + 16t^2, \tag{9}$$

because both velocities are acting simultaneously. It is now easy to substitute any value of t we like in this formula and calculate the distance fallen.

A more interesting variation of the above phenomenon arises when one throws a ball straight up into the air. Let us suppose that it leaves the hand with a velocity of 128 ft/sec. If there were no gravitational attraction to the earth the ball would rise indefinitely at the velocity of 128 ft/sec. But gravity imparts to the ball a downward velocity of $32t$ ft/sec in t seconds. Since the latter velocity is opposite in direction to the former, after t seconds the ball will have a net velocity v given by the formula

$$v = 128 - 32t. \tag{10}$$

This formula should be compared with formula (8).

The formula for the height attained by the ball is also easily obtained. If there were no gravity the ball would attain the height of $128t$ feet in t seconds. However, during these t seconds gravity pulls the ball down a distance of $16t^2$ feet. Hence the height h after t seconds is

$$h = 128t - 16t^2. \tag{11}$$

This formula should be compared with formula (9).

By means of formulas (10) and (11) mathematics can readily answer a number of questions that would be difficult to answer by observation or experimentation. It is a familiar fact that the ball will rise to some definite height and then return to the ground. How high will it go? The ball will rise until its velocity is used up, so to speak, or, in mathematical terms, until its velocity is 0. When will this occur? Let t_1 be the value of t when the velocity is 0. According to formula (10)

$$0 = 128 - 32t_1.$$

By solving this simple equation for t_1 we find that $t_1 = 4$. This means that the ball will reach its highest position after 4 seconds.

To find the maximum height of the ball we have but to substitute this value of t in formula (11). If h_1 is the maximum height attained by the ball then

$$h_1 = 128 \cdot 4 - 16 \cdot 4^2$$

or

$$h_1 = 256.$$

We should note that were the initial velocity larger than 128, t_1 and h_1 would be larger.

Now let us test our intuition a bit. We just saw that it takes the ball 4 seconds to reach its maximum height of 256 feet. How long should it take the ball to reach the ground again? Many people would argue that the ball should reach the ground in less than 4 seconds after attaining its maximum height because the ball falls freely, whereas on its upward flight the action of gravity takes effect slowly in reducing the initial velocity imparted by the hand. Let us see. Let t_2 be the value of t when the ball hits the ground. At that instant the height of the ball above the ground is 0. Hence, from formula (11),

$$0 = 128t_2 - 16t_2^2. \tag{12}$$

We have here a second degree or quadratic equation in the unknown value t_2. But quadratic equations generally, and this one in particular, are readily solved. The easiest way to solve this one is to apply the distributive axiom to write the right-hand side as

$$0 = t_2 (128 - 16t_2).$$

Now the right-hand side is 0 when $t_2 = 0$ or when $128 - 16t_2 = 0$, the latter occurring when $t_2 = 8$.

These values of t_2 are very interesting. The value of 0 for t_2 represents the instant when the ball is just starting out, since time is measured in this problem from the instant the ball leaves the hand. We could have predicted this value of t_2 without any mathematics; yet it is gratifying that equation (12) corresponds so well to the physical facts.

The value of 8 for t_2 is the new information supplied by mathematics. It apparently takes 8 seconds after the ball starts out for it to reach the ground again. Since it takes 4 seconds for the ball to reach its maximum height, it takes only 4 more seconds for the ball to reach the ground. In other words, it takes as much time for the ball to reach its maximum height as it does to reach the ground from there.

Let us test our intuition still further. What velocity does the ball have when it reaches the ground? Would it be the same as that imparted by the hand? More? Less? Let v_2 be the velocity of the ball at the instant t_2 when it returns to the ground. We found that $t_2 = 8$. Hence formula (10) tells us that

$$v_2 = 128 - 32 \cdot 8$$

or

$$v_2 = -128.$$

The minus sign in this answer means merely that the velocity is opposite in direction to the positive velocity the ball has on its upward flight. Formula (10) takes care of these signs automatically. The number 128 answers our question. The ball has exactly the same velocity, except for direction, when it returns to the ground as was given to it originally by the hand.

Our discussion of objects moving vertically has thus far been confined to motion in a vacuum. The primary force causing motion in a vacuum is, of course, the force of gravity or the weight of the object, and this causes a downward acceleration of 32 ft/sec^2. But many centuries before Galileo Archimedes had discovered another principle, which, incidentally, Galileo also considered. Archimedes' principle asserts that an object immersed in a medium such as water is buoyed up by a force equal to the weight (measured in a vacuum) of the water displaced. Hence the net weight of an object immersed in a medium is the force of gravity minus the buoyant force of the medium. Everyone has observed this fact; a person's body feels lighter in water because he is buoyed up by the water. In fact, swimming on the surface of water is possible only because the water offsets the force of gravity and the swimmer's net weight is practically zero.

But now let us apply a little quantitative reasoning to the action of media on masses and see what new information we can deduce thereby. If M is the mass of an object, then, according to formula (3), $W = 32M$ is its weight in a vacuum. Suppose m is the mass of the medium displaced by the object immersed in it. The weight of the displaced medium is then $32m$. According to Archimedes' principle the net weight F of the mass M is

$$F = 32M - 32m. \tag{13}$$

This net weight can be positive or negative depending upon whether $32M$ is larger or smaller than $32m$, that is, upon whether the mass

m of the displaced water is less than or greater than the mass M of the object. If a piece of lead is dropped into water it displaces very little water compared to its mass; hence m is small compared to M. The net weight is practically $32M$ and so the lead sinks rapidly. On the other hand, a balloon filled with hydrogen and placed in air displaces a mass m of air greater than the mass M of the hydrogen. Hence the net weight is negative and the balloon rises.

What we have deduced so far corresponds to what we already know. But let us consider a less familiar situation. An iceberg floats partially immersed in cold water. Can we deduce how much of it is above and how much below the water line? Now cold water is denser than ice because water expands as it is *cooled* from 4° centi-

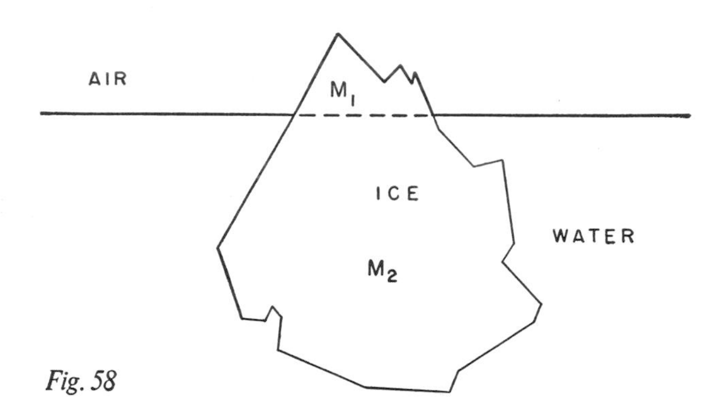

Fig. 58

grade to 0° centigrade. The ratio of the densities is 1.08. This means that whatever mass of ice may be in the water, the mass of the cold water displaced will be 1.08 times the mass of the ice that replaces it.

Let M_1 be the mass of the iceberg above water (fig. 58) and let M_2 be the mass of the iceberg below water. The weight of the iceberg in a vacuum is $32\ (M_1 + M_2)$. In water it is buoyed upward by the weight of the water displaced. But the mass of the displaced water is $1.08M_2$ and its weight is therefore $32\ (1.08M_2)$. Since the iceberg floats, its net weight must be zero. That is,

$$32(M_1+M_2) - 32(1.08M_2) = 0.$$

We may divide both sides of this equation by 32 and then combine the two terms containing M_2. We find that

$$M_1 = .08M_2.$$

In words, the mass of the iceberg above water is about $\frac{1}{12}$ of the mass below water. We can see why a ship striking an iceberg is likely to suffer far more damage than the visible mass would lead one to expect.

The examples of motion presented thus far show that Galileo had hit upon a very valuable method of studying nature. But what we have seen is child's play. The power of his mathematical method has hardly been tested.

13: MOTION ON AN INCLINED PLANE

But where the senses fail us reason must step in.
—GALILEO

IN HIS study of falling bodies Galileo found it difficult to check his mathematical deductions against observational data because the acceleration, and therefore the velocity, of the bodies was so rapid. It must be remembered that Galileo did not have a good clock at his disposal. Though very large mechanical clocks of the size one sees now on towers of public buildings were in use before Galileo's time, the clocks used in homes or such laboratories as existed were water clocks, in which time was measured by means of how much water flowed out of a graduated tube. Obviously the measurements of minutes and certainly of seconds could hardly be made accurately with such clocks.

It occurred to Galileo that the action of gravity could be studied more effectively by slowing up the process. One way to do this would be to cut away part of the earth, since a smaller earth would attract objects more weakly and hence cause less acceleration. But Galileo used a simpler scheme. He noticed, as most of us surely must have, that an object sliding or rolling down a hill moves more slowly than one falling straight down. Presumably, then, the effect of gravity along a hill is weaker than in a free fall. A hill of constant slope is, mathematically, an inclined plane and so Galileo decided to study motion down an inclined plane. Of course such motions are practically important: Trains and automobiles ascend and descend hills. Along the assembly lines of large factories objects are often passed from one operation to another at a lower level by allowing them to slide down inclined planes. Hence, there are good reasons in addition to Galileo's to study the mathematical and physical problem of motion along slopes.

Suppose an object of mass m rests on a smooth inclined plane (fig. 59). Gravity exerts a force on this mass, but this force seeks to pull the mass straight down. The fact that the force of gravity acts in this definite direction is all important for what follows. Since the plane itself prevents motion straight down, the question arises, To what extent is the force of gravity effective in causing the object to slide down the plane? Does the full force of gravity apply or is it in some way dependent upon the steepness of the slope?

Intuition supplies a partial answer at once. Certainly if the inclination of the plane, angle A, were zero, that is, if the plane were horizontal, the object would not slide. As the plane is inclined more and more one expects the force down the plane to be stronger because one observes that the mass slides faster. Of course these two facts are related by the second law of motion, $F = ma$. If the force

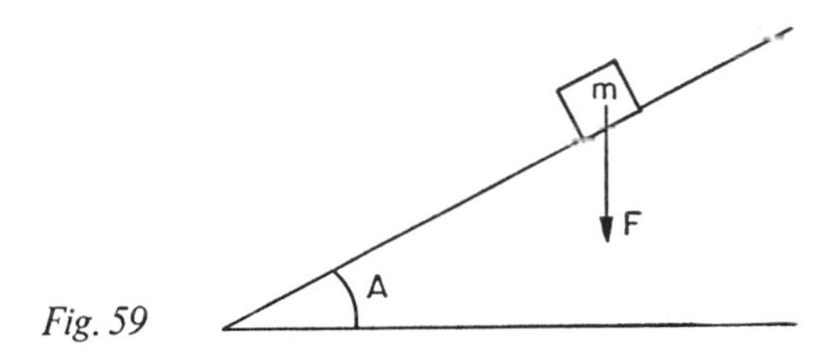

Fig. 59

down the plane is greater then the acceleration is more, and the object gains velocity faster. When the plane is vertical, that is, when the angle of inclination is 90°, and the full force of gravity applies, the object falls freely. Hence the force of gravity down the plane increases with the angle of inclination of the plane. But precisely how does the force vary the inclination?

A mathematician facing some variable quantity, the force down the plane in the present case, which depends upon an angle, immediately thinks of the trigonometric ratios and tries to fit one of these to the problem in hand. In view of what intuition suggests, we should ask which of these ratios is 0 when the angle is 0 and increases to 1 when the angle is 90°. Of the three ratios we have met, sin A meets these conditions. Hence it appears as though the force of gravity straight down multiplied by sin A is the effective component of gravity—this is the technical phrase—along the plane.

This conjecture can be checked experimentally. We place an

object of mass m, and therefore of weight W equal to $32m$, on an inclined plane and connect the object by a rope passed over a pulley at the top (fig. 60) to another object of weight w. Suppose w, to which the full force of gravity applies, is chosen just to balance the force tending to move the weight W down the inclined plane. Then one finds that $w = W \sin A$. In other words, the effective weight of the object on the plane, that is, the effective force of gravity down the plane, is not W but $W \sin A$.

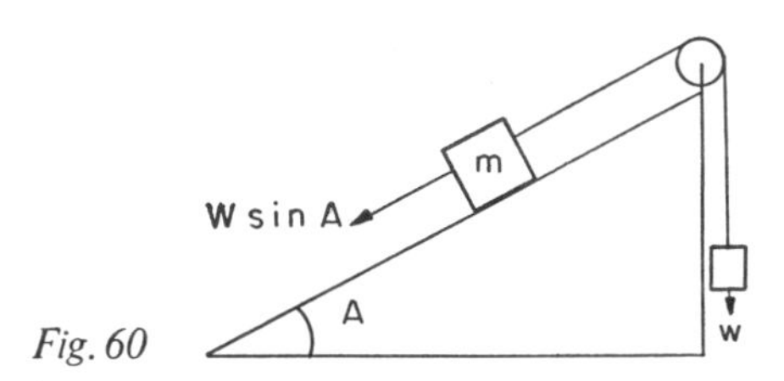

Fig. 60

But now what can be said about the acceleration of objects sliding down inclined planes? If an object of mass m is impelled by a force $W \sin A$, what acceleration would the object gain? The second law of motion says that the force equals the mass times the acceleration. Then

$$W \sin A = ma.$$

But $W = 32m$. Thus we have

$$32m \sin A = ma,$$

or, by dividing both sides of this equation by m, we find that

$$a = 32 \sin A.$$

Hence the action of gravity on a mass resting on an inclined plane of inclination A is to give the mass an acceleration of $32 \sin A$ instead of the usual 32. Galileo was correct in expecting that he could better study the action of gravity by sliding masses down an inclined plane, because on such a plane the acceleration can be made as small as desired by making angle A small enough.

Now that we know the acceleration of objects sliding down inclined planes, what can we deduce about the subsequent motion? For example, if a mass m starts from rest and slides down an inclined plane what velocity will it acquire and how far will it travel in a given time t? Since the acceleration, that is, the gain in velocity each second, is 32 sin A, and the mass starts with zero velocity, the velocity v at any time t is

$$v = (32 \sin A)t,$$

or, as it is usually written,

$$v = 32t \sin A. \tag{1}$$

To obtain the distance traveled in time t we may use the very same argument we used to derive formula (5) of the preceding chapter except that the number 32 sin A now replaces the 32 there. The distance d the object slides in time t is given by

$$d = 16t^2 \sin A. \tag{2}$$

Let us now consider how long it will take an object to slide from a height h to the ground along a slope whose length is l (fig. 61).

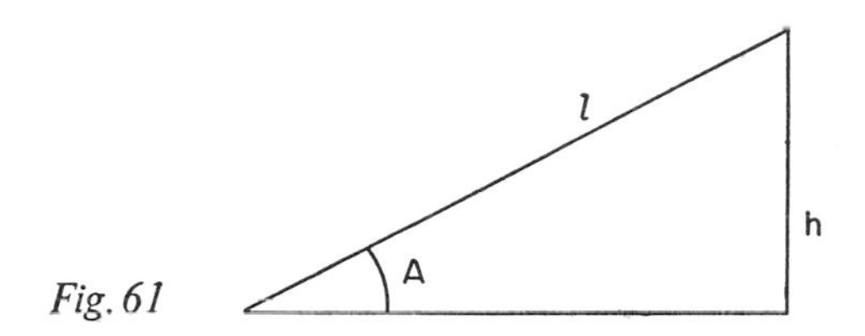

Fig. 61

The time required to travel the distance l is obtainable from formula (2) when $d = l$. Thus, letting t_1 be the time required, we have

$$l = 16t_1^2 \sin A.$$

Dividing both sides of this equation by 16 sin A yields

$$t_1^2 = \frac{l}{16 \sin A}.$$

Then, by taking the square root of both sides and retaining only the positive square root because this root is the physically meaningful one, we have

$$t_1 = \sqrt{\frac{l}{16 \sin A}}. \tag{3}$$

But, as is evident from figure 61,

$$\sin A = \frac{h}{l};$$

by multiplying both sides of this equation by l and dividing by $\sin A$, we obtain

$$l = \frac{h}{\sin A}.$$

If we substitute this quantity in (3) we find that

$$t_1 = \sqrt{\frac{1}{16 \sin A} \cdot \frac{h}{\sin A}} = \sqrt{\frac{1}{16 \sin^2 A} \cdot h},$$

where $\sin^2 A$ means $\sin A \cdot \sin A$. But the square root of a product equals the product of the square roots of the factors. Hence

$$t_1 = \frac{1}{4 \sin A} \sqrt{h}. \qquad (4)$$

This result is by no means surprising. If we consider two inclined planes with the same height h but with different inclinations (fig. 62), then the one with the smaller inclination must be longer.

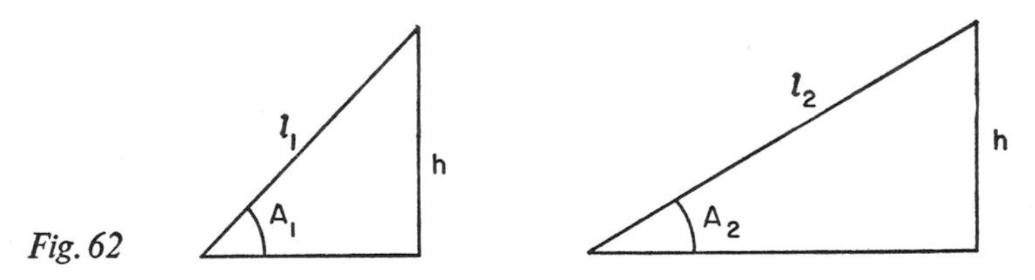

Fig. 62

Moreover, if the inclination A is smaller, $\sin A$ will be smaller and the acceleration of an object sliding down, namely, $32 \sin A$, will be less. Hence the time required for the object to slide down the plane of smaller inclination should be greater. Formula (4) expresses this fact. For small $\sin A$, t_1 is greater than for large $\sin A$ because $\sqrt{h}$ is divided by a smaller quantity. In other words, formula (4) tells us that it takes longer for an object to slide down a gradually inclined plane from a height h than from a steep plane of the same height.

We can in fact make this comparison a little more precise. If l_1 and l_2 are the lengths of the inclined planes with inclinations A_1

and A_2 respectively and with the same height h, then since

$$\sin A_1 = \frac{h}{l_1}$$

it follows from formula (4), by substituting h/l_1 for $\sin A$, that

$$t_1 = \frac{1}{4 \cdot \frac{h}{l_1}} \cdot \sqrt{h} = \frac{l_1}{4h} \cdot \sqrt{h}.$$

If we call t_2 the time to descend the plane of length l_2, then it follows likewise that

$$t_2 = \frac{l_2}{4h} \sqrt{h}.$$

Hence

$$\frac{t_1}{t_2} = \frac{l_1}{l_2}.$$

That is, the times of descent are proportional to the lengths of the inclined planes.

Let us consider next the velocity that an object acquires in sliding down an inclined plane from height h (fig. 61). The velocity of the sliding object at any time t is given by formula (1). The time required to slide the distance l is given by formula (4). If we substitute this value of t_1 in formula (1) we obtain for the velocity v_1 at the time t_1

$$v_1 = 32 \sin A \cdot \frac{1}{4 \sin A} \sqrt{h},$$

or

$$v_1 = 8 \sqrt{h}. \tag{5}$$

Here is an unexpected conclusion. The velocity achieved by the mass in sliding the distance l does not depend upon l or upon the inclination of the plane but solely on the height from which the mass descends. In other words, if an object slides down a long gradual hill or a short steep one, but descends a *vertical* distance of h feet in both cases, the velocity at the bottom is the same. As a matter of fact the velocity would still be the same even if the surface of the hill were curved instead of flat, but our mathematics does not cover this case.

Since the velocity at the bottom depends only upon the vertical distance descended by the object, then perhaps this velocity should be the same as that of a freely falling body which falls straight down the distance of h feet. This conjecture is established at once by mathematics. Formula (7) of the preceding chapter says that the velocity v of an object which starts from rest and falls h feet is

$$v = 8\sqrt{h}\,. \tag{6}$$

Hence the velocity acquired in falling h feet is precisely the same as that given by formula (5).

We should notice, too, that the results obtained for the time and velocity of motion on an inclined plane, formulas (4) and (5), are independent of the mass of the object, just as in the case of freely falling bodies. Hence a ton of lead and a one-pound piece of wood take the same time and attain the same velocity in sliding down the same hill. Of course friction is ignored in these calculations.

Thus far in our discussion of motion along inclined planes we have considered objects that slide down the planes. When Galileo performed his experiments on inclined planes he allowed small spherical balls to roll down the planes because thereby he avoided most of the effect of friction. However, the use of rolling balls raises a question. Suppose one solid disk should slide down a plane and another should roll down (fig. 63). Which one would acquire

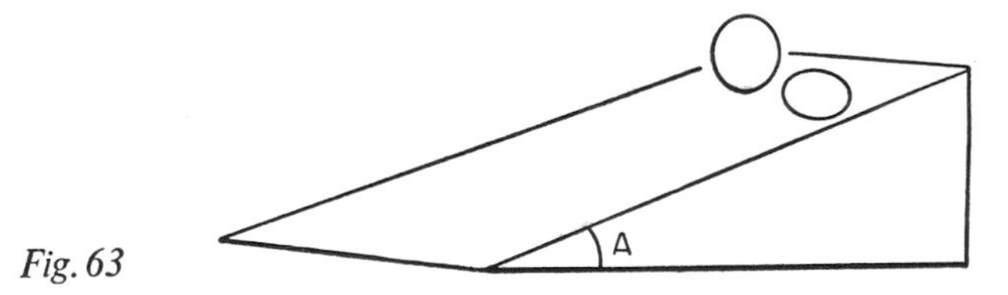

Fig. 63

the greater velocity and therefore reach the bottom sooner? Intuition or experience suggests to most people that the rolling disk would acquire the greater velocity. This is because we generally see objects rolling rapidly but sliding slowly. However, if friction is neglected, as it is in our discussion, the sliding disk will acquire more velocity and reach the bottom sooner than the rolling disk. Moreover, the greater the height of the plane the greater will be the difference in velocity and time of descent. We shall not deduce the formula for the velocity of the rolling disk because this would lead us too far into the physics of motion, but we can explain why the result we have stated is correct. In both cases gravity supplies the force that causes the disks to descend. However, in the case of the rolling disk some of the force is expended in giving the disk the acceleration and velocity of rotation, and only part of the force is expended in giving it acceleration and therefore velocity down the plane. In the case of the sliding disk the entire force of gravity is directed toward causing acceleration down the plane. Of course the rolling ball will still have a rotational velocity when it reaches the bottom but its motion down the plane is nevertheless slower than that of the sliding disk.

For Galileo, inclined planes served as a purely scientific device to study the action of gravity more readily. But today inclined planes, as we have already noted, serve practical purposes. One of the commonest applications of the inclined plane is to raise objects from one level to another. One often sees a heavy object being slid up a flat board from the street level to the floor of a truck. Why is the inclined plane used? We have already learned the answer. An object of mass m and therefore of weight W equal to $32m$ resting on a plane of inclination A exerts a force $W \sin A$ down the plane. To counter this force one need exert only the force $W \sin A$ up the plane, and if one exerts slightly more force than this the object can be set into motion up the plane. Now $W \sin A$ is less than W because $\sin A$ is less than 1 for A less than 90°. Hence the force that must be exerted to raise a mass from one level to another along an inclined plane is less than the force that must be exerted if the object were being lifted straight up. Moreover, if the inclination A of the plane is made small, the force that must be exerted is also small. Hence even a child could push a piano up an inclined plane if the inclination of the plane were made small enough. (Of course we are not considering friction.)

We seem, so to speak, to be able to cheat nature. But we have overlooked something. The more gradual the slope the longer the distance one must travel up that plane to reach a given height h (fig. 62). Perhaps we should investigate not only the force required to move a weight up a slope to a given height h but also the distance involved. We found above that the distance l required to reach height h is $h/\sin A$. The force that must be exerted, we know, is $W \sin A$. A mathematician would immediately notice that the product of these two quantities, $h/\sin A$ and $W \sin A$ is simple, and, being simple-minded, he would look for some meaning and use for this fact.

The product in question is

$$\text{force} \cdot \text{distance} = W \sin A \cdot \frac{h}{\sin A} = Wh. \tag{7}$$

This result says that the force required to pull the object up the plane multiplied into length l of the slope, which is the distance through which the force acts, is independent of the inclination A and the distance l; the product depends only upon the weight W itself and the height h to which the object is raised. But the result states even more. W is the full weight of the object. If this object were raised straight up one would have to exert a force of magnitude W. To raise this weight a distance h, the product of force and distance would be Wh.

Hence to raise a weight W to a height h, the product of force applied and the distance over which it is applied is the same whether one uses a gradual slope over the requisite distance, a steep slope over the corresponding shorter distance, or no slope at all. To put the result in physical terms, one can exert himself rather little over a long distance or exert himself strongly over a short distance. He will do the same amount of work in both cases. Nature seems to care not about the force required but about the total work done. As a consequence physicists have framed a definition of the technical concept of work as the product of force applied and the distance through which the force acts. The work required to do a given job is the same no matter how it is achieved.

To do work one must expend energy. Hence the foregoing result amounts to the statement that one cannot cheat on the energy required to do a given job. Energy is neither lost nor gained in physical processes. The inclined plane therefore furnishes a concrete example of a law of nature known as the conservation of energy.

The fact we have just unearthed would be regarded by most people, except possibly those who are trying to cheat nature, as a pretty one. The mathematical theory of the inclined plane offers other pretty facts about motion. We have already learned that if an object slides down an inclined plane from a point O (fig. 64) and reaches the horizontal it requires more time than if the object falls straight down. Thus the object requires more time to slide the distance OP' than to fall the distance OP. There must then be some point Q on OP' such that the time to descend the distance OQ equals the time required to descend OP. Is there anything mathematically significant about the location of the point Q?

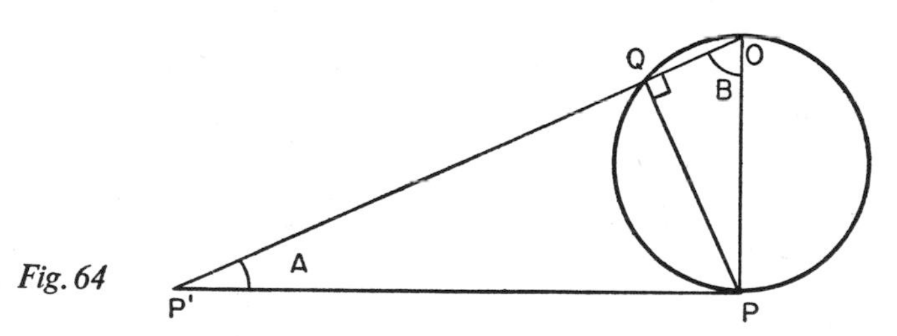

Fig. 64

To answer this at the moment purely academic question, let t_1 be the time required to fall OP. Then, according to formula (5) of chapter 12,

(8) $$OP = 16t_1^2.$$

But according to formula (2) of the present chapter the distance OQ that an object will slide along the plane OP' in time t_1 is

$$OQ = 16t_1^2 \sin A.$$

Figure 64 tells us that

$$\sin A = \frac{OP}{OP'} \text{ and } \cos B = \frac{OP}{OP'} .$$

Since $\sin A = \cos B$, we may write that

(9) $$OQ = 16t_1^2 \cos B.$$

Comparison of results (8) and (9) shows that

$$OQ = OP \cos B,$$

or that

(10 $$\cos B = \frac{OQ}{OP} .$$

This result says that OQ is the side adjacent to angle B in the right triangle whose hypotenuse is OP. Hence OQP must be a right triangle.

If we now consider the circle drawn on *OP* as a diameter then our preceding statement says that *Q* must lie on this circle, because according to a theorem of Euclidean geometry the vertices of all right angles whose sides pass through *O* and *P* lie on a circle with *OP* as diameter. Now *OP′* is *any* inclined plane descending from *O*. Hence on any such inclined plane the point reached by an object in the time t_1 required to fall the distance *OP* straight down lies on the circle with *OP* as diameter.

Now let us consider a set of straight pieces of wire all stemming from a point *O* (fig. 65), the entire structure supported so as to be

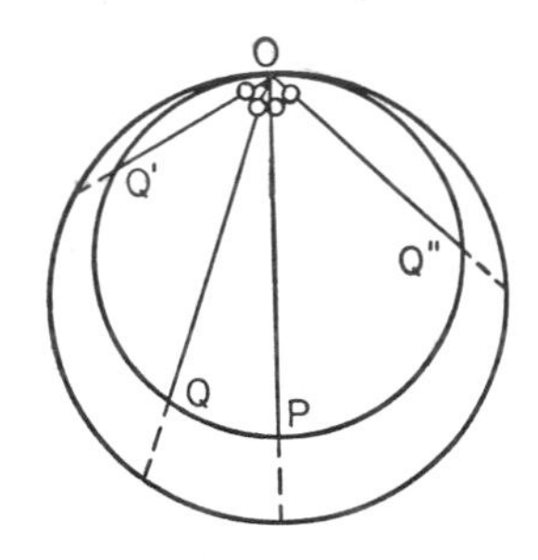

Fig. 65

vertical. A bead is slipped on each wire and held at *O*. All the beads are then released at the same instant and allowed to slide on their respective wires under the action of gravity. What is the subsequent motion of the beads?

At some time *t* the bead that falls straight down may be at *P*. A typical bead on one of the inclined wires will reach *Q*, say on *OQ*, in that same time *t*. We proved above, however, that *Q* lies on a circle with *OP* as diameter. Since *Q* is representative of any one of the beads, the entire set will then lie at time *t* on the circle with *OP* as diameter. As *t* increases *OP* increases, and the circle of beads expands. Hence the beads will lie successively on a series of expanding circles that are readily visualized.

The motion of the beads on the wires is a rather pretty occurrence of circles in mechanical motion. But those people who are more impressed by utility than by beauty may also find something to their liking in this motion. Suppose one wishes to have an object slide along a straight line from a fixed point *O* to the curve *C* (fig. 66) in the least possible time. Along what straight-line path should

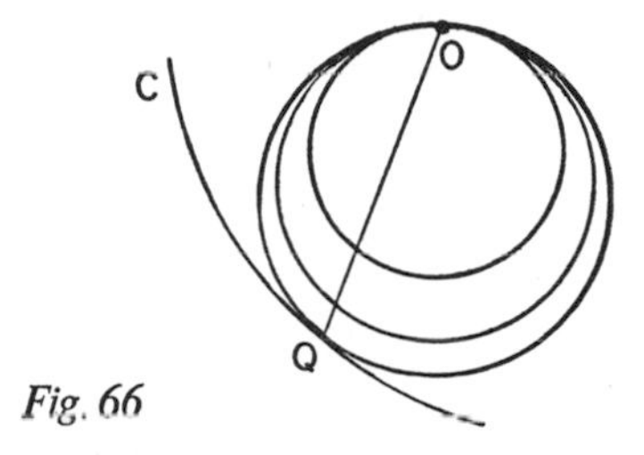

Fig. 66

the object be made to slide? We know from the foregoing study that at any time t objects which slide down the various straight lines emanating from O lie on a circle that has O as its highest point. Consider larger and larger circles through O with O as the highest point on each circle. Each of these circles corresponds to some value of t. The circle which first touches C is the one that corresponds to the smallest value of t for which an object will reach C. Suppose Q is the point on C at which the circle through O touches C. Then OQ is the desired path. Thus if one wished to transport objects from a point O to a hillside having the shape C by sliding them down an inclined plane the straight-line path requiring least time would be OQ.

14: THE MOTION OF PROJECTILES

I value the discovery of a single even insignificant truth more highly than all the argumentation on the highest questions which fails to reach a truth.

—GALILEO

BY analyzing motion along a straight line, whether it be of bodies falling straight down or sliding down inclined planes, Galileo discovered some fundamental physical principles, namely, the first two laws of motion. These laws are in themselves quantitative statements about velocity, force, mass, and acceleration. Moreover, by applying mathematical reasoning, with these laws and the axioms and theorems of mathematics as his basic facts, Galileo made some fundamental discoveries about motion. He was immensely encouraged to proceed with more difficult problems of motion, and in his *Dialogues Concerning Two New Sciences,* a book that had to be secretly transported to Holland for publication because the Church had banned publications by Galileo, he continued his investigations.

His next subject, a new and pressing problem of his day, was the motion of projectiles shot from cannon. He began this study with a somewhat simpler problem, the motion of an object thrown out horizontally but which as it moves out is pulled downward by the action of gravity. Interest in the motion of projectiles has of course by no means diminished since Galileo's time. Bombs dropped from airplanes are examples of objects thrown out horizontally and pulled downward by the action of gravity, and shells shot out from cannon are still a feature of the modern scene. In fact "progress" now enables us to send projectiles thousands of miles whereas in the seventeenth century the range was limited to a few miles. Fortunately, just as diamonds are often found in the most disagreeable surroundings, so some of the brightest scientific ideas can be found in the study of the motion of bombs and shells.

The paths of projectiles are not straight lines but curved. Galileo therefore undertook the study of curvilinear motion. He made several capital discoveries which opened up the mathematical treatment of the subject. If a person rides on a train he shares the motion of the train; in fact, this is why he pays his fare. Likewise, objects on the earth share the motion of the earth and for this reason among others are not left behind by the rotation and revolution of the earth. All of this, of course, seems rather obvious as stated. But now suppose a passenger in a plane suddenly falls out of the plane. He still has the forward velocity that the plane imparted to him while he was a passenger and, were it not for air resistance and the downward pull of gravity, he would continue to travel forward at the same speed as the plane in accordance with Galileo's first law of motion. This fact seems less credible despite the fact that it is verified for us almost daily. Suppose a person rides in an automobile at, say, 60 miles per hour (a trifling speed for modern man, who seems to enjoy going nowhere very fast), and then the brakes are suddenly applied. Of course the automobile's velocity is rapidly decelerated but the passenger's velocity is not. He continues to move forward and, if he ever recovers consciousness, will readily agree that he continued to move at at least 60 miles per hour.

An object dropped from a plane, then, continues to move forward, but it also moves downward under the attraction of gravity. While reflecting on this type of motion Galileo discovered a fundamental principle: the horizontal and downward motions proceed *independently* of each other and the position of the object at any time is determined by considering the forward motion and the downward motion separately. This idea was revolutionary in Galileo's time. The idea that an object can be activated by two forces which do not interfere with each other was contrary to the teachings of Aristotle. He maintained that one force would disturb the other. The predecessors of Galileo, among them Tycho Brahe, accepted Aristotle's teaching unhesitatingly. Yet if we stop to think about Galileo's principle we realize that it must be correct. When a boy throws a ball to a playmate he relies upon the force imparted by his hand and the force of gravity to carry the ball to its goal despite the fact that the boys and the ball are moving with the earth.

Galileo proceeded to use this principle in the study of curvilinear motion. Let us follow his thinking by discussing the motion of an object released from an airplane flying horizontally at 100 feet per second. Those readers who would see more point to the following discussion if the object were a bomb may substitute this word where relevant. We shall consider first the horizontal motion. Since, when released from the plane, the object will have a horizontal velocity of 100 feet per second, and since the first law of motion says that an object in motion and undisturbed by forces will continue indefinitely at the same speed and in the same direction, the object will move horizontally a distance x in time t given by the formula

$$x = 100t. \tag{1}$$

The object will also move downward independently. But the downward motion is due solely to the action of gravity, and we know that a body falling downward from rest falls $16t^2$ feet in t seconds. Hence if we let y be the vertical distance traveled *downward* then

$$y = 16t^2. \tag{2}$$

Equations (1) and (2) taken together describe the entire motion. Thus after one second, $x = 100$ and $y = 16$. The position

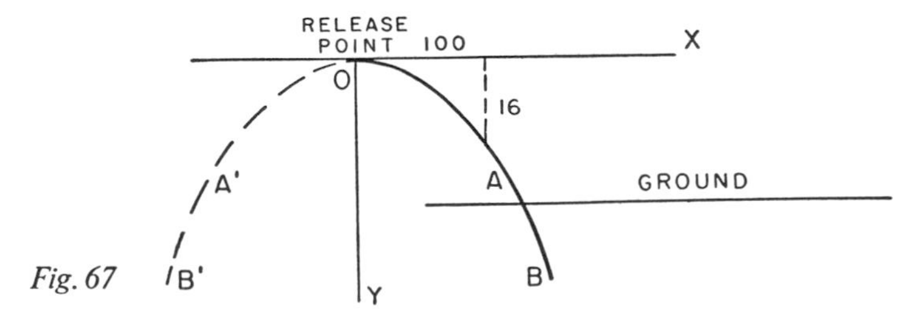

Fig. 67

of the object at this instant is then determined by these values of x and y as shown in figure 67. This scheme of representing a motion and the corresponding curve by two formulas, one describing the horizontal or x-value in terms of the variable t and the second describing the vertical or y-value in terms of t is known in mathematics as the parametric representation of a formula or curve. More often the relationship between y and x is expressed directly in terms of a single formula involving y and x only, as was the case in our study of the equations of curves in chapter 10.

The first question Galileo raised was, What is the shape of the path along which the object travels? To answer this question let us attempt to use our knowledge of equation and curve. Because we are more accustomed to direct relations between y and x rather than parametric relations, let us seek it. At any value of t, the horizontal distance covered by the object is given by formula (1). This says that

$$t = \frac{x}{100} \cdot \tag{3}$$

At the same value of t the vertical distance covered by the object is given by formula (2). We may therefore substitute the value of t given by (3) into formula (2) and obtain

$$y = 16 \left(\frac{x}{100}\right)^2$$

or

$$y = \frac{x^2}{625} \cdot \tag{4}$$

What curve corresponds to this equation?

In our study of curve and equation we found that an equation such as $x^2 = 20y$, (compare formula (10) of chapter 10) represents a parabola which opens upward. Our present formula (4) differs in two respects. First, the number 625 occurs here where the number 20 occurred previously. This is a detail. If in place of the distance of 10 from focus to directrix, which we used in deriving the equation $x^2 = 20y$, we had the distance $625/2$ we would have obtained formula (4) as the equation of the parabola. In other words, any equation of the form

$$y = \frac{x^2}{a} \quad \text{or} \quad ay = x^2 \tag{5}$$

is a parabola. Hence formula (4) also represents a parabola. However, our parabola opens downward because in deriving formula

(4) we agreed to let positive y values be represented by distances downward. Thus the curve represented by formula (4) and shown in figure 67 is a parabola.

There is one further point to be noted about equation (4). Mathematically this equation represents the entire parabola that extends from O along the unending arc OAB and along the unending arc $OA'B'$. However, for our study of the motion of the falling object, only the arc OAB has physical significance. Moreover, the arc OAB continues downward indefinitely, but the physical motion ceases when the object strikes the ground. Hence, of the half of the parabola that lies to the right of the Y-axis only a finite portion is of physical interest. In this situation, as in others we have encountered and shall encounter, mathematics represents the physical phenomenon but must not be identified with the physical phenomenon.

The fact that an object released from an airplane, say, travels along a parabola seems quite acceptable today but in Galileo's time the idea did not win favor. According to Aristotle every object falling downward seeks the center of the earth, whereas an object falling along a parabola, or the arc OAB of figure 67, travels away from the center of the earth insofar as its horizontal motion is concerned. Hence Sagredo, one of the imaginary participants in the dialogues of Galileo's book, objected to Galileo's result.

Whether or not the knowledge that the curve is a parabola is gratifying there are one or two implications that are intriguing. Suppose we consider two airplanes, one flying horizontally at 100 feet per second and the other at 1000 feet per second. Each releases an object that then falls along a parabolic path to the ground, say one mile below (fig. 68). Which of these objects will reach the ground sooner? The answer is that they both reach the ground at the *same* time. The downward motion is governed entirely by formula (2) and this formula does not involve the horizontal velocity. To find how long it takes each object to fall one mile or 5280 feet one has merely to find that value t_1 of t for which

$$5280 = 16t_1^2.$$

Moreover, since, in accordance with our study of motion under the law $d = 16t^2$ in chapter 12, the vertical velocity is given by the

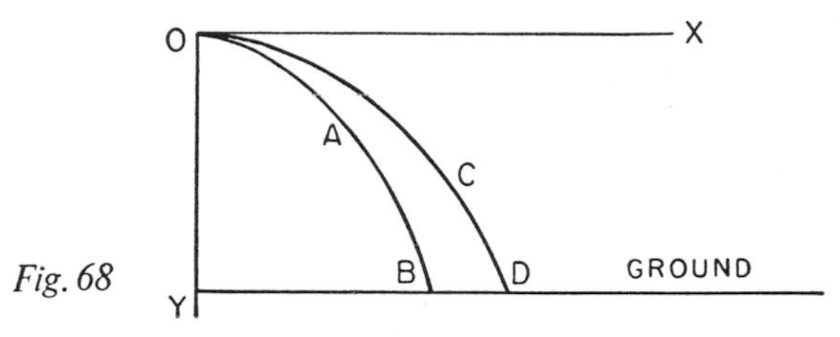

Fig. 68

formula $v = 32t$, the *vertical* velocity with which both objects hit the ground will be the same.

What effect, then, does the horizontal velocity of the plane have on the motion? The object released from the plane with the greater horizontal velocity will reach the ground at a point farther away from the point directly below the point of release (figure 68). The justification of this statement is readily made. According to formula (1) the object released from the plane having the velocity 100 feet per second will travel the horizontal distance $100t_1$ feet in t_1 seconds; the object released with a velocity of 1000 feet per second will travel the horizontal distance $1000t_1$ feet in t_1 seconds. In other words, the second object will travel along the wider parabola *OCD* rather than the path *OAB* of the first object.

We are now prepared to make the next step with Galileo, namely, to consider the motion of projectiles shot from cannon located on the ground. Numerous problems involving such motion were troubling the mathematicians and scientists of Galileo's time. If a shell is fired from a cannon inclined at a given angle to the ground how far will it travel? Another type of question arose in trying to fire a projectile so that it would clear the wall of a fort. One cannot simply aim the projectile at the top of the wall because the projectile will not follow a straight-line path. On the other hand, if the projectile is launched at too high an angle to the ground it might follow the path *OPQ* of figure 69 and still fail to clear the wall. Suppose one wishes to hit a target above the ground, such as

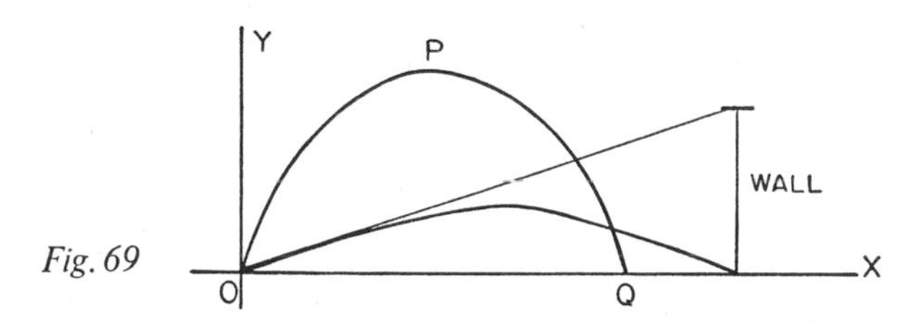

Fig. 69

a tower. At what angle should the projectile be fired? Again one cannot aim it directly at the tower because the projectile will certainly not follow a straight-line path to the tower. What is the effect of muzzle velocity on the projectile? All of these problems and many others cannot be decided by guesswork or intuition. Even experimentation that might be undertaken at great cost and labor would give results for one or two cannon, but these results might not be applicable to others. Let us follow Galileo's mathematical method.

Suppose the projectile leaves the gun at time $t = 0$ with a velocity V, directed upward at an angle A to the ground (fig. 70). This velocity is equivalent to two velocities, one in the horizontal direction and one in the vertical direction, which may be deter-

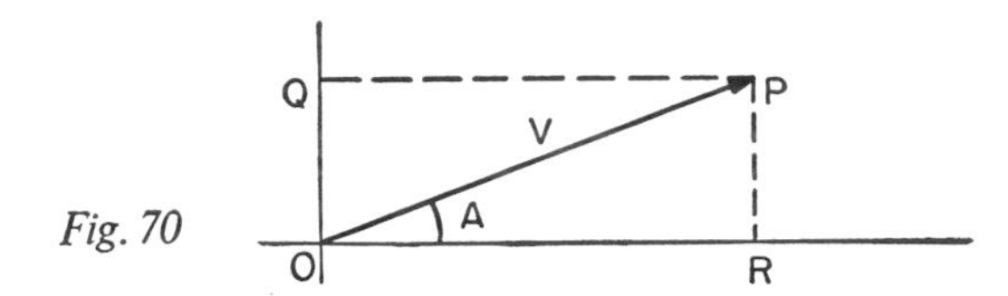

Fig. 70

mined as follows. Let us suppose that the projectile were to travel for one second in the direction OP in which it was fired. Then in that second it will travel some distance which we may as well take to be OP itself, since the unit of length is at our disposal. However, insofar as the motion straight up is concerned the distance covered in one second would be OQ. Hence OQ is the vertical velocity and this is seen to be the same as RP or $V \sin A$. Likewise, insofar as motion in the horizontal direction is concerned, in one second the projectile would travel the distance OR. Hence the horizontal velocity is OR or $V \cos A$. We have, then, that a velocity V in a direction making an angle A with the horizontal is equivalent to a velocity in the vertical direction of amount $V \sin A$ and a velocity in the horizontal direction of amount $V \cos A$.

Now, according to Galileo, the projectile can be regarded as moving independently and simultaneously with the velocity $V \cos A$ in the horizontal direction and the velocity $V \sin A$ in the vertical direction. Let us investigate these two separate motions. First, the horizontal motion. Since no force affects the horizontal motion, Galileo's first law of motion says that the projectile will continue to travel indefinitely at the constant speed $V \cos A$ in the horizontal direction. Thus the velocity v_x in the horizontal direction is

$$v_x = V \cos A. \tag{6}$$

Since this velocity is constant, the horizontal distance x traveled in time t is

$$x = (V \cos A)t = Vt \cos A. \tag{7}$$

The vertical motion is slightly more complicated. If there were no gravitational force pulling the projectile downward, it would continue to travel upward at the constant velocity $V \sin A$. However, as the projectile rises gravity pulls it down. Now we saw in chapter 12 that any object subject to the action of gravity will have an acceleration in the downward direction of 32 feet per second each second and will therefore acquire a downward velocity of $32t$ feet in t seconds. Hence the first statement we may make about the vertical velocity of the projectile is that its net velocity v_y at any time t is

$$v_y = V \sin A - 32t. \tag{8}$$

How far above the ground will the projectile be any time t? Well, the velocity $V \sin A$ will cause the projectile to rise $Vt \sin A$ feet in t seconds. During these t seconds the action of gravity will pull the projectile downward $16t^2$ feet. Hence the net vertical distance y attained by the projectile in t seconds is

$$y = Vt \sin A - 16t^2. \tag{9}$$

In deriving formulas (8) and (9) we have assumed that the motions of rising and falling may be considered separately, even though they take place simultaneously. But each motion is due to a separate and independent force, just as the horizontal and vertical motions are independent, and, consequently, our derivation rests on a sound physical principle.

Formulas (7) and (9) describe the separate motions in the horizontal and vertical directions, each giving a distance in terms of time t. For any given time t we could calculate x and y and thus know where the projectile is at that time. However, we might like to know what the actual path of the projectile is without having to calculate successive x and y values. The answer is easily obtained by algebra. By solving equation (7) for t we obtain

$$t = \frac{x}{V \cos A} \cdot$$

If we substitute this value of t in equation (9) we obtain

$$y = V \sin A \frac{x}{V \cos A} - \frac{16x^2}{V^2 \cos^2 A},$$

or

$$(10) \qquad y = \frac{\sin A}{\cos A} x - \frac{16}{V^2 \cos^2 A} x^2.$$

Now A and V, the angle at which the projectile is fired and the velocity with which it is fired, are constants; hence so are $\sin A$ and $\cos A$. Then equation (10) is of the form

$$(11) \qquad y = ax - bx^2$$

where $a = \sin A/\cos A$ and $b = 16/V^2 \cos^2 A$.

The curve that corresponds to an equation of the form (11) is not immediately indentifiable on the basis of our rather limited study of coordinate geometry. However, a slight extension of the study we made there of curve and equation would show that any equation of the form (11), and therefore equation (10) in particular, is a parabola which lies with relation to the axes as shown in

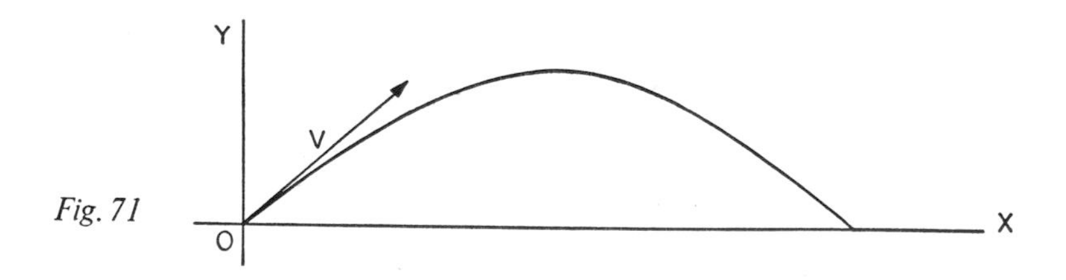

Fig. 71

figure 71. The general conclusion, then, is that the path of a projectile fired at an inclination A and with an initial velocity V in the direction of fire is a parabola.

Let us now proceed to answer some of the questions that troubled Galileo and his contemporaries. How high will a projectile go if fired at a given angle, and how long will it take to reach that height?

To be specific, let us suppose that a projectile is fired at an angle of 53° with a muzzle or initial velocity of 128 feet per second. (These numbers have been chosen merely to make the arithmetic easier.) To answer the question of how high the projectile will go, that is, how large y_1 in figure 72 is, we note first that the vertical motion is governed solely by formulas (8) and (9). However, while these formulas give us the vertical velocity and vertical height

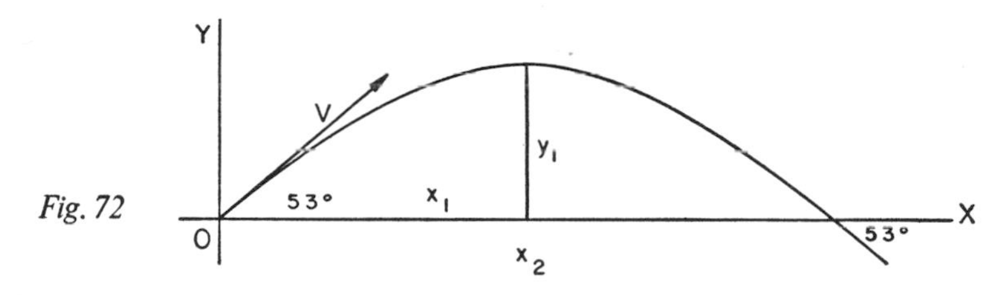

Fig. 72

at any time they do not in themselves tell us the value of t when the projectile is at its maximum height. What fact about the motion does single out this value of t?

The answer is that at the instant when the projectile is at the highest position in its path the *vertical* velocity must be zero, else the projectile would continue to rise. For the value 53° of angle A and the value 128 of V, formula (8), which gives the vertical velocity, reads

$$v_y = 128\ (.8) - 32t,$$

wherein 0.8 represents sin 53°. Let us therefore seek the value t_1 of t when $v_y = 0$. Then

$$0 = 128\ (.8) - 32t_1.$$

This first degree equation is readily solved for t_1 and we find that

$$t_1 = \frac{128(.8)}{32} = 3.2.$$

In words, the projectile reaches its maximum height in 3.2 seconds.

To find the maximum height reached by the projectile we have but to substitute this value of t_1 in formula (9), wherein, of course, we must replace V by 128 and sin A by 0.8. Thus

$$y_1 = 128\ (3.2)\ (.8) - 16\ (3.2)^2 = 163.8.$$

The projectile will reach a maximum height of 163.8 feet.

Having found the time required by the projectile to reach maximum height, as well as the maximum height, we might ask how long it will take the projectile to reach the ground again. Does it take as long to reach the ground as it did to reach the maximum height, longer, or less time? We need not rely upon our unreliable intuition to answer this question. The answer can be obtained from formula (9). At the instant when the projectile reaches the ground the y-value in formula (9) must be zero. Let t_2 be the value of t when y in formula (9) is zero. Then since $V = 128$ and $\sin A = 0.8$ we have

$$0 = 128\ (.8)t_2 - 16t_2^2.$$

To simplify the arithmetic we may divide both sides of this equation by 16 and obtain

$$0 = 8\ (.8)t_2 - t_2^2.$$

We may solve this equation for t_2 by applying the distributive axiom to write

$$0 = t_2(6.4 - t_2).$$

The values of t_2 that make the right side zero are $t_2 = 0$, which, of course, is the value of t when the projectile first starts out, and the value of t_2 for which

$$6.4 - t_2 = 0$$

or

$$t_2 = 6.4.$$

Thus the amount of time that elapses from the instant the projectile starts out until it reaches the ground is 6.4 seconds.

We note that this value of t_2 is twice the value of t_1; that is, it takes twice as long for the projectile to complete its entire up and down motion as it does to reach maximum height. Hence, it must take just as long for the projectile to go from maximum height to the ground as it does to reach the maximum height.

Another significant question one may raise about projectile motion is, What is the velocity with which the projectile strikes the ground? After all, if the velocity is small, the shell may not even explode. This question is easily answered. At the instant the shell strikes the ground the horizontal velocity is still $V \cos A$. The vertical velocity should actually be known to us. We proved in chapter 12 that an object thrown up into the air falls to earth with the same speed as it is thrown up. Insofar as the vertical motion of the projectile is concerned, since it is independent of the horizontal motion, the same conclusion applies. The vertical velocity at the instant the projectile strikes is numerically $V \sin A$ but is directed downward. Since these values of horizontal and vertical velocity are precisely those with which the projectile started out, we have the conclusion that the projectile strikes the ground with precisely the same velocity as it started out.

What is the range of the projectile, that is, how far from the starting point does it strike the ground? Well, the projectile takes 6.4 seconds to travel up and down. While the projectile is going through its vertical motion it is traveling horizontally at the rate of $V \cos A$ feet per second. Since the horizontal distance traveled at any time t is given by formula (7) we may substitute the values of V, $\cos A$, and t which apply to our situation to find how far the projectile travels horizontally in 6.4 seconds. Thus, substituting 128 for V, 0.6 for $\cos A$ (a value gotten from tables), and, of course, 6.4 for t_2, and letting x_2 be the range, we have

$$x_2 = 128\ (.6)\ (6.4),$$

or

$$x_2 = 491.5.$$

We have found the range of the gun for a particular initial velocity and a particular angle of fire. Since in military uses of guns it is often desirable to have the shell reach as far as possible, we should consider the question, How can we obtain the maximum range? Of course it is physically clear that if we increase the muzzle velocity of a gun we will increase the range for any angle of fire. But this solution involves building more powerful guns. Can we, however, choose a favorable angle of fire with a given gun so as to obtain maximum range? Intuitively it is clear that there should be a best angle of fire, for if the projectile is fired almost vertically upward it will of course strike the ground not far from the gun. Also, if it is fired almost horizontally to the ground it will not travel out very far because gravity will pull it down to the ground almost immediately. Presumably at some angle of fire well away from 0° and 90° the projectile will attain maximum range. Let us determine this angle.

Since we wish to study the dependence of range upon angle of fire let us calculate the range of a projectile for *any* angle of fire instead of for the 53° we studied above. We may remember that the procedure for obtaining range was first to find the time t_2 at which the projectile reaches the ground. This value occurs when y in formula (9) is 0. Hence

$$O = Vt_2 \sin A - 16t_2^2.$$

By applying the distributive axiom we have

$$0 = t_2 (V \sin A - 16t_2).$$

Just as in the special case of the 53° angle of fire, there are two values of t_2 for which the last equation is 0, the relevant one for our purposes being the value for which

$$V \sin A - 16t_2 = 0$$

or

$$t_2 = \frac{V \sin A}{16}.$$

The range x_2 is now found by substituting this value of t_2 in formula (7) for the horizontal distance traveled in time t. Then

$$x_2 = V \cos A \cdot \frac{V \sin A}{16}$$

or

$$(12) \qquad x_2 = \frac{V^2}{16} \sin A \cos A.$$

Formula (12) expresses the range x_2 in terms of the initial velocity V and angle of fire A. Since we wish to consider the dependence of range on A, our question now is, For what value of angle A is x_2 a maximum? Evidently, since V is fixed, x_2 will be a maximum when the product $\sin A \cos A$ is a maximum. Hence we are now reduced to a problem in trigonometry, namely, For what angle is the product of the sine and cosine a maximum?

We can answer this question readily. The very definitions of sine and cosine applied to figure 73 state that

$$\sin A \cos A = \frac{a}{c} \cdot \frac{b}{c} = \frac{ab}{c^2}.$$

To consider the possible values of sin A cos A as A varies we can always take right triangles in which the hypotenuse is fixed, for we may recall that the sine and cosine depend only upon the ratios of the sides. Hence the question becomes, When is the product ab a maximum, a and b being the sides of a right triangle of fixed hypotenuse c? But whenever ab is a maximum, a^2b^2 is a maximum and conversely. Moreover, since for any a and b that are under

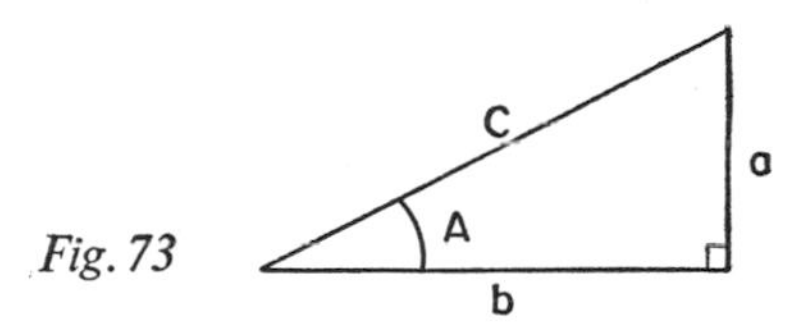

Fig. 73

consideration $a^2 + b^2 = c^2$, we are therefore asking, When is the product a^2b^2 a maximum under the condition $a^2 + b^2 = c^2$, c^2 being fixed?

We answered this question quite some time ago. We can regard a^2 and b^2 as sides of a rectangle of semiperimeter c^2, which is fixed. The quantity a^2b^2 is the area of such a rectangle. Of all such rectangles the area is largest when the rectangle is a square, that is, when $a^2 = b^2$. Hence the product ab is also a maximum when $a = b$. In this case, the right triangle above is isosceles and angle

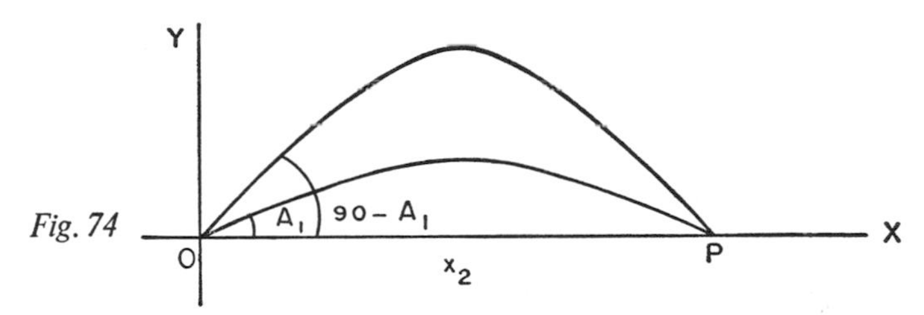

Fig. 74

A is 45°. That is, for any given muzzle velocity the maximum range is obtained by firing at an angle of 45°. This result, a famous one, is Galileo's.

We just found that the maximum range of a gun is attained by firing it at an angle of 45°. If however one wishes to reach a point P (fig. 74) that is not quite as far out as the maximum range then formula (12) has something extra to tell us. If the range x_2 is specified, then formula (12) tells us the angle A at which the projectile should be fired to obtain that range, namely, the angle A

for which the right side equals the required x_2. Now angle A can be calculated but the interesting point that the right side of formula (12) reveals is that there are two values of A that will satisfy the requirement. If A_1 is one value then we shall prove that $90 - A_1$ is another, for

$$\sin (90 - A_1) = \cos A_1 \text{ and } \cos (90 - A_1) = \sin A_1.$$

Hence

$$\frac{V^2}{16} \sin A_1 \cos A_1 = \frac{V^2}{16} \cos (90 - A_1) \sin (90 - A_1)$$
$$= \frac{V^2}{16} \sin (90 - A_1) \cos (90 - A_1).$$

These two values of the angle of fire A that yield the same range are complementary angles; moreover, one is as far below the angle of maximum range as the other is above.

A projectile fired with a fixed initial velocity can not only reach various ranges but it can reach various points in the air by being fired at different angles of inclination. That is, a cannon or gun has a certain coverage of the space in front of it. Let us suppose that an aviator wishes to fly over the gun, perhaps to photograph the area, and hence wishes to go as close as possible but to stay just out of range of the gun so as to avoid being hit. Where can he fly in safety? Figure 75 shows the various paths that a projectile fired at different angles will take. The curve ABC that just envelops all these parabolas evidently marks the boundary of points which can be reached by the gun. Hence the aviator should stay beyond this curve. What is the shape of the curve? We shall not derive its equation here but the interesting fact about it is that it is also a parabola whose focus is at the gun; the equation of this parabola is

$$y = \frac{V^2}{64} - \frac{16x^2}{V^2}.$$

This enveloping curve is in fact called the parabola of surety. No point beyond it can be reached by the gun; any point on it can be reached by just one of the parabolic trajectories; and any point within it can be reached by precisely two of the trajectories. The parabola of surety is of course also of interest to the gunner in command of the gun because it tells him what area is under his coverage.

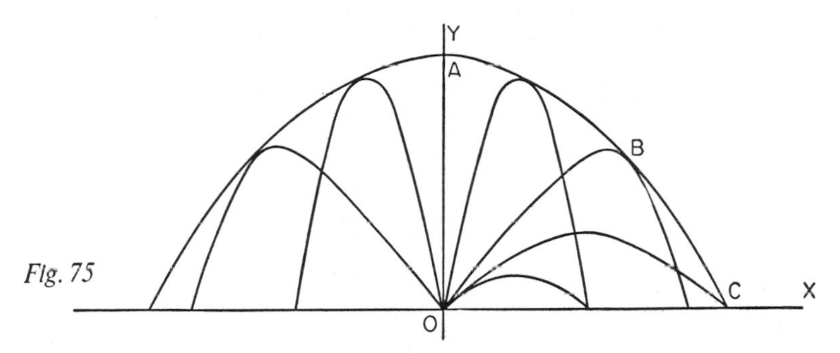

Fig. 75

There are some other facts about projectile motion which are also very pretty—if one may use this word in connection with guns. Suppose a bomb is dropped from an airplane and falls straight down. A gunner on the ground some distance away wishes to hit the bomb somewhere along its path so that it will be exploded in mid-air. How should he aim his gun and when should he fire? The problem is somewhat complicated because the bomb will be continually changing its position while the shell shot from the gun travels toward it and the shell itself does not follow a straight-line path (fig. 76). Intuition or guesswork will not readily furnish the answers, but mathematics offers definite and readily established answers. The gunner should aim his gun directly at the point of release of the bomb and he should fire the instant that the bomb is released. Nature will take care of the rest.

Let us prove that the shell will hit the falling bomb. Let x_1 be the distance between the gunner and the point on the ground directly below the bomb (fig. 76, p. 222). To hit the bomb the shell will certainly have to travel the horizontal distance x_1 as well, of course, as some vertical distance. But the horizontal motion, which is independent of the vertical motion, is governed by the formula

$$x = Vt \cos A.$$

To cover the distance x_1, the time t_1 required is

$$t_1 = \frac{x_1}{V \cos A} \cdot$$

What will the vertical position of the shell be at this time? The vertical motion is given by the formula

$$y = Vt \sin A - 16t^2.$$

Hence at time t_1, the height y_1 of the shell will be

$$V \sin A \cdot \frac{x_1}{V \cos A} - \frac{16x_1^2}{V^2 \cos^2 A},$$

or, since $\sin A/\cos A = \tan A$,

$$x_1 \tan A - \frac{16x_1^2}{V^2 \cos^2 A}. \tag{13}$$

Now let us calculate the position of the falling bomb at time t_1. The initial position of the bomb, point B in figure 76, has a height BD, given by the expression

$$BD = x_1 \tan A$$

where A is the angle of fire and also the angle made by OB with the horizontal. Since the bomb falls straight down, in time t_1 it falls $16t_1^2$ feet. In view of the foregoing value of t_1, the bomb falls

$$\frac{16x_1^2}{V^2 \cos^2 A}$$

in time t_1. Hence the y-value of the position of the bomb after t_1 seconds is BD minus the distance it has fallen or

$$x_1 \tan A - \frac{16x_1^2}{V^2 \cos^2 A}. \tag{14}$$

Comparison of the expressions (13) and (14) shows that the bomb

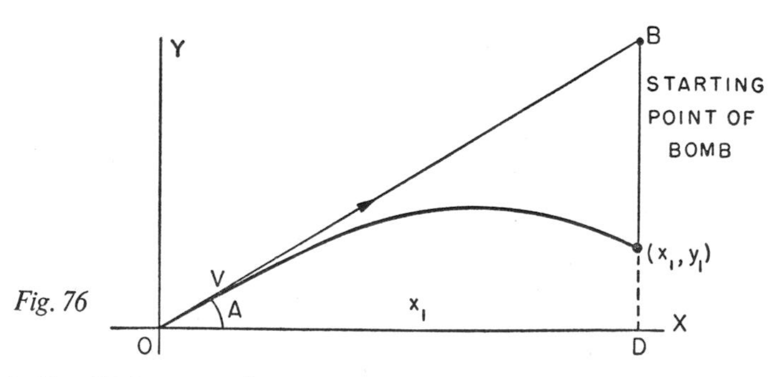

Fig. 76

and shell will be exactly at the same position at the same instant.

The conclusion is doubly surprising. We have said nothing about the muzzle velocity of the gun. Surely this should enter into the problem. But it does not. Regardless of the muzzle velocity of the gun, the shell will hit the bomb, as the above mathematics shows.

The physical explanation of this fact is not hard to find. If the muzzle velocity of the shell is increased it will cover the horizontal distance x_1 much sooner. Since it will also travel faster in the vertical direction it will reach a point on the straight line *BD* higher up and sooner. In this shorter time the bomb will not fall quite so far and hence the shell and bomb will again meet.

It is necessary to remind the reader that the results we have obtained for the motion of a projectile rest upon certain idealizations. We have neglected air resistance, the curvature of the earth, the rotation of the earth, and the variation of the force of gravity with distance from the center of the earth. These factors can be neglected and the mathematics still yield highly accurate results for many practical problems. However, to study the motion of rockets and long-range, high-speed projectiles, these effects cannot be neglected. We shall see later how some of these, at least, are taken into account. For all problems the present idealizations certainly furnish a useful first step and the mathematics employed shows how some simple functions can be used to answer signficant problems.

Our last few chapters have given some indication of how Galileo's plan for describing natural phenomena by means of mathematical functions and for investigating these phenomena by mathematical deductions can operate. The examples of motion we have examined so far, largely Galileo's with a few modern interpolations, are just a handful out of a vast number. They concern motions on and near the earth and in the main show how Galileo constructed a new science of earthly motions based on new physical principles and mathematics. An even greater challenge was being dangled before the eyes of seventeenth-century scientists and mathematicians—the challenge of penetrating somewhat the mysteries of heavenly motions so succinctly and yet so bleakly described by Copernicus and Kepler. But Galileo had already done more than one man's share and in 1642 his infirm body refused to function any longer.

15: FROM PROJECTILE TO PLANET AND SATELLITE

What! You think that Isaac Newton told a lie?
Where do you hope to go when you die?

—ANONYMOUS

TOWARD the end of his *Two New Sciences* Galileo wrote, "The principles which are set forth in this little treatise will, when taken up by speculative minds, lead to many another more remarkable result." To ordinary minds this statement no doubt seemed sheer bombast. Galileo had provided the methodology for the analysis of motions on and near the earth and had applied it successfully. Copernicus and Kepler had previously obtained the laws of motion of the planets and their satellites. The science of motion seemed to be complete. But to the inquiring and restless scientific minds of the seventeenth century these accomplishments served primarily to suggest even broader and deeper physical problems. Confident and proud as Galileo may have been of the significance of his work, he still underestimated its momentous consequences.

From the standpoint of science as a systematic deductive inquiry, Galileo's work, perhaps less spectacular than that of Copernicus and Kepler, nevertheless possessed superior features. Both contributions resulted in mathematical laws of motion, terrestrial motions in the one case and celestial in the other. But Galileo had succeeded in deriving numerous laws from a few physical principles and of course the axioms and theorems of mathematics. Moreover, there was little doubt that many more laws would follow from the same foundation. The Keplerian laws, on the other hand, were not logically related to each other. Each was an independent inference from observations. Though these laws were immensely useful, they did not give promise of leading to new truths. They seemed to be suspended in the same vacuum in which the planets moved.

Galileo's work had the additional advantage of supplying physical insight. The first law of motion and the law that the force of gravitation gives to each mass near the earth a downward acceleration of 32 ft/sec^2 are, once understood, intuitively satisfying principles that explain the vertical rise and fall of bodies, motion on slopes, and projectile motion. Kepler's laws, by contrast, had no physical basis. Why should planets move in ellipses? Why should the line from the sun to any given planet sweep out equal areas in equal intervals of time? And why should the periods of the planets be related to their average distances from the sun in the manner described by the third law? It so happens that the scientific study of magnetism, brilliantly begun by William Gilbert, had attracted attention in the late sixteenth century, and Kepler tried to introduce the idea of a magnetic force which the sun exerted on all the planets, so accounting for their motions. But he failed to relate the behavior of magnets to the precise laws of planetary motion. The question therefore remained open, Was there any deeper physical basis for the motions of the planets being what they were?

Not only was the new astronomical theory devoid of any physical explanation, it was completely isolated from the theory of motion on the earth. This isolation might not bother lesser mortals, but it bothered mathematicians and scientists who believed that *all* the phenomena of the universe were governed by one master plan instituted by the master planner—God. The goal of deriving all the phenomena of nature from a few basic physical laws and the axioms of mathematics had been set by Galileo and was accepted by the seventeenth century. Isaac Newton tells us explicitly that this is the objective of physical science: "But to derive two or three general principles of motion from phenomena, and afterwards to tell us how the properties and actions of all corporeal things follow from these manifest principles, would be a very great step in philosophy. . . ."

Moreover, even if one leaves aside any philosophical convictions as to the design of nature, a comparison of Galileo's work with that of Kepler immediately suggested a rather concrete problem. In studying curvilinear motions on the earth Galileo had found the parabola to be the basic curve. In the heavens, however, Kepler, using the best data available to him, had found the ellipse to be the basic curve. Why this difference? Moreover, since parabola and ellipse are both conic sections there was the provocative suggestion that perhaps some physical law unified these related paths of motion.

On the practical level also, the science of motion continued to present problems. Though the new astronomy had simplified calculations, an accurate method of determining the longitude of a ship at sea was not yet available. In the seventeenth century it seemed that the moon might be the best guidepost because it is almost always visible and its closeness to the earth suggested the possibility of determining its position accurately. Unfortunately, the theoretical knowledge that might have enabled a mariner to relate his observations of the moon's position to his location at sea was not sufficiently precise. The effect of the resistance of air or water on earthly motions was another problem of importance. Galileo, in his calculations, had neglected air resistance, but he had recognized that it would have to be taken into account, as a secondary effect, in many practical problems.

It has often happened in the history of mathematics and science that major problems remained outstanding for quite some time; great minds set to work on them, and yet their efforts succeeded only in revealing the true difficulties in the problems and in generating an atmosphere of despair about anyone's ever resolving them. Then a genius appeared on the scene and, with ideas that seemed remarkably simple once propounded, clarified the entire situation, dispelled the confusion, restored order, and produced a new synthesis that embraced far more even than the phenomena under consideration. The genius who resolved the pressing problems of the latter half of the seventeenth century, and who picked up the torch of science dropped by Galileo, was Isaac Newton.

Newton was born in 1642, the year of Galileo's death. Anyone reading an account of the first twenty-five years of Newton's life might conclude that this was a man destined for insignificance. He was born prematurely, and it was doubtful that he would survive. His father was already dead and his mother, occupied with operating the farm left to her, could pay little attention to him. Newton attended local schools of horribly low educational standards. Except for a strong interest in mechanical contrivances he showed no promise of any sort. He was, in fact, an indifferent student. Only once did he seek to do well, and then simply in order to beat in studies a bully who had kicked him.

At the age of nineteen Newton entered Cambridge University. There he learned the work of Copernicus, Kepler, and Galileo, and he learned some mathematics from one of the outstanding men of the times, Isaac Barrow. But Newton seemed little interested. Because he found himself to be weak in geometry he almost changed his course of study from natural philosophy (science) to law. His three years of study at Cambridge, from 1661 to 1664, ended as unimpressively as they began.

Because a plague was raging in London Newton returned home —to think. His creative faculties blossomed in the years immediately following his graduation from college. In a few years he made first-class contributions to algebra; he formulated and applied the law of gravitation, which proved to be the key to unifying earthly and heavenly motions; he set forth basic procedures in the calculus that entitle him to the honor of being one of the two founders of the subject; and he showed by highly original experiments that white light such as sunlight is actually composed of the entire spectrum of colors from violet to red. The full meaning of these creations and discoveries will be clearer to us later.

Newton returned to Cambridge to secure his master's degree in 1667 and was appointed a Fellow of Trinity College. Two years later, at the age of twenty-seven, Newton, now recognized at least as a serious student of mathematics, was appointed to the Lucasian professorship of mathematics, the post from which Barrow had resigned. His success as a lecturer was inverse to his success in research. At times no one attended. Neither his students nor his colleagues recognized the worth of the original material he presented.

His early publications on the composition of white light and on his philosophy of science were criticized and rejected by many scientists. Newton was hurt and decided he would not subject himself to such humiliation in the future. He broke this resolution to announce further results, and he became embroiled in defending his views against the attacks of others. He resolved this time that he would work only for his private satisfaction and leave his results to be published after his death. Nevertheless, under the urging and financial assistance of the astronomer Edmond Halley, Newton published the classic work *The Mathematical Principles of Natural Philosophy* (1687), which embodies the fruit of his work on gravitation.

Like many deep thinkers Newton was absent-minded. Many stories about this trait have come down to us. One night he left some friends at the dinner table to fetch a bottle of wine from the cellar. On the way he forgot his errand, went to his room, put on his surplice, and ended up in chapel. On another occasion he was leading a horse up a hill but found when he came to remount that he had only the bridle in his hand.

As one might infer from his reluctance to publish and his indifference to popular acclaim, Newton was a very modest man and gave due credit to his predecessors: "If I have seen a little farther than others it is because I have stood on the shoulders of giants." Nor did he feel that his work was of incomparable importance. Well known is his remark, "I do not know what I may appear to the world; but to myself I seem to have been only like a boy playing on the seashore, and diverting myself in now and then finding a smoother pebble or a prettier shell than ordinary, whilst the great ocean of truth lay all undiscovered before me." But Newton is ranked with Archimedes and Gauss as one of the three greatest mathematicians in all history.

Of his many achievements it is Newton's work on gravitation that will concern us here. During the years he spent at home after his college days Newton pondered the problem of unifying the laws of terrestrial and celestial motion, and he began by considering the applicability of Galileo's laws of motion to the behavior of the planets. If the first law of motion applied then the planets, once set into motion and undisturbed by forces, should move in straight lines at a constant speed. Since they did not move in straight lines but rather in ellipses, some force had to be acting to cause the elliptical paths. It seemed clear, too, that this force was directed toward and perhaps exerted by the sun, because the planets were being constantly pulled in toward the sun from what might otherwise have been straight-line motion. This argument, which was made even before Newton's time, suggested that possibly the sun attracted the planets toward it just as the earth attracted objects to itself. That is, presumably the same gravitational force acted in both cases. Hence Newton decided to concentrate on the problem of showing that the same force acted in all cases and, more important, to obtain the precise formula by which this force acted. This formula would, of course, have to be in accord with the laws of motion Galileo had discovered on earth and also account for the elliptical motion of the planets around the sun.

Newton proved that the same force kept the planets in their paths as pulled objects to the earth, and he did so *by showing that the same mathematical law governed all gravitational attractions*. His proof is a classic of scientific reasoning. The popular account, that Newton appreciated the universal action of the law of gravitation when, while dozing under an apple tree, an apple hit him on the head, is obviously nonsense since the fall of an apple involves only the pull of the earth on objects nearby. The mathematician Gauss said that Newton told this story to get rid of stupid persons who asked him how he discovered the law of gravitation. The true account of Newton's thinking is a little deeper.

Newton began by considering the problem of sending a projectile out horizontally from the top of a mountain. Of course, Galileo had worked out just such problems and had shown that the resulting path would be a parabola. Moreover, if the horizontal velocity of the projectile were greater the path would still be a parabola but a wider one so that the projectile would fall farther out horizontally (see fig. 68). But Galileo had considered projectiles that did not travel far, and he had therefore ignored the curvature of the earth. Hence Newton's first thought was that if a projectile were shot out horizontally with enough speed it might follow the curved path *VD* (fig. 77, p. 230). But would it not move off into space and escape from the earth altogether? No, because the earth would continue to attract it. In what direction would the earth pull the projectile? Galileo had always considered that gravity pulls all objects straight down. But straight down for an object that is circling the earth means toward the center of the earth. Hence, projectiles shot out from the mountaintop would be pulled in toward the earth. If shot out with greater velocity a projectile might take the path *VE* and if shot out with a sufficiently large velocity, might travel around the earth and perhaps continue to circle it indefinitely. And so Newton argues in his *Mathematical Principles,* "And after the same manner that a projectile, by the force of gravity, may be made to revolve in an orbit, and go round the whole earth, the moon also, either by the force of gravity, if it is endowed with gravity, or by any other force, that impels it toward the earth, may be continually drawn aside towards the earth, out of the rectilinear way which by its innate force [inertia] it would pursue; and would be made to revolve in the orbit which it now describes; . . ."

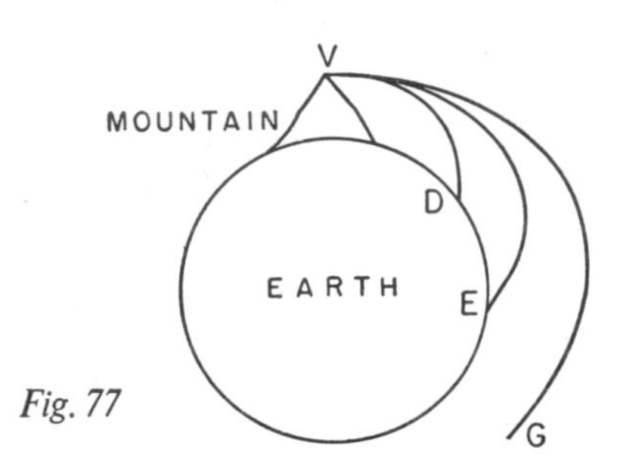

Fig. 77

If the earth by its gravitational attraction could cause the moon to "circle" the earth, then so might the sun by its gravitational attraction cause the planets to "circle" the sun. Hence Newton had some basis for entertaining the exciting prospect that the same force which pulled objects near the earth to it also kept the moon moving around the earth and the planets moving around the sun.

All of Newton's reasoning thus far was qualitative and speculative. To make progress one now had to become quantitative. Referring still to the force acting on the moon Newton says next, "If this force were too small, it would not sufficiently turn the moon out of a rectilinear [straight line] course; if it were too great, it would turn it too much, and draw the moon from its orbit toward the earth. It is necessary that the force be of a just quantity, and it belongs to the mathematicians to find the force that may serve exactly to retain a body in a given orbit with a given velocity; . . ."

Newton was a mathematician as well as a physicist, and so he undertook the task of finding the precise mathematical law of attraction between any two objects. Since the gravitational force with which the earth attracts an object depends upon its mass, it was reasonable that the force of attraction between any two objects should depend directly upon both masses. That is, if either mass should be doubled the force should be doubled. Mathematically, this suggests that if F is the force of gravitation and m and M are the two masses, F might vary with the *product* of the two masses. But F might also depend upon the distance between the two masses. Galileo had shown that all bodies fall to earth with the same acceleration, the acceleration being caused, of course, by the force of attraction of the earth. But Newton realized that the bodies Galileo had considered were so near the surface of the earth that the slight differences in distance from the surface did not matter. If, however, an object were very far from the earth then most likely the force of attraction would be weaker. The open question seemed to be, How much weaker? Here Newton adopted a conjecture that had been at least voiced by others before him, namely, that the force becomes weaker as the square of the distance r between the two masses increases. All of this is more understandable in symbols. What Newton conjectured is that

$$F = \frac{GmM}{r^2}, \tag{1}$$

where G is a constant that depends upon the units of mass, distance, and force employed. We shall say more about G later. What matters now is that Newton conjectured that formula (1) applied to *any* two masses. Of course this conjecture had to be established.

Scientific and mathematical conjectures are established by supplying the appropriate evidence. In the case of a mathematical theorem the appropriate evidence is the deductive proof; in the case of a scientific law the evidence might be experimental verification of the law itself or of some consequence of the law. However, before one seeks this confirming evidence, which may be difficult to obtain, it is advisable to test the conjecture by applying it to the simplest situation to which it is supposed to apply. Thus, a possible theorem that is supposed to apply to all polygons might be tested by noting what it says for triangles and then seeing whether the affirmation is true for triangles. Certainly if the theorem does not apply to triangles it cannot apply to all polygons. After a conjecture stands the simple tests it is then more reasonable to expend a great deal of effort in finding a deductive proof or in setting up an elaborate experimental test.

Now formula (1) was supposed by Newton to apply to any two masses in the universe. Hence in particular it should apply to the earth and an object near the earth. Galileo's information about the attraction exerted by the earth might make it easy to test formula (1) by applying it to this special case. Let us see what the formula has to say about objects falling to earth.

We first write formula (1) as

$$F = m\frac{GM}{r^2} \cdot \tag{2}$$

But Newton's second law of motion says that the relation of any force to the acceleration it gives a mass m is

$$F = ma. \tag{3}$$

Hence the quantity GM/r^2 in (2) must be the acceleration that the force of attraction exerted by the mass M imparts to mass m. That is,

$$a = \frac{GM}{r^2} \cdot \tag{4}$$

Now let M in formula (2) be the mass of the earth and let m be the mass of any object near the surface. The quantity r in the formula is supposed to be the distance between the two masses; however there is some question as to how much this distance should be since the extent of the earth is considerable. Newton conjectured that a sphere of uniform mass acted as though its mass were concentrated at the center. Let us suppose this to be the case. Then since objects *near the surface of the earth* are all practically 4000 miles from the center, r in formulas (2) and (4) would be practically the same for all such objects. Since G is always the same and M is the constant mass of the earth, then the entire quantity on the right side of formula (4) is constant and the acceleration of all objects falling to earth from points nearby should be the same. This is precisely what Galileo had discovered. As a matter of fact the quantity a in this case is just 32 ft/sec^2 and F becomes the weight W of the object so that (2) becomes $W = m \cdot 32$. Thus Newton's conjecture passed its first test.

The next most reasonable test of formula (1) might be to measure the force of attraction that the earth exerts on objects far enough away so that the quantity r in the formula might be sufficiently larger than 4000 miles to cause a decrease in the gravitational force and hence in the acceleration imparted to bodies. Robert Hooke tried such experiments by measuring the acceleration exerted on bodies at high altitudes. However, these bodies were still so close to the earth that their distance from the center did not differ significantly from 4000 miles. Hence he did not detect any difference in the acceleration caused by the earth's attraction. In any case the crucial point about formula (1) or (2) was whether it applied to any two heavenly bodies. Newton decided to apply it to the earth and moon.

Newton's plan was a bit lengthy and so we shall survey it before plunging into the details. If the moon were not attracted by the earth then, since no other apparent force is acting on it, the first law of motion says that it would travel in a straight line. Newton therefore argued that the force of attraction the earth exerted on the moon must impart the very acceleration to that body that causes it to "fall" toward the earth instead of following a straight-line path. Thus if mP (fig. 78)is the distance the moon would travel in one minute were the earth not present, PQ is the distance the

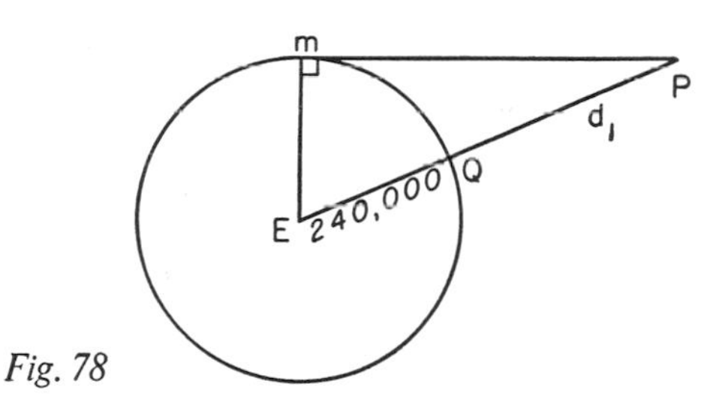

Fig. 78

moon "falls" toward the earth in that minute because the earth's attractive force gives the moon an acceleration acting along PQ. Newton's plan was to calculate this acceleration on the assumption that the law of gravitation applied. He then proposed to compare the calculated value with the value derived from observation. If the two values agreed, the evidence for the law of gravitation would be immensely strong.

Newton's first task, then, was to calculate the acceleration that the earth imparted to the moon. This value is determined by formula (4) when r is 240,000 miles, the distance from the earth to the moon. Let us call the value of the acceleration a_1. Then

$$a_1 = \frac{GM}{(240{,}000)^2}. \tag{5}$$

When $r = 4000$ formula (4) yields, as we saw above, the acceleration that the earth imparts to objects nearby. But this acceleration is 32 ft/sec^2. Hence

$$32 = \frac{GM}{(4000)^2}. \tag{6}$$

If we now divide equation (5) by equation (6) we obtain

$$\frac{a_1}{32} = \frac{(4000)^2}{(240{,}000)^2} = \frac{4^2\,(1000)^2}{(240)^2\,(1000)^2} = \frac{4^2}{(240)^2} = \frac{1}{(60)^2}. \tag{7}$$

Thus

$$a_1 = \frac{1}{(60)^2} \cdot 32. \tag{8}$$

The quantity a_1 then gives the acceleration of the moon.

Now Newton faced the problem of comparing this calculated value with a value of the same acceleration based on some observational evidence. Here Newton had to be somewhat roundabout. We found in chapter 12 that a body that falls to earth with a constant acceleration of 32 ft/sec^2 falls in t seconds the distance d given by the formula

$$d = 16t^2 = \frac{1}{2} \cdot 32t^2.$$

The very same reasoning would show that if a body falls with an acceleration of a ft/sec^2, it would fall a distance d given by

$$d = \frac{1}{2}at^2. \tag{9}$$

In one minute, or 60 seconds, the body would fall the distance

$$d_1 = \frac{1}{2}a(60)^2. \tag{10}$$

Simple algebra applied to this formula tells us that

$$a = \frac{2d_1}{(60)^2} \cdot \qquad (11)$$

If therefore Newton could find the distance d_1 fallen by the moon in one minute, he could then calculate the acceleration of the moon.

But the calculation of d_1 was child's play to Newton. Let the arc *mQ* in figure 78 be the distance the moon travels along its actual path, which for simplicity Newton assumed to be circular, in one minute. The angle *E* which the moon's path subtends is that part of 360° which one minute is of the time the moon takes to traverse the entire circle, that is, one month. Now the lunar month is 27 days, 7 hours, and 43 minutes, or 39,343 minutes. Hence

$$\measuredangle E = \frac{1}{39{,}343} \cdot 360 .$$

Since angle *E* is known and the distance *Em* in figure 78 is 240,000 miles, the distance *EP* can be calculated because

$$\cos E = \frac{Em}{EP}$$

and therefore

$$EP = \frac{Em}{\cos E} \cdot \qquad (12)$$

Now the distance *PQ*, which is the distance d_1 of formula (11), is *EP* — *EQ*. But *EQ* is also 240,000 miles, and *EP* can be calculated from formula (12). Hence *PQ* or d_1 can be calculated and formula (11) used to obtain the acceleration of the moon. The result agrees excellently with the value of a_1 given by formula (8).

Newton performed the foregoing reasoning and calculations in 1666 when he was twenty-two years old, but he did not publish the work until he reported it in his *Mathematical Principles of Natural Philosophy* in 1687. There he says "the force by which the moon is retained in its orbit is that very same force which we commonly call gravity. . . ." There has been some uncertainty as to why Newton delayed announcing this extraordinarily significant evidence for his law of gravitation. The explanation is that he used the fact, as we did above, that the earth's gravitational force acts as though the mass of the earth were concentrated at the center, but he was not certain that this was correct. However, he proved this result in 1685 and had no reason to hesitate any longer.

The argument by which Newton established his law of gravitation should be carefully noted. He conjectured that the law was correct and showed first that it accounted for the constant acceleration of objects falling to the surface of the earth. He then showed that the law led to the correct prediction of the acceleration of the moon. This evidence was so striking that Newton was willing to generalize and conclude that the law applied everywhere. Newton now intended to use the law as a physical axiom along with the two laws of motion, and from these laws and the axioms and theorems of mathematics he expected to derive a number of new physical laws. If all of these derived laws continued to check with experiment and observation then the assumption that the law of gravitation applied universally would be amply justified.

We should note once again how the ways of science differ from those of mathematics. The mathematician starts with axioms about number and geometrical figures that immediately carry conviction. They seem to be self-evident truths. From these axioms theorems are derived by deductive arguments, which are unassailable, and thus the mathematician arrives at new truths. In effect, the scientist works backward. He adopts physical axioms such as the laws of motion and the law of gravitation. There may be little evidence in favor of these axioms. Certainly they are not self-evident. He then proceeds to derive new conclusions from these axioms, often with the help of mathematics and even deductive arguments. If the conclusions continue to agree with observation and experimentation he infers that the physical axioms are correct, or, at least he acquires more and more confidence in them as the evidence for them increases. He must not be astonished, however, if some day some old or new deduction no longer agrees with physical evidence. If this should happen he must revise his physical axioms.

We intend to see in the next chapter how a number of significant mathematical deductions from the law of gravitation supported Newton's bold conjecture. But before entering upon these matters we should note how the reasoning that led Newton to the law of gravitation has attained new significance in our time. Newton considered the motion of projectiles shot from the top of a high mountain with great velocities and concluded that these would circle the earth. He then passed to the consideration of the moon's motion around the earth and of planetary motion around the sun. Today scientists have created satellites that circle the earth and have launched them in almost the same manner as Newton thought projectiles might be made to circle the earth. Scientists do not launch the satellites by shooting them from the top of a mountain. Accessible mountaintops are not high enough to clear other obstacles in the satellites' paths and moreover would be so low that the satellites would encounter serious air resistance. Instead, scientists use rockets to raise a satellite to a height of several hundred miles at which point a mechanism turns the satellite to a horizontal direction. Additional rockets now give the satellite sufficient velocity to have it pursue an elliptical path around the earth. The upward motion need not be vertical, for a projectile shot out from the earth at an angle will rise until its vertical velocity is exhausted and then, if its horizontal velocity is sufficient, will travel in an elliptical path.

Newton applied this very reasoning to answer the question of how the planets originated. He argued that they must have been shot out from the sun at some angle and with sufficient velocity to reach their present distances from the sun. Then, under the gravitational attraction of the sun, they were forced to travel in the elliptical orbits they now pursue. Hence the mathematics of projectile motion led among other things to a theory about the origin of the planets that is still accepted.

16: DEDUCTIONS FROM THE LAW OF GRAVITATION

Who by a vigor of mind almost divine, the motions and figures of planets, the paths of comets, and the tides of the seas, first demonstrated.

—NEWTON'S EPITAPH

IN scientific work the proof of the pudding is in the eating. By applying his mathematical law of gravitation to the motion of objects near the surface of the earth and to the motion of the moon Newton found that the deductions agreed with known facts. Hence he felt confident that gravitation was a universal force, and that he had the correct mathematical description of its operation. But the full test of the law, as Newton realized, would come from applying it to a variety of terrestrial and celestial phenomena. If the deductions from this law and any other securely established physical and mathematical principles continued to agree with observations and experiments, then the law would be amply tested and could be accepted as a basic physical principle. Moreover, still further deductions might then produce reliable new knowledge. Let us see, therefore, what conclusions can be drawn from the law and to what extent they agree with knowledge obtained in other ways.

The law of gravitation itself says that any two masses in the universe attract each other with a force which, quantitatively, is given by the formula

$$F = \frac{GmM}{r^2} . \tag{1}$$

In this formula F is the force of attraction, m and M are the masses of the two bodies, r is the distance between the masses (for simple shapes r is the distance between the geometrical centers), and G is a constant under all conditions. Let us note that formula (1) expresses the force F as a function of three variables m, M, and r. Hence this formula is more complex than those we encountered in straight-line motion or projectile motion.

One of the first implications of the law is to generalize the notion of weight and to clarify the relationship between weight and mass. Galileo had introduced the concept of mass as the resistance an object offers to change in the speed and the direction of its motion, and he had also recognized that weight is the force with which the earth attracts an object. As already noted, these two properties of matter are often and understandably confused. But formula (1) provides some insight.

Let m be the mass of some object and let M be the mass of the earth; then r is the distance from the center of the earth to the object. And since F is the force with which the earth attracts the object, F is the weight of the object. Let us write formula (1) in the for present purposes more convenient form

$$\frac{F}{m} = \frac{GM}{r^2}. \tag{2}$$

Then, insofar as the attraction of the earth on all objects at the distance r from the center is concerned, the quantities G, M, and r are always the same. Hence formula (2) shows that the ratio of F to m is constant, or, in other words, the ratio of the weight to the mass of all objects at any fixed distance r units from the center of earth is the same.

Previously Galileo had noted this fact for objects near the surface of the earth. In this case, $r = 4000$ miles, and, as we pointed out in discussing formula (4) of the preceding chapter, the quantity GM/r^2 amounts to 32 ft/sec^2. Hence, if we use the designation W for the force of gravity when it applies to objects near the surface of the earth, we may write formula (2) in this special case as $W/m = 32$, and this is the relation of weight to mass that we used in preceding chapters.

But formula (1) gives us much more insight into the concept of weight. It shows that if r, the distance between the two masses is increased, then F decreases and so the weight of the object decreases. Specifically, if an object of mass m is taken 4000 miles above the surface of the earth, the distance of the object from the center of the earth is doubled. If r, which represents the original distance of the object from the earth's center, is now replaced by $2r$, the denominator in (1) becomes $(2r)^2$ or $4r^2$. That is, the new force or weight is ¼ of the original weight. Hence the weight of an object can be varied appreciably by varying its distance from the center of the earth whereas the mass of the object remains constant. We can now detect experimentally the difference in the weight of a mass at the foot and at the top of a mountain. Thus the variation of weight with distance from the earth's center, as predicted by formula (1), is confirmed experimentally.

Let us explore further the distinction between weight and mass. We normally consider the weight of an object to be its weight near the surface of the earth. But suppose the same object were taken to the moon. Would it have weight there? For the moment let us try to answer this question on the very reasonable assumption that all other masses in the universe are so far away that only the object and the moon are taken into account. Now the moon will attract the object. Hence if we agree to call the weight of the object the force with which the moon attracts it, then the object will have weight on the moon. How much weight?

Formula (1) should supply the answer if we let M be the mass of the moon, m the mass of the object, and r the distance between the two masses, which distance would be the radius of the moon. But now let us see how much we know about these quantities. We found in an earlier chapter that the moon's radius is 1080 miles. Hence we know the value of r. We must also know the mass of the moon, but we don't. Let us therefore see if we can get around that obstacle. We shall denote by F_1 the force with which the moon attracts the object. Then

$$F_1 = \frac{GmM}{(1080)^2}. \tag{3}$$

Let E be the mass of the earth and let F_2 denote the force with which the earth attracts the mass m when it is near the surface of the earth. Then

$$F_2 = \frac{GmE}{(4000)^2}. \tag{4}$$

We can at least compare F_1 and F_2. To do this we divide equation (3) by (4). Then

$$\frac{F_1}{F_2} = \frac{(4000)^2}{(1080)^2} \cdot \frac{M}{E} = (13.4)\frac{M}{E}. \tag{5}$$

Hence the ratio of weight of the object on the moon to the weight of the object on the earth is approximately (13.4) times the ratio of the mass of the moon to the mass of the earth.

We still do not know the mass M of the moon and the mass E of the earth. For present purposes we could make a rough estimate of the ratio of these two masses by assuming that both masses are proportional to their volumes. The ratio of the volumes of two spheres is the cube of the ratio of their radii. Since the radius of the moon is to the radius of the earth as about 1 to 4, the ratio of the two volumes is $\frac{1}{64}$. However, a more accurate determination of the ratio of the two masses has been made by rather advanced methods, and we shall therefore accept the result. The correct ratio is $\frac{1}{82}$. If we therefore substitute this value in formula (5) we see that

$$\frac{F_1}{F_2} = (13.4) \cdot \frac{1}{82} = \frac{1}{6}.$$

Thus an object of mass m weighs $\frac{1}{6}$ as much on the moon as it does on the earth.

Now let us consider the weight of an object in the light of the fact that it is actually attracted both by the earth and the moon. If by the weight of the object we now mean the net gravitational pull of the earth and moon on the object, then as the object travels from the earth to the moon the weight decreases on two accounts. The distance from the center of the earth is increasing and so the earth's pull is decreasing, while the distance to center of the moon is decreasing so that the moon's pull is increasing. At some point in its trip to the moon the object will have zero weight. As a matter of fact, this point is about 24,000 miles from the center of the moon along the line joining that center to the earth's center.

The foregoing considerations of weight, which were introduced to make clear the distinction between mass and weight, seem to have led to fanciful speculations. But this is not at all the case. Plans for space ships being seriously made today must take into account the change in weight of an object as it leaves the earth.

One of the rather simple and yet spectacular deductions that can be made on the basis of the law of gravitation is to calculate the mass of the earth. To do this we shall first make an auxiliary calculation. Suppose that an object of mass m is suspended at C on a string AC (figure 79a). Of course gravity pulls straight down on this mass. Nearby a one-ton mass of lead is placed and it pulls the mass m to B. The position of the lead is adjusted until it is on the same level and one yard from B. Hence the mass m remains suspended at B and the string now has the position AB.

Let us analyze the forces acting on the mass m at B. The force of gravity continues to pull it straight down and the lead pulls it to the left. The combined action of these two forces is some force that acts in an oblique direction somewhere between straight down and horizontally to the left. This resultant force must in fact lie along the line AB because the tension in the string just opposes the resultant and keeps the mass m motionless at B. It is helpful to make a separate diagram of the forces acting on the mass m. Thus the line PQ (fig. 79b) represents the force of gravity and the line PR represents the attractive force of the lead. It is known from ex-

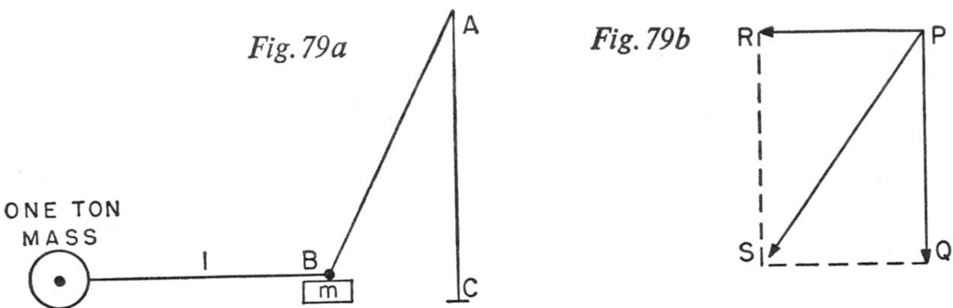

Fig. 79a *Fig. 79b*

perience that forces combine like velocities; that is, the combined effect of the forces PQ and PR is the diagonal of the rectangle determined by PQ and PR. This resultant force PS, as noted above, must have the same slope as the line AB. Hence angle $SPQ =$ angle A. But

$$\tan SPQ = \frac{QS}{PQ} = \frac{PR}{PQ}. \tag{6}$$

Since angle A can be measured we also know angle SPQ. It turns out to be a very small angle, 0.000,000,45°. From tables one can find that $\tan SPQ = 8 \cdot 10^{-9}$, that is, 8 divided by one billion. Hence we know the ratio of the force PR to the force PQ.

Let us now use the law of gravitation to calculate the same ratio of the two forces PQ and PR. The force of gravitation PQ acting straight down on the mass m is given by

$$PQ = \frac{GmM}{(4000 \cdot 1760)^2}, \tag{7}$$

where M is the mass of the earth, and the number in the parentheses is the distance from the mass m to the center of the earth in yards. The force of attraction PR of the one-ton mass of lead on the mass m one yard away is given by

$$PR = \frac{Gm \cdot 1}{1^2}. \tag{8}$$

The ratio of PR to PQ is then obtained by dividing equation (8) by equation (7). The result is

$$\frac{PR}{PQ} = \frac{(4000 \cdot 1760)^2}{M}.$$

But we have already found the ratio of PR to PQ to be $8 \cdot 10^{-9}$. Hence

$$8 \cdot 10^{-9} = \frac{(4000 \cdot 1760)^2}{M}.$$

Then

$$M = \frac{(4000 \cdot 1760)^2}{8 \cdot 10^{-9}} = 6.5(10)^{21}.$$

The quantity $6.5(10)^{21}$ is tons of mass, since our unit of mass has been tons.

Thus we have the remarkable result, the mass of the earth. We should emphasize first of all that this result is a quantity of mass and not weight. We cannot speak of the weight of the earth because weight is a force that the earth exerts on *other* masses. However, just to get some feeling for the magnitude of the result we note first that in our common, as opposed to scientific, system of measuring mass and weight, a ton of mass will also weigh one ton. Hence one can try to imagine a weight of $6.5(10)^{21}$ tons and thereby get some indication of the "weight" of the earth.

The calculation of the mass of the earth was made without knowing the value of the gravitational constant G. There is of course no reason why one should not measure G, provided we agree on a system of units for measuring mass, force, and distance. A common system (not used above) is to measure mass in grams, distance in centimeters, and time in seconds. Force is then measured in dynes. Since we are accustomed to weights that are forces measured in pounds, we should note that 980 dynes equal 0.0022 pounds of force. There is unfortunately much confusion about units because many systems are in use and because the pound (and the gram) are used as units of mass as well as of force. At any rate, if we now take two known masses measured in grams at a known distance apart measured in centimeters and measure the force of attraction between the masses in dynes, we shall know every quantity except G in formula (1), and we can then calculate G. In the system of units, grams, centimeters, and seconds, G turns out $6.67 \cdot 10^{-8}$. This is a very small quantity so that for ordinary masses separated even by small distances the gravitational force is insensible and negligible for most purposes.

With this value of G we can measure the mass of the earth by another method. We know from our discussion of the law of gravitation in the preceding chapter that the quantity GM/r^2 is the acceleration that the mass M imparts to some other mass r units distant. If M is the mass of the earth then the acceleration it imparts to objects on the surface is 32 ft/sec² or 980 cm/sec². In this case r is 4000 miles or $6.38 \cdot 10^8$ centimeters. Hence

$$980 = \frac{GM}{(6.38 \cdot 10^8)^2}. \qquad (9)$$

Above, we found the value of G. Hence if we substitute this known value in formula (9) and perform the necessary arithmetic we find that

$$M = 5.98 \cdot 10^{27} \text{ grams},$$

which is the same quantity in other units as the one obtained above, because one gram equals 0.0022 pounds of mass.

From this information we can proceed to another deduction that supplies information about the interior of the earth. The earth is approximately a sphere, and its radius is known. Its volume, therefore, given by the formula $V = 4\pi r^3/3$, turns out to be $1.09 \cdot 10^{27}$ cubic centimeters. The density of any substance is the mass divided by the volume. Thus the more mass for the same volume the denser the material, and this is what we mean when we say that the density of metals, for example, is greater than the density of wood. In the case of the earth, the density is now calculable, for we have the mass M and the volume V. Then

$$\frac{M}{V} = \frac{5.98 \cdot 10^{27}}{1.09 \cdot 10^{27}} = 5.5.$$

The density of water, which is the standard, is 1. Since the material found on the earth's surface is water or other material not nearly as dense as 5.5, we may conclude that the interior of the earth must contain heavy minerals such as lead or iron.

Having obtained the mass of the earth we may be emboldened to tackle a more daring calculation, the mass of the sun. The mathematics is hardly more involved but since the sun is much farther away, it will take a little longer to get its mass. The basic mathematical facts we shall use are the law of gravitation, that is, formula (1), and the second law of motion, which says that any force acting on a mass m equals the mass times the acceleration created by the force. To keep clearly in mind the masses involved we shall let S be the mass of the sun and E the mass of the earth. Then, the gravitational force with which the sun attracts the earth must equal the mass of the earth times the acceleration which the sun imparts to the earth. In symbols

$$\frac{GSE}{R^2} = EA, \tag{10}$$

where R is the distance between the earth and the sun and A is the acceleration of the earth's motion.

We see that we can divide both sides of equation (10) by E and obtain

$$\frac{GS}{R^2} = A. \tag{11}$$

Now G is the gravitational constant, which is known, and R is 93,000,000 miles. Hence if we knew A, the acceleration of the earth in its motion around the sun, we could then solve equation (11) for S, the desired mass of the sun.

It is possible to find the acceleration A of the earth with the very same method by which we found the acceleration of the moon in the preceding chapter. A brief review of the method employed there in the sequence of steps from formula (9) to formula (12) would show that all of the information about the moon's motion that we required there we also possess about the earth's motion; also, whereas we dealt there with the acceleration of the moon's motion in its path about the earth, we have now but to think in terms of the acceleration of the earth in its path around the sun.

The method just referred to for finding the acceleration of the earth is certainly feasible but lengthy. We may be confident that when the mathematician faces an arduous task his characteristic laziness will drive him to seek an easy way out. And there is one. A little reasoning will replace a lot of labor.

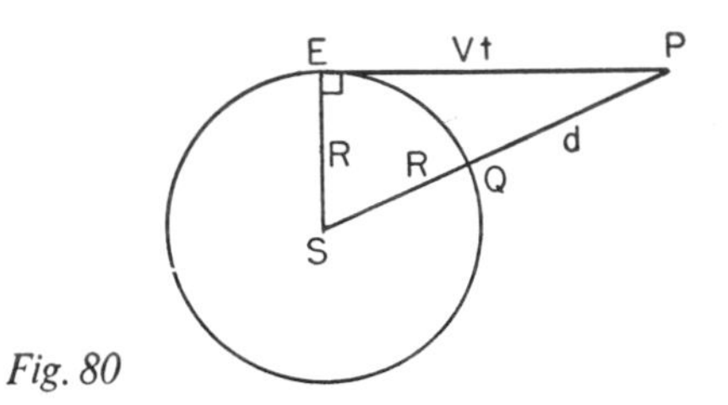

Fig. 80

The distance *EP* (fig. 80) is the distance the earth would travel in time t if it were to continue to travel in a straight line with its velocity V; thus $EP = Vt$. Also PQ is the distance d the earth "falls" toward the sun in that time t. Since *SEP* is a right triangle

$$(R + d)^2 = R^2 + (Vt)^2$$

or

$$R^2 + 2dR + d^2 = R^2 + V^2t^2.$$

We subtract R^2 from both sides and obtain

$$2dR + d^2 = V^2t^2.$$

By applying the distributive axiom we may write

$$(12) \qquad 2d\left(R + \frac{d}{2}\right) = V^2t^2.$$

However, the distance any body "falls" in time t and with constant acceleration A is given by the now familiar formula

$$(13) \qquad d = \frac{At^2}{2}\,.$$

If we substitute this value of d in the first factor on the left side of (12) we obtain

$$At^2\left(R + \frac{d}{2}\right) = V^2t^2.$$

We now divide both sides of this equation by t^2. Then

$$A\left(R + \frac{d}{2}\right) = V^2. \tag{14}$$

Equation (14) holds no matter what d may be because d is the distance the earth falls in time t for any time t. Hence we may let t become smaller and smaller, in which case EP and PQ become smaller and smaller. Then equation (14) holds when EP becomes zero and d likewise becomes zero. In this final or limiting case equation (14) becomes

$$AR = V^2$$

or

$$A = \frac{V^2}{R}.\text{*} \tag{15}$$

We have finally found the expression for the acceleration of the earth and can substitute this value on the right side of formula (11). Thus we have

$$\frac{GS}{R^2} = \frac{V^2}{R}$$

or

$$S = \frac{RV^2}{G}. \tag{16}$$

Now V, the velocity of the earth, is known. It is the circumference of the earth's path around the sun divided by the time the earth takes to travel this distance, namely, one year. To calculate S in grams, says, since the most convenient value of G is dependent upon the centimeter-grams-second system, we would have to ex-

* In deriving formula (15) we used a new process, called a limiting process. The full significance of this process will be clearer when we study the calculus. We might note here, however, that the quantity A in (14) is the average acceleration during the time t while the quantity A in (15) is the acceleration at the instant the earth is at the point E in figure 80.

press R in centimeters and V in centimeters per second and, of course, use the known value of G. Such a lengthy calculation is abhorrent to mathematicians and can be shortened by another little bit of reasoning that takes advantage of an earlier result.

The reasoning that led to formula (16) for the mass of the sun in terms of the radius and velocity of the earth can be applied to the earth and moon. Since the relationship between the latter two bodies is the same as that between the sun and earth we can write down at once

$$E = \frac{rv^2}{G}, \tag{17}$$

where r is the radius of the moon's path around the earth and v is the moon's velocity. If we now divide equation (16) by equation (17) we have

$$\frac{S}{E} = \frac{RV^2}{rv^2}. \tag{18}$$

Formula (18) is a gain over formula (16) in that we can use any convenient units for R and V provided we use the same units in numerator and denominator. If we take the radius of the earth's path to be 93,000,000 miles and remember that the circumference of this path is traversed in one year, or 365 days approximately, we have

$$RV^2 = 93{,}000{,}000\frac{(2\pi \cdot 93{,}000{,}000)^2}{(365)^2}. \tag{19}$$

The radius of the moon's path is about 240,000 miles and the average time of one revolution around the earth is 27⅓ days. Hence

$$rv^2 = 240{,}000\,\frac{(2\,\pi \cdot 240{,}000)^2}{(27\frac{1}{3})^2}. \tag{20}$$

Since we are interested in RV^2/rv^2 we can ignore the $2\,\pi$ in (19) and (20) and we can cancel numerous zeros in this quotient. The rest of the arithmetic yields

$$\frac{S}{E} = 340{,}000.$$

A more accurate calculation would yield the value of 333,000. Thus we have found the ratio of the mass of the sun to the mass of the earth. Since we know the mass of the earth we can now find the mass of the sun. More stupendous than this mass is the ease with which we have calculated it.

We have made a number of deductions from the law of gravitation and the laws of motion that show how the laws can be used to obtain information about masses and weights of objects. But one of the major questions that Newton and others had set out to answer is thus far untouched. What does the law of gravitation have to say about the paths of planets? Kepler had inferred from observation three famous laws of planetary motion, and these still seemed to be unrelated to the basic physical laws of motion and gravitation. Could the Keplerian laws be embraced in the theory of Galileo and Newton? Here indeed was a test of the power and applicability of these basic physical laws.

In a famous series of mathematical deductions Newton showed that all three of Kepler's laws were necessary consequences of the laws of motion and gravitation. To follow Newton's arguments would require a knowledge of the calculus (the reason for this will be clearer when we discuss that subject), but if we make the assumption that the planetary paths are circular, then we can see at once that the third Keplerian law follows from the basic laws of motion and gravitation.

Let m be the mass of any planet; let M be the mass of the sun; and let R be the distance between them. Then the now familiar law of gravitation says that the force F exerted by the sun on the planet is

$$F = \frac{GmM}{R^2}. \tag{21}$$

We also know that the sun's force causes any planet to depart from straight-line motion and "fall" toward the sun with some acceleration. This acceleration A is given by formula (15), that is,

$$A = \frac{V^2}{R}. \tag{22}$$

It is true that formula (15) was derived to obtain the acceleration which the sun exerts on the earth, and in that derivation V was the velocity of the earth and R the earth's distance from the sun. But the derivation applies equally well to any planet if we now understand V to be the velocity of that planet and R its distance from the sun. On the basis of this expression for the acceleration with which the sun attracts any planet, we may also assert, by the second law of motion, that the force F with which the sun attracts that planet is

$$F = m\frac{V^2}{R}.$$

The velocity V of any planet is the circumference of its path divided by the time T of revolution around the sun; that is, $V = 2\pi R/T$. Hence

$$F = m\frac{4\pi^2R^2}{RT^2} = m\frac{4\pi^2R}{T^2}. \tag{23}$$

Now formulas (21) and (23) give two different expressions for the force with which the sun attracts any planet. Hence we may equate these two expressions and obtain

$$\frac{GmM}{R^2} = \frac{m4\pi^2R}{T^2}.$$

We may divide both sides of this equation by m and so eliminate that quantity. By multiplying both sides by T^2 and then multiplying both sides of the resulting equation by R^2/GM we get

$$T^2 = \frac{4\pi^2}{GM}R^3. \tag{24}$$

The quantity $4\pi^2/GM$ is the same no matter what planet is being considered, because G is a constant, M is the mass of the sun, and $4\pi^2$ is a constant. Hence formula (24) says that T^2 is some constant, say K, times R^3; in symbols

$$T^2 = KR^3. \tag{25}$$

Thus the square of the time of revolution of any planet is a constant (the same for all planets) times the cube of that planet's distance from the sun. Formula (25) is, then, Kepler's third law of planetary motion. We now see that is a mathematical consequence of the two laws of motion and the law of gravitation.

With arguments such as we have just examined Newton showed in a matter of minutes that all three famous laws, which Kepler had obtained only after years of observation and trial and error, must hold. Moreover, the laws of planetary motion, which theretofore seemed to have no relationship to earthly motions and the law of gravitation, were shown to be logically inescapable. In this sense Newton "explained" the elliptical motion of the planets. This fact was as much a consequence of basic physical laws as the straight-line motion of objects falling to earth from rest or of projectiles following parabolic paths. Newton's original conjecture that the parabolic motion of projectiles should be intimately related to the elliptical motion of the planets was gloriously established. We should note, too, that since the Keplerian laws do fit observations their derivation from the law of gravitation constituted superb evidence for the correctness of that law. At the same time the triumph of the heliocentric theory was assured.

We have now seen a segment of the accomplishments of Galileo and Newton and may therefore appreciate the nature of their plan to seek basic, mathematically formulated physical principles and to use these in conjunction with the theorems of mathematics proper to derive new knowledge. Newton very modestly describes this plan in the preface to the first edition of his *Mathematical Principles of Natural Philosophy:*

> *I offer this work as the mathematical principles of philosophy* [*science*]*; for all the difficulty in philosophy seems to consist in this—from the phenomena of motions to investigate the forces of nature, and then from these forces to demonstrate the other phenomena; and to this end the general propositions in the first and second Books are directed. In the third Book I give an example of this in the explication of the system of the world for by propositions mathematically demonstrated in the former Books, in the third I derive from the celestial phenomena the forces of gravity with which bodies tend to the sun and the several planets. Then, from these forces, by other propositions which are also mathematical, we deduce the motions of the planets, the comets, the moon and the sea.*

That the vision of Galileo and Newton was extraordinary may be judged from the evidence we have already examined, to say nothing about some additional evidence we shall examine in later chapters.

There are many points about the work of Galileo and Newton that are worthy of attention. Their work offers, first of all, a comprehensive, systematic, rationally connected account of terrestrial and celestial motions. Thereby they established the existence of *universal* mathematical laws. These laws describe the behavior of a speck of dust and the most distant star. No corner of the universe is outside their range. Thus the evidence for mathematical design of the universe was immensely strengthened. Whereas the Greeks asserted the rationality of nature, Galileo and Newton proved it. Moreover, the unvarying adherence of natural phenomena to the pattern of these laws speaks for the uniformity and invariability of nature and opposes the medieval belief in an active Providence to whose will the universe was continually subject.

A second point is that the work of Galileo and Newton swept away the last traces of mysticism and superstition that had always been associated with the heavens. The heliocentric theory of Copernicus and Kepler had classed the earth among the other planets, so that there was good reason to believe that the heavens were made of the stuff of earth rather than, as Greek and medieval philosophers had maintained, of some light, perfect, indestructible substance. But the heliocentric theory could have been regarded by everyone and was regarded by many as a mathematical contrivance convenient for the calculations of the paths of the moon and planets but not physically true. Moreover, aside from its isolation from the main body of scientific knowledge, the heliocentric theory created difficulties in accounting for the phenomena of motion readily observed here on earth (chapter 8), and hence encountered legitimate objections.

The work of Galileo and Newton resolved all the difficulties in the theory of motion and incorporated the theory of the heavenly motions in the very same physical theory that treated motions on earth. There could be no doubt now not only that the other planets were no different from the planet earth but that the substance of the other planets could be identified with the rock and clay beneath man's feet, for this affirmation is the very essence of the law of gravitation. The identification of the stuff of the heavenly bodies with the crust of earth wiped out libraries of speculation and dogma on this subject. Today the astronomer looks at stars so distant that their light must impinge on a photographic plate for hours to make a sensible impression. Yet he is certain that this light comes from bodies like the sun.

There is a third and deeper feature of the work of Galileo and Newton that has disturbing implications. On the one hand their decision to seek the measurable properties of matter and to relate these by mathematical formulas or, as Newton put it, to set forth the exact manner in which "all things had been ordered in measure, number and weight," seemed all wise. The nature of matter itself is incredibly complex, as modern research in atomic theory is beginning to make us aware. Galileo and Newton avoided all discussion of this subject and concentrated on inertia or mass, velocity, acceleration, force, and other measurable properties of matter. This concentration on measurable properties in itself disturbed no one. People thought at that time that they understood what matter is and graciously allowed Galileo and Newton to confine themselves to certain quantitative aspects of matter.

But there is a fairy tale about a boy who ate so much pie that he became all pie, and fairy tales often reflect reality even though they deal with fancies. Through the work of Galileo and Newton science began to absorb so much mathematics that for the first time the danger became evident that it was going to contain little else. When the concept of a force of gravity was first broached—even Copernicus used the term *gravity* to describe the pull of the earth on objects near it, and Kepler spoke of a universal force acting between any two pieces of matter—people thought they understood physically what the force was. Everyone is conscious of force when he exerts it himself in lifting or pushing something. So, too, the earth, though inanimate, could by some mechanism exert a force on bodies near it. But Newton's law of gravitation asserts that the force of gravitation acts between the sun and planets hundreds of millions of miles away. This force presumably acts instantaneously and constantly through empty space of unimaginable extent. Nor can one suspend or block the action of gravity. When Venus, for example, comes between the earth and the sun, the sun's gravitational attraction of the earth does not cease. Light can be blocked by opaque bodies and deviated by lenses. Electric and magnetic forces can be controlled. Gravitation alone, immutable and independent of all other forces, reigns supreme and aloof. Further, the extension of the simple concept of the earth's attraction of objects near it to enormous masses and distances makes belief in the existence of such a force more difficult than belief in the most outlandish astrological superstitions.

What did Newton know about the force of gravitation and why did he advance such an incredible hypothesis? Newton knew one fact—a mathematical formula. This formula served him so well in the deduction of hundreds of physical laws that he unhesitatingly asserted the existence of the force. But he admitted that the physical nature of the force of gravity was a mystery to him, and he framed no hypothesis concerning it. He even attacked the vague physical explanations that had characterized science of the late medieval period, saying: "To tell us that every species of things is endowed with an occult specific quality by which it acts and produces manifest effects is to tell us nothing." Since he could not explain the action of gravity he said, "These forces being unknown, philosophers have hitherto attempted the search of Nature in vain; but I hope the principles here laid down will afford some light either to this or some truer method of philosophy."

Newton was insistent that though the causes of gravitation were not known he should nevertheless propose the mathematical formulation and utilize it, leaving the causes to be found perhaps by others. Insofar as the work in his *Mathematical Principles* is concerned he says, "For I here design only to give a mathematical notion of these forces, without considering their physical causes and seats." Again he writes, "But our purpose is only to trace out the quantity and properties of this force from the phenomena, and to apply what we discover in some simple cases as principles, by which, in a mathematical way, we may estimate the effects thereof in more involved cases; for it would be endless and impossible to bring every particular to direct and immediate observation. We said, *in a mathematical way* [italics are Newton's], to avoid all questions about the nature or quality of this force, which we would not be understood to determine by any hypothesis; . . ." With prophetic insight he adhered strictly to the mathematical formulation of the action of gravity and to the mathematical consequences of this formulation.

Newton emphasized even more than Galileo that science seeks to understand how the world acts and not why. Like Galileo he wished to know *how* the Almighty had fashioned the universe but was not presumptuous enough to inquire toward what end; nor did he hope to fathom the mechanism behind many phenomena.

In other writings Newton did attempt to make some material explanation of the action of gravity. He believed that a force must be transmitted from one body to another either through direct contact or through a medium, but he rejected his own speculations as vague and unripe and left the problem to others. Failing to give such an explanation, he could only describe gravitational effects as action at a distance. His contemporaries, Leibniz, Huygens, and John Bernoulli, and successors such as Leonhard Euler were more seriously disturbed by the introduction of such an inexplicable force but were no more successful in furnishing an explanation. During the seventeenth and eighteenth centuries scientists regarded gravitation as an "uncommon unintelligibility." Lesser scientists adopted the phrase "action at a distance," as though the phrase somehow contained meaning; gradually the endless repetition of the phrase lulled people into accepting a familiar phrase as an explanation. By the nineteenth century gravitation was accepted as "common unintelligibility."

The mathematical law of gravitation has, however, been so effective that the phenomenon has been accepted as an integral part of physical science. The significance, then, of Galileo's and Newton's decisions to seek the quantitative aspects of physical phenomena is actually to replace physical explanation by mathematical description. The physical behavior of nature was less and less sought after, and mathematical laws became the goal of science. In the Newtonian age mathematics mounted the steed of science and took the reins in its hands. One might almost say that since then mathematics has been replacing physics. Let us not be surprised, then, if we find as we follow the application of mathematics to nature that the reality of the physical universe seems to be hidden more and more in a maze of mathematical laws.

17: MORE LIGHT ON LIGHT

My design in this Book is not to explain the Properties of Light by Hypotheses, but to propose and prove them by Reason and Experiments: In order to which I shall premise the following Definitions and Axioms.

—NEWTON, IN HIS *Opticks*

MORE pervasive than motion is the phenomenon of light that permits man to see motions on the earth and in the heavens. Since ancient times scientists have undertaken to investigate the nature of light, its behavior as it passes through a medium such as air or water, and the associated process of vision. The progress they have made is peculiar. As to the physical nature of light we know no more today than men did several thousands of years ago. Yet man's knowledge of the behavior of light and his power to put the various phenomena of light to use has grown enormously, especially in the last few centuries.

The progress that has been made already in the field of light is similar to the developments in gravitation. Though, as we have noted in the preceding chapter, the nature of the force of gravitation is also a mystery, our mathematical knowledge of this phenomenon is sufficiently powerful to enable man to predict the paths of moving objects on the earth and in the heavens with remarkable preciseness. Some observations of the behavior of the gravitational force led, as we saw, to mathematical laws, and from these, most important new knowledge was derived by purely mathematical processes. The same procedure has been pursued in the investigation of light, and our present concern will be to see how some extremely simple mathematics permits us to acquire mastery of this phenomenon despite our woefully inadequate physical understanding.

Theories as to the nature of light and vision were formulated by many Greeks, among them Plato and Aristotle. A modicum of mathematical knowledge about the behavior of light also dates back to Greek times. Euclid, we saw, knew the law of reflection, namely, that light is reflected from a smooth surface in accordance with the law that the angle of incidence equals the angle of reflection. The Alexandrian Greek, Heron, observed the fact we used in an earlier chapter, that when light travels from one point to another by way of reflection from a mirror, it takes the shortest path. These two mathematical facts were about the only two precise ones the ancient world knew about light.

Though some interest in the phenomenon of light was shown by the Arabians and by the Europeans of the late medieval period and the Renaissance, it was not till the seventeenth century that the subject of light became a major scientific pursuit. The reason for this—the invention of the telescope and microscope—has already been mentioned.

The telescope itself, according to somewhat debatable history, was invented by several Dutchmen, Hans Lippershey and Zacharias Jansen, spectacle makers, and independently by James Metius, brother of the minor mathematician Adrian Metius. Like many fundamental inventions this one was the result of an accident that happened to the right people, people who were alert enough to note what had happened and sufficiently informed to understand what was taking place and to capitalize on it. Story has it that the first two men held two pairs of spectacles some distance apart and looked through both. They noticed that a distant church spire appeared very close. Instead of dismissing this phenomenon as a mere curiosity or as a peculiarity of the particular pair of spectacles they experimented further and designed a usable telescope. When Galileo visited Venice in 1609 he heard of this instrument and immediately made one himself by putting together two lenses of the right shapes, one at each end of a long tube.

We have already mentioned the astonishing discoveries Galileo made with his telescope. The hope of improving observations of the conjunctions of the heavenly bodies to aid ships at sea in determining longitude and the use of the telescope as a spy glass for warfare and navigation whetted other interests in optics. These uses of the telescope and the apparent potential value of the microscope stirred even the most indifferent minds.

Almost every great scientist and mathematician of the seventeenth century became interested in optics, the branch of physics that treats of light. Kepler, Descartes, Galileo, Newton, Huygens, and Leibniz made contributions. Newton and Huygens contributed the outstanding theories, Newton insisting that light consisted of streams of minute particles, while Huygens maintained that light consisted of a series of waves in an all pervasive ether. As we shall see later, neither was to say the last word on the subject.

One of the first definitive results, early in the century, was the discovery by the Dutch mathematician Willebrord Snell and René Descartes of the correct law of refraction (chapter 7). This rather simple law had eluded Greek, Arabian, and Renaissance scientists, who had all sought to explain not only the bending of light as it passes say from air to water but also the spectacular phenomenon of the rainbow. Among physical discoveries, most striking was Newton's decomposition of white light—sunlight—into the colors of the spectrum. Newton achieved this decomposition by passing sunlight through a prism. To prove that the colors were in the sunlight and were not due to the effect of the prism he passed the refracted beam through a second prism placed so as to reverse the original effect, and the many colored beams of light recombined to emerge from this second prism as white light. Another impressive result, which settled the question of whether light traveled with a finite velocity, was Römer's measurement of the velocity. As to subsequent progress we shall learn more as we proceed with our own account.

Experimentation in the field of light, as in any other scientific field, yields information, indeed basic information. But to resort to experimentation to obtain all the information one desires is not only a costly and time-consuming procedure; it yields just disconnected bits of information. As Galileo and Newton had foreseen and shown mathematics not only formulates the experimental results compactly but often leads to the prediction of new results that are impossible to obtain experimentally or that an investigator would not think of seeking experimentally. In the field of light, mathematics has proved to be an enormous boon. Let us see what it can accomplish.

We shall utilize the mathematical law of reflection as a basic principle. In addition we shall use the physical fact that in homogeneous media, that is, media that possess the same physical properties throughout, light travels in straight lines. Now the atmosphere is not strictly homogeneous because the density of air decreases with altitude and the moisture content or the humidity also varies with altitude and from place to place. However, for the transmission of light over short distances along the surface of the earth the air may be regarded as practically homogeneous, and straight-line propagation of light may be assumed.

The basic device in any optical instrument such as spectacles, the telescope, or the microscope is a lens through which light passes, or a mirror which serves to reflect light. The surfaces of lenses or mirrors are usually spherical because these are easiest to construct.

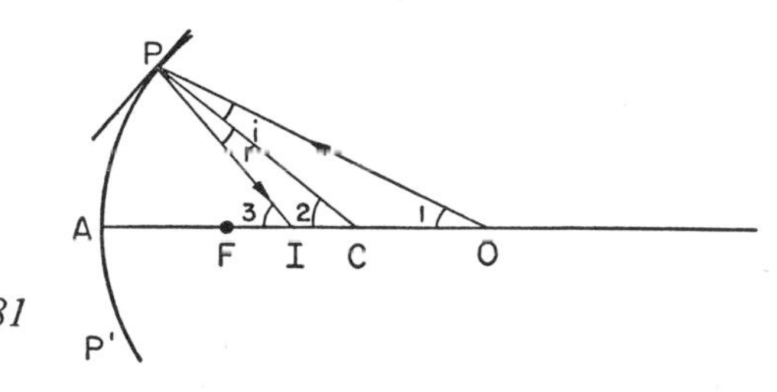

Fig. 81

Let us begin by considering a spherical mirror of which PAP' (fig. 81) is a cross section, and let us suppose that the surface is silvered so that light rays entering from the right are reflected to the right.

Let the point O be an object from which light rays emanate and let OP be a typical light ray that strikes the mirror at P. Of course this light ray is reflected in accordance with the law that the angle of incidence i equals the angle of reflection r. To locate these angles we observe first that if C is the center of the spherical surface, and therefore the center of the circular arc PAP', then CP is perpendicular to the circle, that is perpendicular to the tangent at P. The reason is the familiar theorem of geometry that a radius of a circle is perpendicular to the tangent at the point of intersection of radius and tangent. Let PI be the reflected ray. Hence i and r are the angles shown in figure 81. We shall also utilize the angles 1, 2, and 3 indicated in the figure.

Our object now will be to get some relationship among OA, CA, and IA. Let us begin by considering triangle OPC. Since an exterior angle of a triangle is equal to the sum of the two remote interior angles, and angle 2 is an exterior angle of this triangle,

$$\measuredangle 2 = \measuredangle 1 + \measuredangle i$$

or

(1) $$\measuredangle i = \measuredangle 2 - \measuredangle 1.$$

Likewise, if we consider triangle *CPI* of which angle 3 is an exterior angle, we obtain

$$\measuredangle 3 = \measuredangle 2 + \measuredangle r$$

or

(2) $$\measuredangle \mathrm{r} = \measuredangle 3 - \measuredangle 2.$$

But angle $i =$ angle r. Hence we may equate the right sides of equations (1) and (2) and write

$$\measuredangle 2 - \measuredangle 1 = \measuredangle 3 - \measuredangle 2$$

or, by adding angle 1 and angle 2 to both sides,

(3) $$\measuredangle 1 + \measuredangle 3 = 2 \measuredangle 2.$$

We note that angles 1, 2, and 3 are associated, respectively, with the lengths *OA, CA,* and *IA,* which we seek to relate, and with the arc *AP*. Hence we have moved closer to our objective by obtaining equation (3).

Let us note next that angle 2 is a central angle of a circle (fig. 81) and that it subtends arc AP. Clearly the size of *AP* must vary with the size of angle 2, and one might even be tempted to say that the length of arc *AP* could be used to specify the size of angle 2 instead of the usual specification of the size of an angle by the number of degrees in it. However, the same angle 2 subtends a large arc, *A′P′*, in a circle of larger radius (fig. 82). Hence the size of the subtended arc does not determine the size of the angle. But the ratio of the arc length to the radius, that is, AP/CA or $A'P'/CA'$ is the same in the two circles. Hence this ratio can serve as a measure of the size of the central angle. This method of denoting the size of an angle, which amounts to specifying how many radii the subtended arc contains, is called the radian measure of an angle and is occasionally convenient, just as it is often convenient to measure a length in yards instead of feet.

Let us now return to equation (3) and figure 81. The size of angle 2, as noted in the preceding paragraph, is AP/CA. We should like a similar measure of angles 1 and 3. However, neither angle is a central angle. But we can use as an *approximate* measure of angle 1, the ratio of the arc PA to OA. This approximation is not bad because angle 1 is smaller than angle 2 and correspondingly the denominator OA is larger than CA. Likewise an approximate measure of angle 3 is arc PA divided by IA, and here

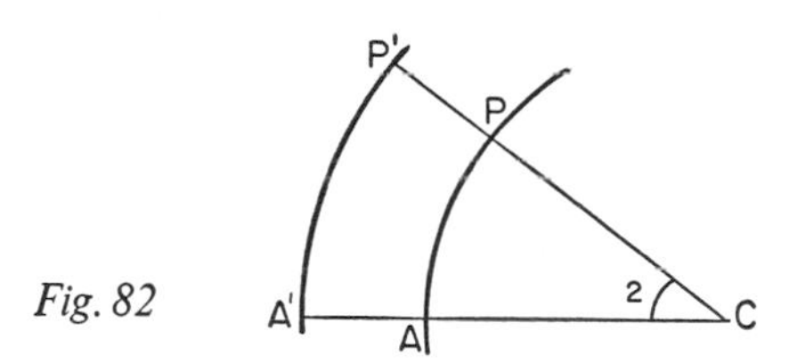

Fig. 82

too the approximation is reasonable because angle 3 is larger than angle 2 and correspondingly IA is smaller than CA. We shall comment on these approximations later; at present let us proceed to use them.

With these measures of the respective angles we may write equation (3) as

$$\frac{\overset{\frown}{PA}}{OA} + \frac{\overset{\frown}{PA}}{IA} = 2\,\frac{\overset{\frown}{PA}}{CA}.$$

Since arc AP occurs in each numerator we may divide the equation by this quantity and eliminate it. Then

$$\frac{1}{OA} + \frac{1}{IA} = \frac{2}{CA}. \tag{4}$$

It is customary to denote the distance OA by p and to denote the distance IA by q. CA is the radius of the circle, and so we shall denote it by R. Then equation (4) reads

$$\frac{1}{p} + \frac{1}{q} = \frac{2}{R}. \tag{5}$$

This result is in itself surprising. We started with an object placed at O and considered *any* ray that leaves O, strikes the mirror, and is reflected to some point on the line OCA. Of course the distance OA or p is fixed, as is the distance CA or R. But one would expect that the distance IA or q would depend upon which ray OP we happened to start with. But the value of q, according to equation (5), depends only on p and R and is independent of the ray OP with which one starts. This fact means that *all* rays which leave O will be reflected by the mirror to the *same* point I on the axis OCA. Then we shall see at I a sharp *image* of the object at O because all the light from O that strikes the mirror will be re-collected at I. Hence we have learned thus far that a spherical mirror will form an image of an object.

Equation (5) can tell us much more. Mathematically, equation (5) represents a function relating three variables, p, q, and R. If R is kept fixed, that is, if we consider a fixed spherical surface, then only p and q can vary, and by selecting a value of p we can calculate the corresponding value of q. Of course this functional relation is different from those we encountered in straight-line motion, projectile motion, or the law of gravitation. But by studying this function and by deducing consequences from it we can derive new physical knowledge just as we did in the case of the other functions studied.

Formula (5) was derived under the limitation that the object O lies to the right of the center C. Hence we shall restrict our deductions to this circumstance. Let us consider first what happens when p is very large; this means that the object is at the right and very far from the mirror. Since we can write formula (5) as

$$\frac{1}{q}=\frac{2}{R}-\frac{1}{p}, \tag{6}$$

we see that the larger p is the smaller $1/p$ is and the closer $1/q$ comes the value $2/R$, which means the q approaches $R/2$. To use the suggestive but imprecise language of mathematicians and photographers, when the object is at infinity—in symbols, $p=\infty$ —then $1/p=0$ and $q=R/2$. Of course this language must be understood to mean that when the object is so far away that for all practical purposes the incoming rays are parallel to the axis of the mirror, then the image will be practically at the point F (fig. 83) halfway between C and A. This point F is called the focus of the mirror.

Let us consider next what happens to the position of the image as the object moves in from infinity toward the left. As the object moves in then of course p decreases and so $1/p$ increases; hence the quantity on the right side of (6) decreases because $1/p$ is being subtracted from $2/R$. Then $1/q$ decreases and q increases. This means that as the object moves to the left, the image, which was at F when p was at infinity, moves to the right. In particular, when the object reaches C, so that $p = R$, then, by (6), $1/q = 1/R$ and $q = R$. In other words, when the object reaches C, the image also reaches C and object and image are at the same spot. This conclusion seems strange, but if we examine the meaning of it physically we see at once what is going on. If the object is at C, then all rays from the object travel along radii to the mirror and

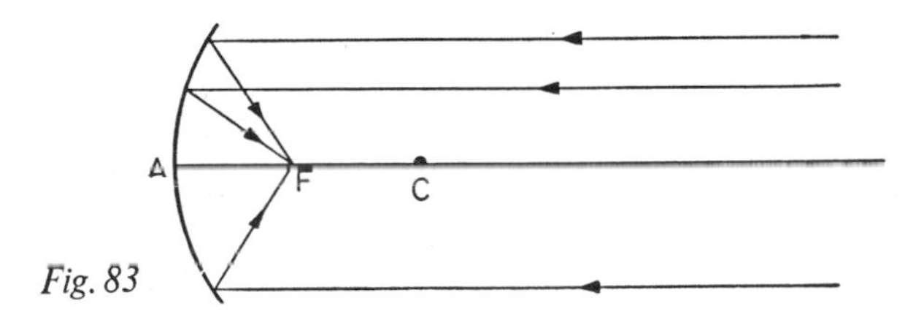

Fig. 83

strike the mirror perpendicularly. The angle of incidence is now zero; hence so is the angle of reflection, and the reflected rays return along the radius. The image will therefore form at C.

Thus far we have reasoned on the basis of formula (5) under the condition imposed by the derivation, namely, the object lies to the right of C. Ignoring for the present the approximations made in the derivation of this formula, we may assert that the conclusions established are unquestionable because the entire process is deductive, and the original formula correctly describes the physical situation. We have been operating with formula (5) as mathematicians, but now let us operate as scientists.

Before proceeding we might explain what we propose to do by recalling what Newton did with the law of gravitation. Newton adopted the law on the basis of two pieces of evidence. It gave results for motions on the earth that agreed with what Galileo had found, and it gave results for the motion of the moon that agreed with known data. Thereupon Newton generalized and concluded that the law applied universally. The mirror law rests initially on a somewhat different foundation. It was derived deductively from other established physical and mathematical facts. However, just as Newton assumed that his law applied to new physical situations, so we shall suppose that the mirror law applies to physical situations not covered by its derivation. Very often a mathematical formula that fits some phenomena of nature will apply far beyond the range of conditions under which it was originally derived, and through exploration and the insight gained thereby one finds that the proof of the formula can be generalized to cover new situations or that one has a basic law of nature so broadly applicable that it may be accepted as a physical axiom. At the very least some exploration may be physically instructive or suggestive.

Let us see, therefore, what we can learn by assuming that the mirror law applies to all positions of object and image. Let u be a value of p and let v be a value of q which together satisfy equation (5). If we examine equation (5) we see that we can substitute the value of v for p and u for q and still satisfy the equation, simply because

$$\frac{1}{q}+\frac{1}{p}=\frac{1}{p}+\frac{1}{q}\ .$$

However, interchanging the values of p and q means that the object is now at I in figure 81 and the image is at O. In other words, object and image will occupy positions that image and object occupied previously.

This result is deduced on the assumption that formula (5) represents the physical situation when the object is to the left of C. One could now test experimentally whether object and image do interchange positions and thus acquire a new physical fact through a mathematical lead, or alternatively one could now try to rederive formula (5) under the condition that the object lies to the left of C. Actually the derivation under the new condition can be carried through in the same way as we derived (5) in the first place, by merely changing the notations in figure 81. Hence we now know that formula (5) holds when the object is anywhere to the right of F, because previously the images all lay between C and F when the object was to the right of C.

However one establishes that we may interchange object and image, it now follows from what we found earlier that if the object moves from C to F, the image will move from C to the right toward infinity. But let us persist and ask what happens to the image when the object moves still farther to the left from F toward A (fig. 84). The image can no longer move farther to the right because it has already reached infinity. The poor image has no place to go.

Let us consult formula (5) or, for present purposes, the more convenient form, (6). If the object is between F and A then let us suppose, for concreteness, that $p = R/4$. Then $1/p = 4/R$. If we substitute this value of $1/p$ in (6) we find that

$$\frac{1}{q} = \frac{2}{R} - \frac{4}{R}.$$

Hence

$$\frac{1}{q} = -\frac{2}{R}$$

and

$$q = -\frac{R}{2}.$$

Now the first thing to note about this value of q is that it is negative.

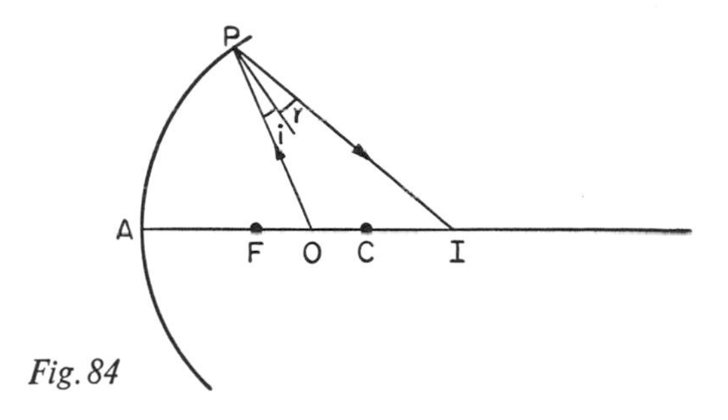

Fig. 84

tive. What clue does this furnish to the solution of our problem? Well, up to now we have regarded the distances p and q as positive and these positive values were always measured from A toward the right. Hence a negative q value should be a distance along the line CA but measured to the *left* of A. In other words, the mathematics tells us that when the object is between F and A the image will appear to the left of A, *behind the mirror.*

When the object was at F the image was at infinity to the right of the mirror. But as the object moves to some point between F and

A, the image suddenly appears to the left of A (fig. 85). How did it get there? Infinity is a mysterious place where strange things happen. Somehow the two "ends" of the line AC at the extreme right and extreme left join at infinity, and the image slides from the right side around to the left and so sidles up behind the mirror. Of course the reader is warned that lines do not behave this way all the time. The magic is done with mirrors.

Well, the image does appear precisely where mathematics says it should. It appears to the left of the mirror and a person looking

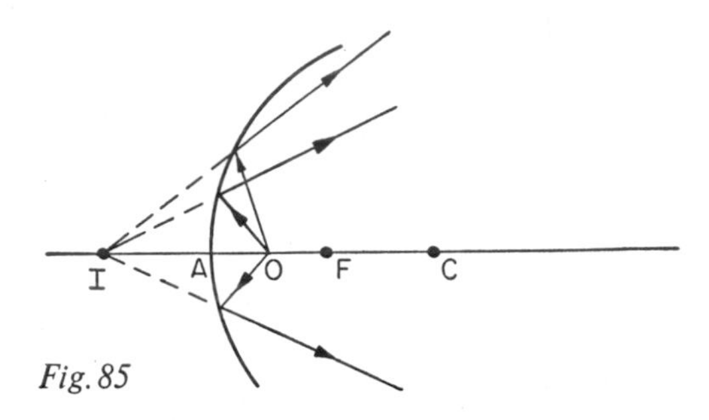

Fig. 85

into the mirror will see the image behind the mirror. This image is called virtual, whereas the ones we discussed earlier are called real. Both types of images are clearly visible; however, in the case of real images the reflected light does actually concentrate at the image, whereas in the case of the virtual image, no light emanating from the object ever gets behind the mirror to concentrate at the image. The light rays merely *appear to come* from the location of the image. These facts can be verified. A camera placed in front of the mirror and focused on a real image will receive the image on the film, whereas one placed behind the mirror and focused on a virtual image will not receive any picture, just as a camera placed behind an ordinary plane mirror will not obtain a picture.

The prediction we made mathematically as to the existence of a virtual image was made on the *assumption* that formula (5) continues to hold even when the object lies to the left of F. Though we have no reason to believe it does hold in this case, apparently it still predicts what does happen. We can understand now what the distinguished physicist Henrich Hertz meant when he said that mathematical formulas are cleverer than we are.

Thus far we have determined the location of the image for all positions of the object along the line CA from infinity at the right up to the point A. Let us now ask a seemingly meaningless question: What happens to the image if the object passes through the mirror and stops somewhere to the left of A? Now the object is behind the mirror and cannot produce any image. But the mirror law still has something to say. Now p is negative and if we substitute a negative value of p in formula (6) we get a positive value of q. This result should mean that if the object is to the left of A, there is an image to the right. Since the object is behind the mirror, how can there be an image? Well, we ought to do some thinking for ourselves and not depend upon mathematical formulas to tell us everything. To produce an image when the object is behind the mirror, simply reverse the silvering on the mirror! In other words, instead of silvering the glass to produce the concave mirror we considered up to now, we should silver the glass to produce a convex mirror. The mirror law then tells us that if the object is anywhere to the left of A along CA, that is, if p is any negative quantity, then the image will be to the right of A. In fact object and image should be interchanged in figure 85. No doubt the reader has seen his virtual image in many a convex mirror. In the form of silvered globes they are often used for garden decorations.

We can readily show by means of equation (6) that when p is negative the image will always lie between F and A. If p is negative, $-1/p$ is a positive quantity and this quantity is added to $2/R$. Hence $1/q$ is positive and larger than $2/R$ and so q is less than $R/2$. This result means that the image will be between F and A.

If we had any lingering doubts as to the physical applicability of the mirror law we could make one more rather simple test. A useful principle to keep in mind when one is testing a general law is to apply it to extreme but simple cases. Thus a theorem about regular polygons of n sides certainly holds for very large values of n, but the theorem makes no sense when n is infinite because infinity is not a number. However, a regular polygon of more and more sides approaches a circle in shape and there is good reason to expect that the theorem, perhaps with appropriate changes in wording, should apply to circles. If it does apply there is some heuristic evidence in favor of the original theorem.

To test the mirror law in an extreme case we shall consider what happens when the radius R of the sphere is very large. Now the larger the radius of a sphere the less rounded the surface; and, to use loose language, when the radius becomes infinite the sphere becomes a plane. Let R become infinite in formula (5). Then the right side approaches 0. In the limiting situation in which R is infinite the sphere is a plane and the mirror law becomes

$$\frac{1}{p}+\frac{1}{q}=0.$$

It follows algebraically that

$$p=-q.$$

The meaning of this statement is that in a plane mirror the image appears as far "into" the mirror as the object is in front of the mirror (fig. 86). This is the case, and we used this fact in discussing the law of reflection in chapter 6.

We could continue to draw conclusions from the mirror law, and

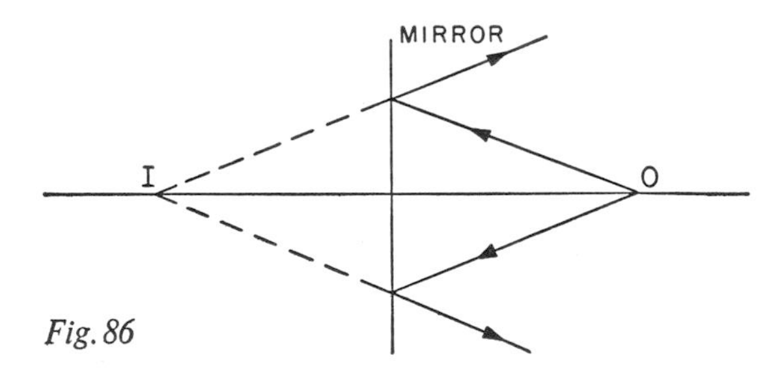

Fig. 86

we could show that this same law applies to the formation of images by thin lenses such as are used in spectacles and in other optical instruments. But our intention in presenting and discussing equation (5) was not to study the optics of mirrors and lenses, important as this subject may be. Rather it was to see how functions, formula (5), can aid us in studying natural phenomena.

One point concerning the derivation of formula (5) remains to be discussed. In this derivation we used some approximate expressions for angles 1 and 3 that certainly should introduce some error in the final formula. Nevertheless the final formula proved to be highly useful and, in fact, it is used in the design of mirrors and lenses. This type of approximation must be distinguished first of all from the kind that is involved in the usual mathematical idealization. Mathematics may solve a practical problem concerning spheres, for example. Of course no physical sphere is perfect; hence, when the mathematical result derived for perfect spheres is applied some deviation from the exact mathematical result must be expected. But in the derivation of the mirror law we consciously used an erroneous mathematical expression in the idealized geometrical account itself. Here we deliberately committed a distinct mathematical error. Yet the result proved to be useful nevertheless. What we did here is characteristic of what is done daily in applied mathematics. A mathematician, a physicist, or an engineer seeks to solve a physical problem by mathematical means. He encounters difficulties in solving the problem, but finds that by using approximate expressions he can get a mathematical answer. Should he trust his result and apply it, knowing that it is only approximate?

The scientist has two courses at his disposal. He can try to determine the amount of the error that his approximations introduce and then decide whether this error will be serious in the applications he has in mind. Since actual instruments can be constructed with only a certain degree of accuracy, the mathematical error in the theory may be far less than the error that necessarily enters into the physical construction. If the scientist cannot or will not take the trouble to determine the amount of error introduced by the mathematical approximation, he may nevertheless apply the resulting formula and see whether the device constructed thereby works satisfactorily. If it does, he continues to use the approximate mathematical law for new applications. Sometimes, as in designing steel beams for bridges and buildings, the engineer overdesigns; that is, he makes the beams far stronger than the formulas dictate and thereby allows for approximations in the theory.

In the case of the mirror law we could show that the result is reasonably accurate for small mirror surfaces. Strictly, the various rays from the object do *not* converge to the image point. But for small mirrors angles 1 and 3 are small, and the reflected rays come close to passing through the point I of figure 81. Hence the mirror law is useful for such cases.

Suppose, however, that for some application one had to use a large mirror, that is, suppose the arc *PAP′* of figure 81 should be large, and one wished to produce a sharp image at some given point *I* of an object located at point *O*. What kind of mirror will do this? The mathematicians knew the answer to this question long before the question was raised. The surface that will concentrate rays from *O* at a given point *I* is a part of an ellipsoidal surface (surface of a football) that has the points *O* and *I* as foci (fig. 87). If one passes a plane through this surface and through the points *O* and *I* he obtains an ellipse; the arc PAP′ of figure 87 is part of such an ellipse. Of course there are many such ellipsoidal surfaces, and other considerations might determine which of these to choose, just as there are many spherical mirrors of differing radii that would serve more or less well to concentrate rays from a point

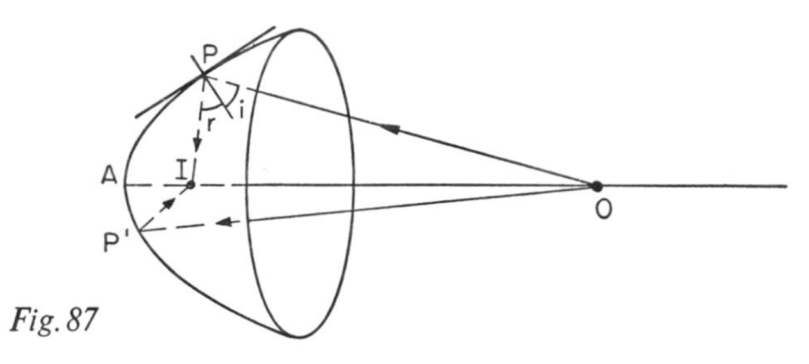

Fig. 87

O at a point *I*. The proof that any ray from *O* will, after striking the ellipsoidal surface, be reflected to *I* rests upon a geometrical property of the ellipse, namely, that the lines *OP* and *IP* make equal angles with the tangent to the surface at *P* (in the plane *OPI*); this property is precisely what is needed to make the incident ray *OP* take the path *PI* after reflection at *P*.

When the object *O* is very far away—for example, when it is a star—the rays coming toward the mirror are practically parallel. In this case the mathematician regards the object as being "infinitely far away." The question then arises, What surface would concentrate these parallel rays at one point *I?* Here too the answer to the question was well known long before the question became important for science. One can show that when one focus of an ellipse remains fixed and the other moves off to infinity, the ellipse becomes a parabola. Thus if the point *I* of figure 87 remains fixed and the point *O* moves off indefinitely far to the right, the ellipse of which *PAP′* is an arc becomes a parabola. The fact that a parabola is a limiting case of the ellipse suggests a theorem about the parabola that is the key to the problem of reflecting incoming parallel rays to one point. We noted in connection with the ellipse that the lines *OP* and *IP* make equal angles with the tangent at *P*. When the point *O* moves off to infinity the line *OP* becomes parallel to the axis of the parabola (fig. 88). Hence it seems likely that in the case of the parabola the line *OP* and the line *IP* make equal angles with the tangent at *P*. This geometrical fact can be proved. It follows, then, that a light ray coming in along the line *OP* will be reflected to *I*, because the light ray obeys the law that the angle of incidence equals the angle of reflection. All of this mathematics tells us, then, that a paraboloidal surface—that is, the surface gen-

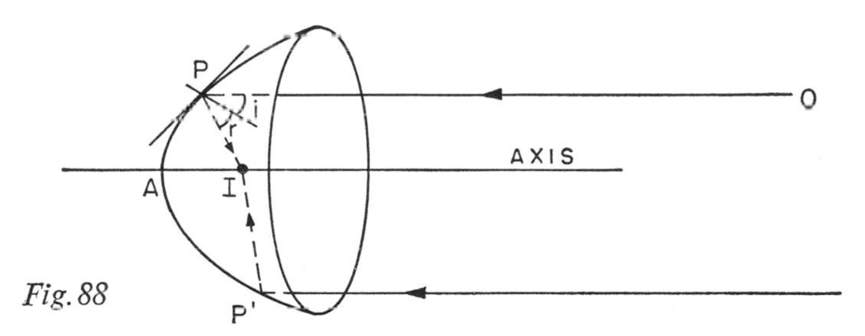

Fig. 88

erated by revolving a parabola about its axis of symmetry—will serve to concentrate parallel rays at one point.

Because paraboloidal surfaces will reflect to one point a family of incoming parallel rays, paraboloidal mirrors are used in large reflecting telescopes to obtain sharp images of the sun or stars. Archimedes used a large curved mirror, presumably paraboloidal, to focus the sun's rays on the Roman fleet besieging his native Syracuse in 212 B.C. The resulting heat at the focus (the word means burning-place) is supposed to have set fire to the ships.

We learned in connection with spherical mirrors that we may interchange the positions of object and image. This fact applies to the paraboloidal mirror as well. Hence the rays from a source of light placed at *I* will be reflected out along lines parallel to the axis of the paraboloid, and the reflected beam will be strong in that one direction. This property of paraboloids, which we proved on other grounds in our work on coordinate geometry, is used in automobile headlights, searchlights, radio antennas and other devices. The concentration of the light in one direction, that is, along the axis of the paraoloid, is remarkably effective. For instance, an automobile headlight with a small bulb at the focus of a paraboloidal mirror produces a beam on the road 6000 times as strong as the source alone would produce in that direction.

When the bulb is exactly at the focus of the paraboloid the beam of the automobile headlight shines directly ahead. Sometimes we wish to have this beam illuminate a wider area, for example, the countryside and the road. At other times we wish to illuminate the road, which means that the beam should point somewhat downward. These objectives and many others can be accomplished by utilizing the mathematical properties of paraboloidal mirrors and the laws of behavior of light rays.

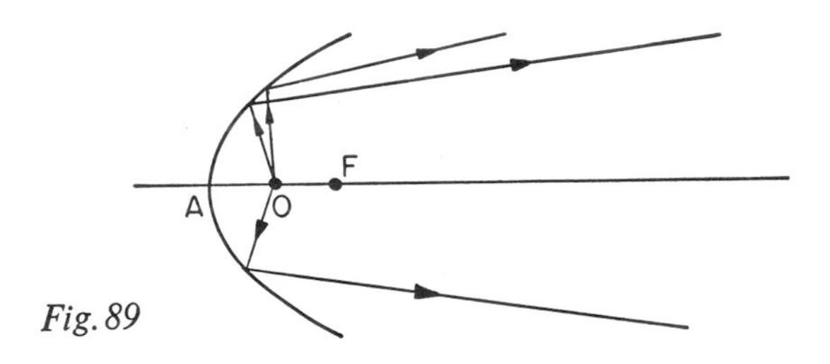

Fig. 89

If the light source is placed on the axis of a paraboloidal surface but somewhat behind the focus (fig. 89), then by tracing the course of a few rays, which, of course, continue to satisfy the law of reflection, we see that the beam is spread out.

If the light source is placed above the focus, then the beam is directed downward (fig. 90). Alternatively, if the light source is placed below the focus, the beam is directed upward (fig. 91).

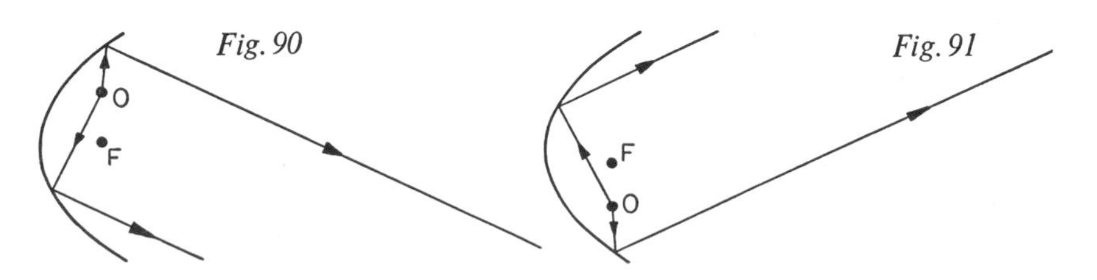

Fig. 90 *Fig. 91*

The emerging rays are not parallel in these last three cases but are so nearly so that a powerful beam is obtained. In practice, to change the direction of the auto headlight's beam, we shift from one light filament to another by means of a switch. By shaping the glass surface that seals the beam off from the air, so that the glass serves as a lens, further beam-shaping can be achieved.

Those who are indifferent to nature and more concerned about themselves should note that the eye is an optical system (fig. 92). Its main component is a lens that focuses incoming rays on the retina. Muscle action can change the focus of the lens so that light from far or from nearby sources will focus on the retina.

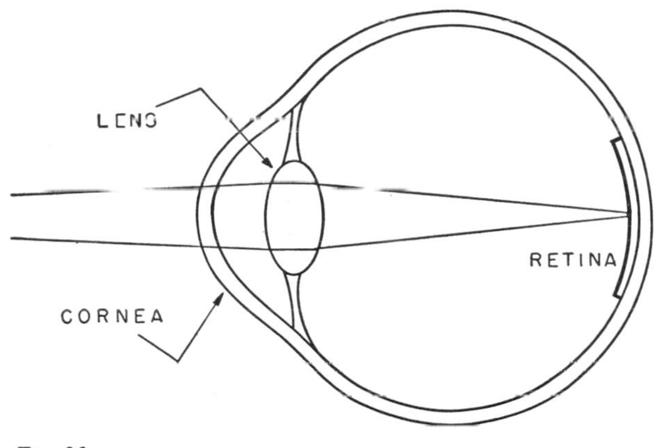

Fig. 92 HUMAN EYE VERY MUCH SIMPLIFIED

The knowledge of the behavior of light gained by the mathematical techniques described in this chapter not only enables us to understand the action of the eye but enables opticians to supply corrective lenses, and enables physicians to diagnose illnesses of the eye and to perform helpful operations. Hence the science of medicine is a direct beneficiary.

When one recalls the various instruments that use mirrors and lenses to redirect or concentrate light—the human eye, spectacles, cameras, telescopes, microscopes, binoculars, flashlights, headlights, searchlights, and specialized medical instruments—one begins to see the vast importance of the mathematics of optics in the study and mastery of nature.

18: THE MATHEMATICS OF OSCILLATORY MOTION

Ut tensio, sic vis.

—ROBERT HOOKE

THE motions of projectiles, of planets, and of light have an obvious importance or attraction, and one would therefore hardly question why men should study these motions. On the other hand, the motion of an object suspended from a spring and bobbing up and down repeatedly seems to offer no more attraction than to while away an idle hour. Of course an idle mind readily becomes the devil's playground, but we shall tempt the devil.

Let us be clear first as to what is different about the motion of objects on springs as compared with the motions we have previously examined. A projectile shot from a cannon traverses its course to the target and stops there. It does not then reverse its motion and return to the cannon. In fact it is apparently so pleased to have completed its motion that it announces its arrival at its destination with a bang. Some motions are never completed but continue indefinitely in one direction. Thus the earth, after completing one revolution around the sun, does not stop and then move in the opposite direction. Projectiles and planets move, so to speak, along one-way streets. On the other hand, the object bobbing up and down in response to the force exerted by the spring travels back and forth endlessly. The object goes nowhere in this oscillatory motion; yet the study of such motions has carried man far in his scientific travels.

Historically, the investigation of oscillatory motions was motivated by the desire to improve methods of telling time. The primary standard of time is, of course, the motion of the earth around the sun, but this natural clock is not particularly useful in the daily affairs of man. In the seventeenth century the need to measure small periods of time accurately for the purpose of telling longitude at sea caused scientists to search for increasingly accurate clocks. The search resulted in some major successes that were at least as valuable for the advancement of mathematics and the study of other phenomena of nature, such as light and sound, as they were for the specific problem of measuring time.

Scientists naturally concentrated on any physical phenomena that seemed to be periodic or repetitive and might therefore be related to the periodic motion of the planets. Two phenomena recommended themselves for closer investigation, the motion of an object or bob, as it is called, on a spring, and the motion of a pendulum. The first of these attracted the attention of Robert Hooke (1635–1703), a contemporary of Newton, a professor of mathematics and mechanics at Gresham College in England, and a noted experimentalist.

Let us consider with Hooke the motion of a bob on a spring. The bob is attached to the lower end of the spring (fig. 93) whose upper end is attached to a fixed support. Gravity will pull the bob

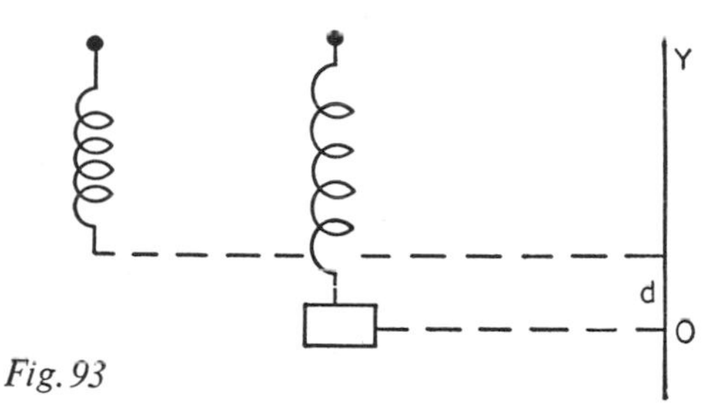

Fig. 93

down until the tension in the spring offsets the force of gravity. Unless disturbed the bob will then remain in a fixed position, which is called its rest or equilibrium position. Since we intend to disturb the bob and set it in motion let us agree to introduce a variable y that will represent its displacement from the rest position. When the displacement is upward we shall consider y to be positive and when downward, negative. Also, we shall ignore the resistance of air. While this resistance does dampen the motion its effect is secondary.

Hooke made a capital discovery about the motion of bobs on springs, but before we describe and utilize it fully let us see what mere observation of such motions can tell us. Let us suppose that the bob is pulled down below its rest position and then released.

We know from experience that the bob will move up past the rest position to some highest point or maximum displacement, then move downward to some lowest point or minimum displacement, and repeat this behavior endlessly, if the damping effect of air and (strictly) any internal energy losses in the spring are negligible. The motion is best exhibited by a graph (fig. 94) that shows how the displacement varies with time from the instant the bob passes the rest position on the way up. Of course the bob moves up and down along a vertical path whereas the graph is spread out horizontally because time values are graphed along the horizontal axis. The graph shows, of course, that the motion of the bob is periodic and that the behavior from O to P is repeated. The interval OP is therefore called the *period,* which is, then, the time required by the bob to go through one complete oscillation.

To undertake a mathematical study of the motion of the bob the first fact one would want is the formula that represents the graph, that is, the formula that relates the displacement and time of motion of the bob. But no one of the functions we have had occasion to use seems to fit this periodic behavior. We are therefore faced with the unfortunate necessity of extending our mathematical tools. Let us examine the graph more closely to see just what is involved.

According to the graph the bob executes one complete oscillation—that is, it moves from the rest position upward to its highest position, back to rest position, down to the lowest position, and then back up to the rest position—in one period. Let us suppose that the period is one second and that the maximum and minimum displacements are 1 and -1, respectively. We note also that during one period the graph naturally divides itself into four quarters, namely, the interval from 0 to $\frac{1}{4}$ second during which the bob rises to its maximum displacement, the interval from $\frac{1}{4}$ to $\frac{1}{2}$ second during which the bob goes through the same sequence of displacements but in *reverse* order, and the two quarters from $\frac{1}{2}$ to 1 second during which the bob repeats the motion of the first half except that the displacements are now downward or negative. Every period thereafter the graph repeats exactly the pattern of the first period.

These observations show that the shape of the entire graph is determined by what happens in the first quarter of a period, since the displacements after that merely repeat in different order or with different sign what happens in this quarter. Hence let us concentrate on the shape of the graph during the first quarter. We notice from the graph that we are dealing with a function, that is, a relation between displacement and time, wherein the displacement starts with the value 0 when $t = 0$, rises rapidly for small values of t, and then rises more slowly to the value 1 when $t = \frac{1}{4}$. To the mathematician these observations on the behavior of the displacement suggest the behavior of sin A when A varies from 0° to 90°. Indeed, if one were to plot the values of angle A as abscissas and the values of sin A as ordinates the graph would then have the same shape as the graph in figure 94 as t varies from 0 to $\frac{1}{4}$.

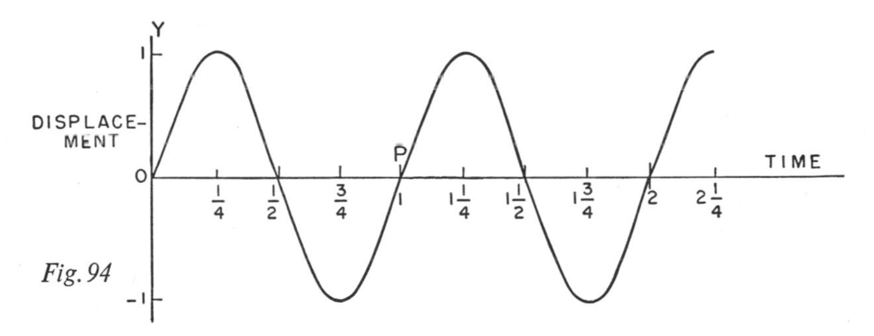

Fig. 94

There are, however, several differences. In the first place the value of sin A ends abruptly with the value 1 when A reaches 90°. There is no functional behavior corresponding to the second, third, and fourth quarters of the displacement-time relationship. This defect is easily remedied. We shall create a new function that will extend the meaning of sin A. Let us agree to have sin A vary from 1 to 0 as A increases from 90° to 180°, and, moreover, to have this function decrease in the reverse order from that in which it increases as A varies from 0° to 90°. Thus we require that sin 100° = sin 80°, sin 110° = sin 70°, and so forth, until A reaches the value of 180°. In the interval from 180° to 360° we shall agree to have sin A take on the same values as it does from 0° to 180° except that we shall now require that sin A be negative. Thus we require that sin 190° = —sin 10°, sin 200° = —sin 20°, and so forth.

We can consider angles larger than 360° merely by regarding such angles as generated by a rotation. Thus the minute hand of a clock rotates through 360° in one hour. As it continues to rotate we may regard the additional angle through which it rotates as added to 360°. In an hour and a half the minute hand sweeps through 540°, in two hours through 720°, and so forth. We shall now require that in each 360° interval, beyond the value of 360° for A, the function sin A is to repeat its behavior in the interval from 0° to 360°. Thus, for example, sin 370° = sin 10°, sin 380° = sin 20°, and sin 540° = sin 180° The function $y = \sin A$, which we have just created, is pictured in figure 95.

Fig. 95

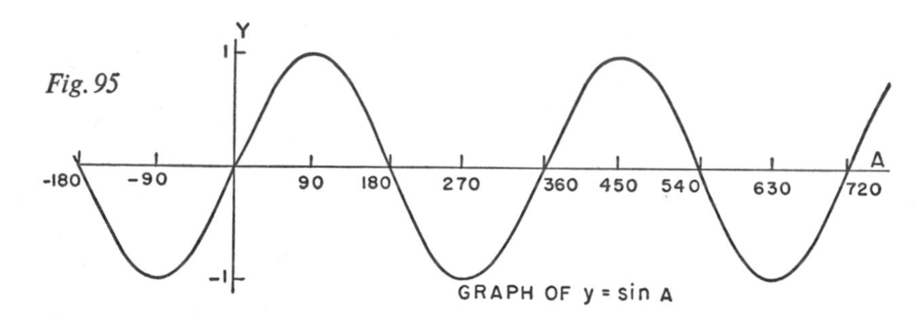

There are occasions when one wishes to use negative angles as well as positive angles. Of course, a negative angle means no more than a rotation of some number of degrees in the direction opposite to whatever direction is called positive. It is conventional to call the direction in which the minute hand of a clock rotates negative; accordingly, the angle formed by the initial and final positions of the minute hand is called negative. Correspondingly, the angle generated by a counterclockwise rotation is called positive. Though minute hands of clocks do not usually rotate in what we have called the positive direction, other devices such as wheels or gears on machines do. Of course the convention that the negative direction be clockwise and the positive be counterclockwise need not be adhered to any more than that the upward direction must always be called positive and the downward be called negative.

Since negative angles are used and since there will be occasion to utilize sin A when A is negative, the function sin A is defined for negative angles as shown in figure 95 so as to fit the agreement that the function repeats its behavior in each 360° interval. Thus $\sin(-90°) = \sin 270°$.

There is, however, a difficulty in attempting to employ the function $y = \sin A$ to describe the motion of the bob on the spring. The bob repeats its behavior each second, while the function $y = \sin A$ repeats its behavior every 360°. This difficulty is readily circumvented. To "speed up" the behavior of $y = \sin A$ we have but to make the values of A change more rapidly while the values of $\sin A$ continue their usual course. A simple maneuver will do the trick. Let us consider $y = \sin 2A$. As A varies from 0° to 180°, $2A$ varies from 0° to 360° and $\sin 2A$ must go through the full range or cycle of values from 0 to 1, to 0, to −1, and then to 0. Thus the function $y = \sin 2A$, as shown in figure 96, repeats its behavior every 180°. If we use the term *frequency* to mean the number of times that the function $y = \sin 2A$ repeats its behavior in 360°, we may say that the frequency of this function is 2.

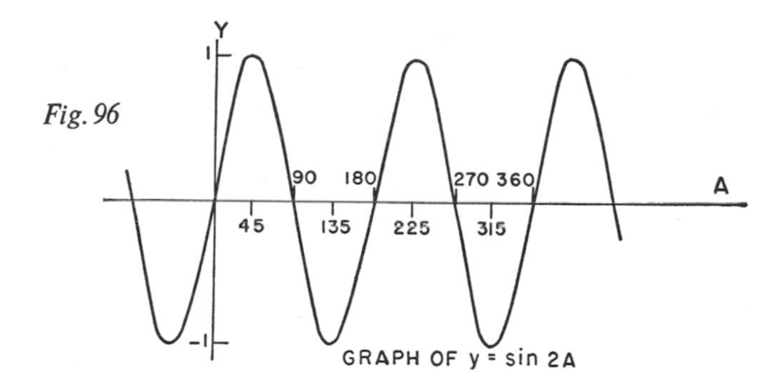

Fig. 96

We now have a mathematical device that we can carry further. The function $y = \sin 3A$ will have a frequency of 3, or repeat its behavior every 120°. Obviously we can produce a function that repeats its behavior as often as we please merely by using the proper coefficient of the angle A. This coefficient will be the frequency with which the sine function repeats its behavior in each 360° period.

While we are on the subject of creating new trigonometric functions we shall anticipate another need. In discussing the motion of the bob on the spring we supposed that its maximum displacement, or the *amplitude* of the motion, was 1. Let us suppose that each displacement were one-half of what is shown in figure 94. Could a trigonometric function still represent this motion? This question is readily answered. The function $y = (\frac{1}{2}) \sin A$ will have exactly one-half the y-values of the function $y = \sin A$ for the same values of A, and the function $y = (\frac{1}{3}) \sin A$ will have exactly one-third the y-values of $y = \sin A$ for the same values of A. Evidently we can modify the displacement values by any factor we please merely by putting that factor in front of $\sin A$.

By the same argument we can decrease or increase the displacement values of $y = \sin 2A$ by any constant factor merely by placing that factor in front of $\sin 2A$. Thus $y = (\frac{1}{3}) \sin 2A$ will rise and fall one-third as much as $y = \sin 2A$ in each 180° period (fig. 97).

We have therefore arrived at the following stage. The function

$$y = D \sin FA \tag{1}$$

will have a frequency F in 360°, that is, it will repeat its behavior F times in each 360° interval, and it will have an amplitude or maximum displacement D. It seems clear, then, that by choosing F and D properly we can produce a formula that will describe a sinusoidal, that is, sine-like, function with any frequency and amplitude.

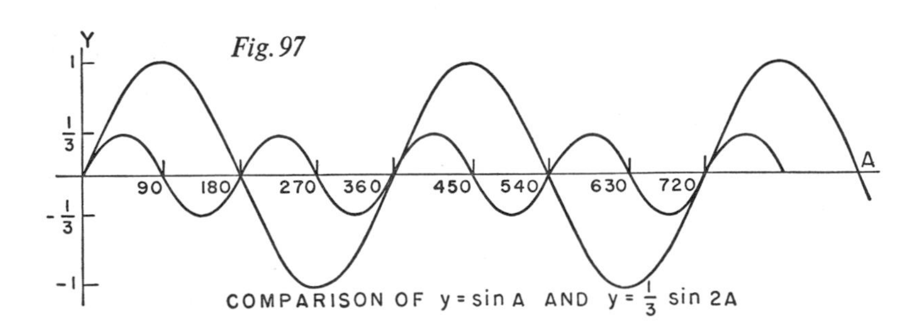

There is one more detail to be covered before we can describe mathematically the motion of a bob on a spring. We had occasion to point out in the preceding chapter that the size of an angle can be described in terms of radians instead of degrees. An angle can always be thought of as a central angle in a circle, and the number of radians in the angle is the subtended arc divided by the radius,

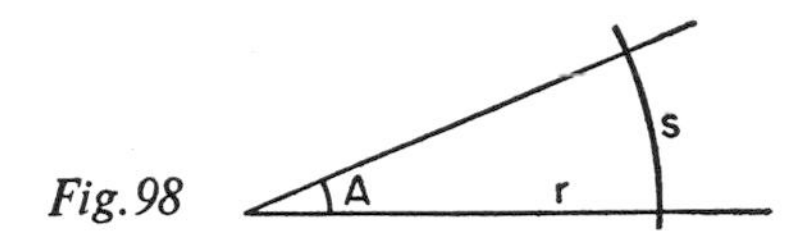

Fig. 98

that is, s divided by r (fig. 98). In this *radian* system of measuring angles, the sizes of angles usually denoted as varying from 0° to 360° can vary from 0 to 2π because the subtended arc can vary from 0 to 2π times the radius. Thus a 90° angle, which is one-quarter of a complete rotation, would have the size $\pi/2$ in the radian system of measuring angles. The use of radians instead of degrees involves no more difficult an idea than the use of yards instead of feet to measure length. Whereas in the latter case we remember the conversion relationship, that 1 yard is equivalent to 3 feet, in the case of radian measure we need only remember that an angle of 2π radians is equal in size to one of 360°.

Though we could continue to work with angles measured in degrees we shall adopt the radian measure because it will ultimately be more convenient. At the moment the adoption of radian measure will serve no more purpose than conformity to scientific usage. We shall therefore restate the conclusion arrived at in connection with formula (1) as follows. The function

$$y = D \sin FA \tag{2}$$

has a frequency F in 2π radians and an amplitude D.

The function (2) relates two variables, the size of angle A and a variable y that has a specific numerical value for each value of A. But this statement is not exactly right. The fact that the numbers chosen for A were originally suggested by the sizes of angles, and the fact that the y-values were originally suggested by ratios associated with angles, are *not* contained in formula (2). The formula merely relates two variables, A and y. Given a numerical value of A we can find the numerical value of y. The situation here is not different from that involving a simple formula such as $d = 16t^2$. We often say that this formula relates the distance in feet that a body falls to the number of seconds it has been falling. Actually the formula itself does no such thing. As a mathematical formula it relates two variables, d and t. That d can represent the distance a body falls when t represents the number of seconds the body falls is a physical interpretation and application of the mathematical formula. There is, however, no reason why $d = 16t^2$ cannot represent the number of apples that t pigs eat each month.

We are therefore free to use formula (2) to represent some other physical phenomena having little or nothing to do with angles. In particular, we can let A stand for the time that the bob on the spring is in motion and let y stand for the displacement of the bob, and it may very well be that for the proper choice of F and D, the formula can represent the motion of the bob. To emphasize that the values of A in formula (2) may, in some applications at least, represent time values, we shall write

$$y = D \sin Ft \tag{3}$$

and now say that this function has a frequency F in an interval of 2π units of t and an amplitude D.

Up to this point we have been constructing the mathematical equipment needed to study the motion of bobs on springs. To recapitulate, it seemed from mere observation of the motion that some sort of periodic function would be needed to represent it, and we have therefore built up such functions by extending some concepts of trigonometry. Now let us see what all this mathematics will do for us in studying the motion of bobs on springs.

Hooke's great discovery in connection with the action of springs may be explained as follows. We all know from experience that if we stretch or compress a spring the spring exerts a force that tends to restore the normal length. Suppose d is the increase or decrease in length of the spring resulting from extension or contraction. Hooke found that the restoring force the spring exerts is proportional to d; that is, the force is a constant, k, say, times d. This is the meaning of the quotation from Hooke that heads this chapter. The value of the constant k depends upon the elasticity of the spring. Let us apply Hooke's law to the motion of a bob on a spring.

Suppose that the bob is attached to the lower end of the vertical spring shown in figure 93. Since gravity pulls the bob downward the spring will be extended by some amount, say d. The force that gravity exerts on the bob is of course the weight of the bob, and this is $32m$, where m is the mass of the bob. The restoring force, according to Hooke's law, is kd. The bob will settle in what we have called the rest position, which is fixed by the condition that the restoring force will just offset the pull of gravity. Hence

$$32m = kd. \tag{4}$$

But now suppose the bob is pulled further downward so that the new position of the bob is y units below the rest position (fig. 99).

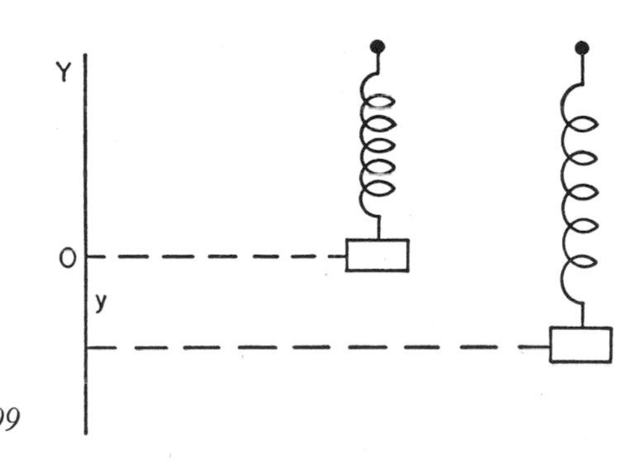

Fig. 99

Then the total extension of the spring is now $d - y$, the minus entering because y-values below the rest position are negative. The restoring force exerted by the spring is now $k(d - y)$. Since the weight of the mass acts downward, the net upward force is

$$kd - ky - 32m. \tag{5}$$

In view of equation (4) the net upward force acting on the mass m is $-ky$. When the bob is released it will begin to move under the action of the net force, $-ky$, exerted by the spring. Now Newton's second law of motion says that the force equals the mass times the acceleration created by the force. Hence

$$ma = -ky, \tag{6}$$

or

$$a = -\frac{ky}{m}. \tag{7}$$

We now have a mathematical law governing the motion of the bob. Let us therefore utilize the law to see what we can establish about this motion. When the motion begins, the bob is at some point below the rest position to which it has been pulled. At this position the velocity is zero and y is negative. Hence, according to equation (7), the acceleration is positive and the bob will begin to acquire positive velocity as it moves upward. This velocity will continue to increase because the acceleration remains positive until $y = 0$ when the bob is at the rest position. The acceleration is now zero but this does not mean that the velocity is zero. In fact the velocity, as we have just noted, has been increasing up to this point and hence the bob is moving very rapidly as it passes the rest position. Since the velocity there is not zero, the bob will continue to move upward past the rest position. But y is now positive and a is accordingly negative. The bob will lose velocity and in fact the larger y becomes the more negative is the acceleration. Hence the velocity will ultimately become zero and the bob will stop moving upward.

Will the motion then cease? No, because acceleration is still present and moreover negative. Hence the velocity will decrease below the zero value and become more and more negative. Physically, the negative velocity means that the bob will reverse direction and move downward. The y-values will now decrease from that of the highest position and the acceleration, though still negative, will likewise decrease in numerical value. Nevertheless, since acceleration is present and negative, the velocity will continue to become more and more negative. When the bob reaches the rest position on the way downward the velocity is highly negative and hence the bob is moving rapidly downward. But as y becomes negative, that is, as the bob falls below the rest position, the acceleration becomes positive. This means that positive increases are being made to a negative velocity and so the velocity becomes less and less negative and ultimately reaches the value zero. At this point the bob will stop moving downward. However, the bob will not remain motionless because the acceleration, now positive and large, will add positive increments to the velocity and so the velocity will increase from the zero value to positive values and the bob will move upward. The behavior of the bob from this position onward repeats the original behavior.

This analysis of formula (7) reveals several features of the motion. The motion is slowest, because the velocity is close to zero, when the bob is near its lowest or highest positions. The motion is fastest when the bob passes through the rest position on the way upward and on the way downward. Moreover, the entire downward motion has all the features of the upward motion except, of course, for the direction. Finally, the entire upward and downward motion repeats itself because when the bob reaches the lowest position after the first oscillation, the acceleration, velocity, and position are precisely what they were to start with.

This tedious analysis serves one purpose at the moment. Near the beginning of this chapter we graphed the relationship between displacement and time of the bob moving up and down at the lower end of a spring, and we surmised from the appearance of the graph that the motion might be represented by a sinusoidal function. The more careful analysis based on formula (7) shows that the rate at which the displacement of the bob changes does indeed follow the rate at which the y-values of a function such as $y = D \sin Ft$ change. Of course familiarity with such sinusoidal functions is presupposed; this familiarity naturally comes through experience with such functions. The identification of the two functional relationships—the relationship between the displacement of the bob and time of motion on the one hand and $y = D \sin Ft$ on the other—is not mathematically demonstrated by our argument. In a later chapter we shall show how this identification can be established rigorously, but let us now rely upon the above evidence, which in fact is pretty much what Hooke used, to conclude that the bob does pursue a sinusoidal behavior. However, we still do not know what D and F of formula (3) should be to make the formula fit the motion of the bob. As to D, this is the maximum displacement, and its value is determined by how far below the rest position the bob is pulled before it is released. Hence we know D.

What will the frequency F of the motion be? Let us see what determines this frequency. One would expect that the elasticity or stiffness of the spring would be involved. Some springs stretch readily, others have to be pulled hard to stretch them. This property of the spring is represented in the constant k. If the spring is very stiff then k is large. According to formula (7) the larger k is, the greater the acceleration of the mass. Greater acceleration means that the bob will go through the cycles of changes in velocity and therefore in displacement more rapidly; that is, there will be more oscillations per second or the frequency will be larger. On the other hand, if a bob of larger mass is attached to a given spring, then formula (7) says that the acceleration will be less. Hence, for a given spring stiffness, the larger the mass the less the frequency. These considerations suggest that the frequency F of the oscillating mass is determined by k and m, and, moreover, that F must be large when k is large and that F must be small when m is large. To the mathematician these considerations suggest that a possible formula for F is k/m.

However, there is no reason why the formula could not as well be k^2/m^2 or k^3/m^3 or some other one that fits the foregoing considerations. With the mathematics presently at our disposal we could not determine the exact formula for F. An experimental physicist could, however, proceed as follows. He would first determine the value of k. That is, using a known mass he would observe how much the spring is stretched when the mass is attached. This increase in length is d. Then, knowing m and d, he would use formula (4) to calculate k. He would now set the mass in motion and observe what F is for the known k and m. He would find that

$$F = \sqrt{\frac{k}{m}}, \tag{8}$$

where F is the frequency or number of oscillations in 2π seconds. To verify experimentally that (8) is the correct formula he could try several masses on the same spring and also try several different springs.

In a later chapter we shall say more about the determination of F. Let us now, however, accept formula (8) on the basis of experimentation and recognize that we now have determined the D and F which we need to apply formula (3) to the motion of the bob on the spring. The formula that relates the displacement and time of the bob's motion is then

$$y = D \sin \sqrt{\frac{k}{m}}\, t, \tag{9}$$

if the time t is measured from the instant the bob moves upward from the rest position.

Since $\sqrt{k/m}$ is the frequency in 2π seconds, the frequency f in one second is of course

$$f = \frac{1}{2\pi}\sqrt{\frac{k}{m}}. \tag{10}$$

One more quantity is generally desired in connection with the motion of bobs on springs, namely, the period, or the time required to make one complete oscillation. If a man runs 10 feet in 1 second he requires 1/10 of a second to run 1 foot. Likewise, if a bob makes f oscillations in one second it requires $1/f$ seconds to make one oscillation. Hence the period T is given by the formula

$$T = \frac{1}{f} = \frac{2\pi}{\sqrt{\frac{k}{m}}} = 2\pi\sqrt{\frac{m}{k}}. \tag{11}$$

The remarkable fact, which Hooke observed and emphasized, about formula (11) is that the period is independent of the amplitude. In other words, whether one pulls the bob down a small or large distance from its rest position and then releases it, the period of motion is the same. This result is not in accord with most people's intuitions. When the bob is pulled down a great distance D, the restoring force, according to formula (6), is $-kD$, wherefore the bob will shoot up rapidly and indeed move rapidly thereafter. But this does not mean that the period will be shorter. Actually, since the amplitude D of the motion will be larger, the bob will have to cover more distance in each oscillation and, though it moves rapidly, its period of oscillation will be the same as when the amplitude is small.

We have devoted much space to the derivation of formulas (9), (10), and (11), and it would now seem appropriate that we utilize these formulas to derive some new and useful or at least interesting information about the motion of bobs on springs. Indeed we could do so, but we shall refrain from capitalizing on these accomplishments until we have introduced a closely related phenomenon, the motion of a pendulum. Let us regard the work done thus far as an investment. We have learned some facts about a new type of mathematical function, the sinusoidal function, and we have seen that this function can represent the motion of bobs on springs. Perhaps this information can be turned to good advantage later. As a matter of fact our position is not greatly different from Hooke's. He learned, largely by experimentation, a great deal about the action of springs and sought to apply this knowledge to the construction of an accurate clock. Although he did not succeed in producing the first successful spring-regulated clock, his contributions did ultimately make this device possible.

While Hooke was studying the action of springs, Galileo and Huygens were exploring another device that had attracted Galileo's attention when he was a youth. Galileo had observed the swinging of lamps suspended from long cords in churches. By using his pulse beat to time the swings he found that the time required for a complete back and forth motion, that is, a complete oscillation, was the same whether the lamps swung through wide arcs or narrow ones. This observation excited Galileo's interest in the pendulum, though at the time he did not envisage its use in a clock. What did interest him was that the pendulum seemed to have some special properties. Moreover, because its motion is not hindered as much by air resistance as is the motion of objects sliding or rolling down planes, observations of pendulum motion might agree far better with theoretical studies of idealized situations. Hence Galileo proceeded to study the pendulum.

Galileo's contemporaries thought that the pendulum was kept in its regular motion by the action of the air. But Galileo dismissed this idea because, he said, "in that case, the air must needs have considerable judgment and little else to do but kill time by pushing to and fro a pendent weight with perfect regularity." Just as he ignored the presence of air in his study of straight-line and projectile motion, Galileo decided to study the action of pendulums in a vacuum.

Following Galileo, we shall suppose that a bob is hung at one end of a length of string, and the other end is held fixed. The bob is set in motion by pulling it to one side and then releasing it. What does mathematics tell us about the subsequent motion?

We shall denote by A (fig. 100) the angle the string makes with the vertical when the bob is in some arbitrary position during its motion. If the bob is to the right of the vertical then A will be

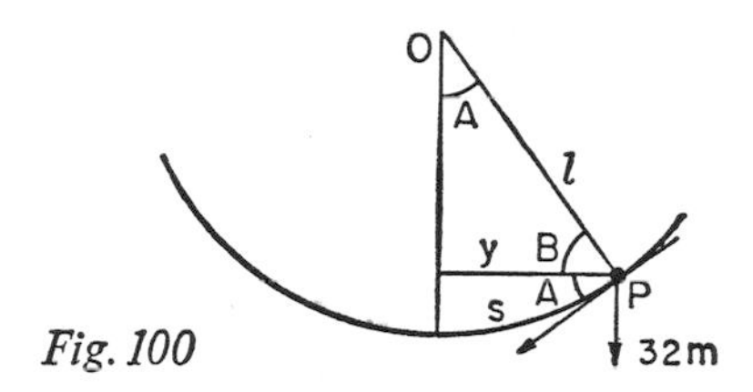

Fig. 100

understood to have positive values, and if to the left, negative values. The bob moves along the arc of a circle, and the direction of its motion at any position P is along the tangent at that position. The tangent at P is perpendicular to the string OP because OP is a radius of the circle. Hence the two angles marked A in figure 100 are equal because both are complements of angle B.

What force acts on the bob? If m is the mass of the bob, the force of gravity pulls the bob straight down with a force of $32m$, which is, of course the weight of the bob. But the bob is not free to fall straight down because the string resists somewhat the pull of gravity. The bob can, however, move in the direction of the tangent, which, as noted, makes an angle A with the horizontal. We may then regard the bob as moving along an inclined plane that makes an angle A with the horizontal. In our study of the motion of an object sliding down an inclined plane we learned that the effectiveness of the force of gravity along the direction of the plane is not $32m$ but $32m \sin A$. Hence this latter quantity is the force acting on the bob and causing it to move along the circle.

There is, however, a matter of sign. When the bob is to the right of its rest position, the force $32m \sin A$ acts to the left; when the bob is to the left of the rest position, the force causes it to move to the right. Thus the force is opposite to the displacement. Hence we must represent the force as $-32m \sin A$. Since this force is, according to Newton's second law, the mass times the acceleration of the bob, we have

$$ma = -32m \sin A$$

or

$$a = -32 \sin A.$$

If we refer to figure 100 we see that

$$\sin A = \frac{y}{l},$$

where y is the *horizontal* displacement of the bob. Hence

$$(12) \qquad a = \frac{-32y}{l}.$$

Equation (12), though correct, is not quite what we should like. It says that the acceleration of the bob is a constant, $32/l$, times the horizontal displacement of the bob. But the bob does not move horizontally. Rather, it moves along the arc of a circle and the distance it actually moves along this arc is the length s shown in figure 100. To relate the acceleration to the actual distance moved we shall consciously make an approximation of the type we made in our study of light and, at the moment, hope for the best insofar as the usefulness of our result is concerned. The quantity y is half the chord of the arc $2s$. When the arc is small the chord is a good approximation to the arc. Hence if we restrict the theory to oscillations of small amplitude, we can with some reasonableness replace y by s in (12) and obtain

$$(13) \qquad a = -\frac{32}{l}s.$$

This formula gives the acceleration acting on the bob in terms of the *actual* displacement of the bob from the rest position, that is, from the vertical.

We are now going to take advantage, in a big way, of the abstractness of mathematics. We have seen, for example, that the same geometric theorem can apply to many quite different physical situations. The same is true of formulas. What acceleration dictated the motion of the bob on the spring? Formula (7) gives the answer. Now formula (13) governs the motion of the bob on the pendulum, and the two formulas are identical except for notation. Just as k and m were constants in the case of the motion of the first bob, so 32 and l are constants in the motion of the second one. Moreover, just as y in formula (7) is the actual displacement of the bob on the spring, so s in formula (13) is the actual displacement of the second bob. Since the same mathematical formula governs the acceleration in the two cases, and since the acceleration determines the motion, we may conclude that the mathematical deductions from that formula apply to both cases. Hence, in view of formula (9), the relationship between the displacement and time in the case of the motion of the bob on the pendulum is

$$(14) \qquad s = D \sin \sqrt{\frac{32}{l}}\, t,$$

where D is now the maximum displacement of the bob, that is, the largest value of s, and t is time measured from the instant the bob starts to move to the right from the lowest position. It follows from formula (10) that the frequency of the bob, that is, the number of oscillations or complete swings it makes in one second is

$$(15) \qquad f = \frac{1}{2\pi}\sqrt{\frac{32}{l}},$$

and from formula (11) we see that the period of the pendulum is

$$(16) \qquad T = 2\pi\sqrt{\frac{l}{32}}.$$

Before we discuss formulas (14) to (16) let us emphasize an aspect of the reasoning that led to them. Because the same formula for acceleration acts in two different situations, and the acceleration determines the final formula for the motion, we conclude that the same final formula must apply in the two different situations. The abstractness of the mathematical formulas pays unexpected dividends, no doubt as compensation for our willingness to think abstractly.

Formula (16) shows exactly what Galileo had observed, namely, that the period of the pendulum is independent of the amplitude or maximum displacement of the bob. Thus, whether the bob is pulled aside to make an angle of 30° with the vertical or so as to make an angle of 5° with the vertical and then released, the period of the subsequent swings is the same. We must remember, however, and we shall return to this point later, that, in the case of the pendulum, formulas (14), (15), and (16) are only approximately correct because we replaced a chord of a circle by an arc, and this approximation is reasonably accurate only for small amplitudes of oscillation. However, if the pendulum can be kept in motion at about the same amplitude, as is done in pendulum clocks by making a weight that falls gradually under the action of gravity apply a force to a mainspring, then the period of the pendulum will be practically constant, though not given exactly by formula (16).

Another deduction from (16), which Galileo also made, is that the period is independent of the mass of the pendulum. Whether the bob is made of lead or cork the period will be the same.

Formula (16) reveals the immense usefulness of the pendulum. By knowing the acceleration due to gravity, the quantity 32 in the formula, and the length l of the string to which the bob is attached, one can predict the period of the pendulum and hence measure time. Moreover, by fixing l, one can make the period whatever he wishes. Many years after Galileo studied the pendulum to learn some facts about its motion he designed a practical pendulum clock and had his students construct one. About fifteen years later, Huygens, working independently, constructed one that gained wide acceptance. Of course pendulum clocks do not work well on rolling ships and so the search for a reliable spring clock continued long after Galileo's and Huygen's time.

In our treatment of the pendulum, as well as in that of the motion of the bob on a spring, we used the fact that the force of gravity acting on a mass m is $32m$. The quantity 32 is of course the acceleration due to gravity at points near the surface of the earth. We know, however, that the acceleration of masses under the pull of gravitation is not always 32. In fact the acceleration is, more generally, the quantity GM/r^2—formula (2) of chapter 16—where G is the gravitational constant, M is the mass of the earth, and r is the distance of the mass being accelerated from the center of the earth. Then the acceleration varies with distance from the center of the earth. Hence, if a pendulum clock is adjusted so that it beats seconds at one location, it will not be accurate at another location if the acceleration due to gravity is different there. The pendulum clock may have to be readjusted if it is moved from one location to another.

It is an ill wind, however, that blows no good. The fact that the period of the pendulum depends upon the acceleration due to gravity was turned to excellent advantage by seventeenth- and eighteenth-century mathematicians and scientists. To follow their work let us first rewrite formula (16). In our derivation of this result we used 32 for the acceleration due to gravity. Let us now reconsider the derivation with the quantity 32 replaced by the symbol g, which is the customary symbol for the acceleration due to the direct action of gravity, as opposed to the symbol a, which is used for any acceleration, whether due to gravity or some other force. Then the derivation would continue to hold as long as g is constant, which it is in any one location, and formula (16) would read

$$T = 2\pi\sqrt{\frac{l}{g}}. \tag{17}$$

Now suppose that by adjusting l, the period T of a pendulum is made to be one second at a location where g is, say, 32. If the pendulum is moved to another location and if the period is now measured, then formula (17) may be used to calculate g. Hence the acceleration due to gravity can be measured at a variety of locations. This method is very accurate and is used today for precise determinations of g.

As the pendulum is moved from either pole toward the equator it is found that g decreases from the value of 32.257 ft/sec^2 at the poles to 32.089 ft/sec^2 at the equator. This difference, small as it may seem, must be explained.

Newton had considered the shape of the earth. On the assumption that it was homogeneous throughout, that is, it possessed the same density all the way down to the center, and that it was subject to the attraction of the sun and to rotation, he proved that the earth's shape is not a sphere but an ellipsoid of revolution, roughly a sphere flattened somewhat at the top and bottom. Hence an object on the surface of the earth is farthest from the center at the equator and closest to the center at the poles. Since r, the distance from the center of the earth, decreases the closer one is to the poles, the greater g must be as one moves from the equator to either pole. Thus far at least the argument agrees with the measured values of g. However, according to the calculations of Newton and Huygens and the more accurate later calculations of the mathematician Clairaut, the flattening of the earth at the poles is not sufficient to account for the observed variation in g; that is, the distance from the center of the earth to either pole is not sufficiently less than the distance to the equator to account for the larger value of g at the poles.

The full explanation of the observed variation in g was given by Huygens. An object whirled at the end of a string tends to fly off into space and follow the straight-line path that the first law of motion says all objects seek to follow. That the object continues to move in a circle is due to the fact that the hand keeps pulling the object into the center. The force the hand exerts is called a centripetal (center-seeking) force. Objects on the surface of the earth are whirled around by the rotation of the earth and hence are subject to a centripetal force. Where does this force come from? When we examined the motion of the moon around the earth we pointed out that the earth pulls the moon toward the earth and thus prevents the moon from moving along a straight-line path. In other words, the gravitational attraction of the earth supplies the centripetal force. Let us now consider an object on the earth. If the earth were not rotating, it would still of course, exert its gravitational force on the object; and this force would be the weight of the object. But since the earth rotates and keeps objects rotating along with it, the gravitational force must be used partly to supply the centripetal force. Hence the weight of the object is reduced. Since the weight of an object near the surface is gm, a reduction in weight must mean a reduction in g compared to the value on a stationary earth.

Why, then, does the measured g vary from the equator to the poles? An object on the equator is whirled on a circle whose radius is the radius of the earth. As one goes north or south from the equator the circle on which an object is whirled by the rotation of the earth shrinks in size; this circle is in fact the circle of latitude at the object's position. At the poles the circle shrinks to zero. Let us take into account now that an object traveling with velocity v on a circle of radius r is subject to the centripetal acceleration of v^2/r—see chapter 16, formulas (15) and (22). The velocity of an object rotating with the earth is the circumference of the circle on which it travels divided by the time, that is $2\pi r/T$. Hence

$$(18) \qquad a = \frac{v^2}{r} = \frac{4\pi^2 r^2}{T^2 r} = \frac{4\pi^2 r}{T^2}.$$

The time of revolution is 24 hours, or 86,400 seconds, no matter where the object is. But the radius of the path depends upon the latitude of the object. Thus $r = 4000$ miles at the equator and $r = 0$ at the poles. Evidently the centripetal acceleration is greatest at the equator and decreases to zero at the poles. The centripetal force exerted by the earth on a mass m is just m times the centripetal acceleration. Since the centripetal force must be subtracted from the total gravitational force obviously the net gravitational force, that is, mg, should be less at the equator than at the poles. Thus the values of g calculated by means of the formula for the period of a pendulum confirm the rotation of the earth, or may be regarded as independent evidence for the rotation.

The value of g varies from place to place on the earth for still another reason. The earth's mass would attract any object as if the mass were concentrated at the center, but only if the earth's mass were uniformly distributed. But large deposits in one area of iron ore, oil, or other highly dense substances disturb the uniform distributions of the earth's mass and cause the value of g to vary. Such variations in g can be calculated through the action of a pendulum, as noted above, and departures from the expected values at a given locale indicate to geologists information as to the local matter in the earth. Detection of such variations in g is one of the methods of geophysical prospecting.

There is a postscript to the story of pendulum motion, written by Huygens. We pointed out above that the period of a pendulum is not strictly independent of the amplitude of swing. Huygens thereupon set for himself the problem of determining the curve along which a bob might swing, such that the period would be exactly the same whether the bob traversed a long or a short arc. It so happens that the curve Huygens found to be the solution of this problem was already being intensively studied by many mathematicians, largely as an intellectual challenge.

The curve itself is actually before our eyes daily, though we do not take the trouble to see it, much less investigate its mathematical properties. Suppose one fixes his attention on a point on the circumference of an automobile tire. The path *of that point* (fig. 101, p. 296) as the tire rolls along the ground is known as a cycloid. The

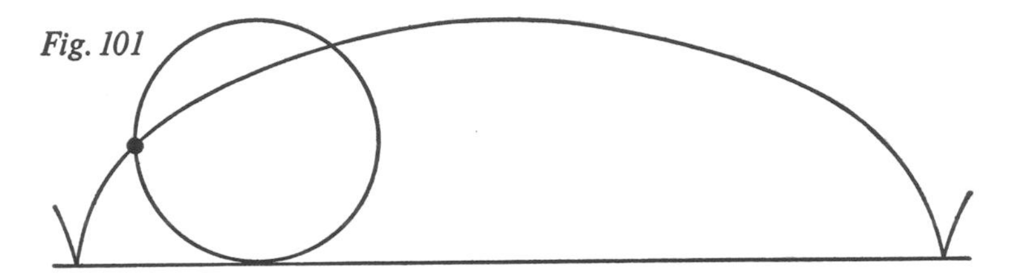

Fig. 101

figure shows what is called one arch of the cycloid; the full curve consists of repeated arches.

Let us visualize this curve as inverted (fig. 102), and let an object slide along this curve under the action of gravity. Huygens proved that no matter at what point A the mass starts it will reach the lowest point O in the same time and will go on to reach the symmetrically located point A′ in that same time.

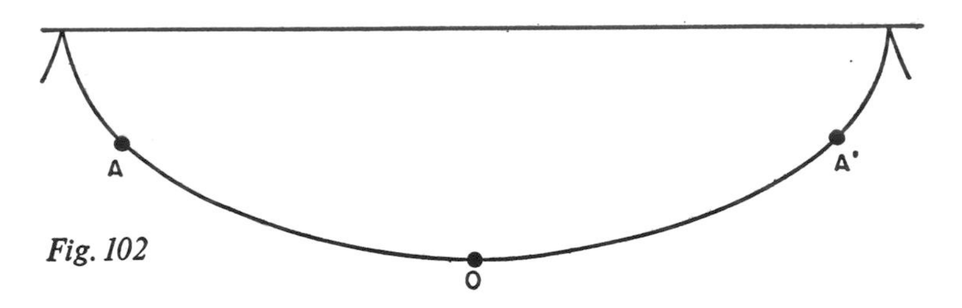

Fig. 102

This mathematical property of the cycloid is of considerable interest in itself, but Huygens, concerned with designing a clock, was not interested in sliding objects. If he could get the bob of a pendulum to follow such a curve under the action of gravity, then he would have the essential device for the truly accurate clock he was seeking. Huygens found the answer to this question through

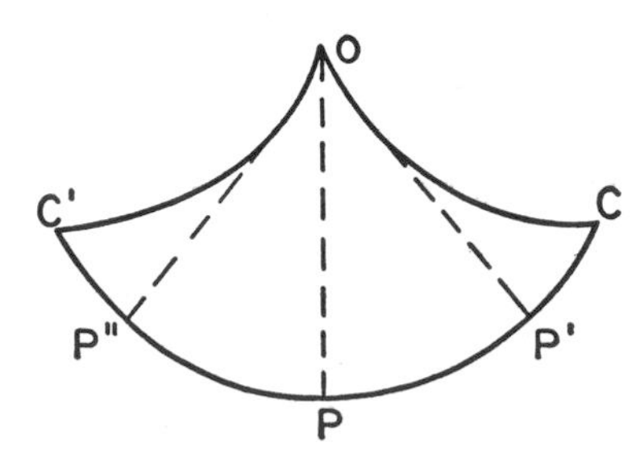

Fig. 103

further mathematical work on the cycloid. Let OC and OC' be identical half cycloidal arches (fig. 103), and OP be a string with a bob at P. As the bob is swung to the right the string will wind itself partly around the cycloidal arc OC and the rest of the string will be pulled taut by the pull of gravity on the bob. Huygens proved that if the length of OP is twice the height of either cycloid OC or OC' then the locus PP' of the bob will also be an arc of a cycloid. Likewise, as the bob swings to the left the path PP'' will be an arc of the same cycloid with P as the lowest point. Mathematically speaking, the arc OC is called the evolute of the cycloidal arc $PP'C$ and OC' is the evolute of the cycloidal arc $PP''C'$.

Hence Huygens had found the way to make the bob swing along the arc of a cycloid and had designed the perfect pendulum. The cycloid is often referred to as the isochronous curve, or the tautochronous curve, because the period of swing of a pendulum following a cycloidal path is truly independent of the amplitude of the swing.

19: OSCILLATIONS OF THE AIR

I have never been able to fully understand why some combinations of tones are more pleasing than others, or why certain combinations not only fail to please but are even highly offensive.

—GALILEO

SINCE the phenomenon of sound almost constantly engages our attention, it is not surprising that mathematicians and scientists include this subject in their investigations. Though the use of sound in social conversations has been reduced in our time to desultory remarks that hapless people must resort to when television sets break down, other uses and sources of sound are no doubt more widespread today. The phonograph, radio, television set, and motion pictures add to the sounds of the winds, thunder, and the oceans. The layman who considers the welter of sounds which constantly besiege him would no doubt reject the suggestion that there is some basic law and order in all this variety, and would question the profitableness of undertaking the scientific study of such confusion. But there is a mathematical science of sounds, and the mathematical tool that proves to be the key to the analysis of sound is the very same function which represents the motion of masses on springs and of pendulums.

Physically, all sounds are motions of air molecules. However, the sound of a tin can bouncing along the street and many of the pathetic sounds emanating from radio and television sets are created by disorderly air motions. Hence we label them for what they are—sheer noise—and dismiss them as beneath notice. On the other hand, those vocal and instrumental sounds that are intelligible must possess some property which enables the human ear to identify them. Without being too clear or precise at the moment as to what this property is we can at least recognize that when we utter a sound, we repeat it a sufficient number of times to make possible the identification. For example, when one pronounces "t" or "a" or some combination of letters he actually gives forth this sound many times, though perhaps in a short interval of time. Thus he is repeating some basic pattern. In other words, intelligible sounds must be periodic. Such sounds are technically called musical sounds. In this sense of the term even the individual sounds of such a distasteful word as "mathematics" are musical sounds.

Musical sounds, then, contain some basic patterns and these patterns might be representable and studied mathematically. The search for mathematical properties of musical sounds was undertaken long ago by the Pythagoreans. Sounds are usually distinguished by what is called pitch; that is, they can vary from low, deep sounds to high, piercing sounds. The first fact the Pythagoreans discovered was that the pitch of sounds emitted by a plucked string varies inversely with the length of the string; that is, the shorter the string the higher the pitch of the sound. They found further that the sounds emitted by two equally taut strings are harmonious when the lengths of these strings bear simple numerical ratios to each other. Thus two strings, one of which is one-half the length of the other, give forth a harmonious combination of sounds. The musical interval between these sounds is called an octave. Later we shall see that the frequency, or number of vibrations per second, of the higher-pitched sound is twice that of the lower one. Another pleasing combination of sounds is given off by strings whose lengths are in the ratio of 2 to 3. In this case the musical interval is called the fifth, and the higher-pitched sound makes 3 vibrations in the same time as the lower one makes 2. A third basic interval is the fourth, and this interval describes two sounds given off by strings whose lengths are in the ratio of 3 to 4, or whose frequencies are in the ratio of 4 to 3.

Though the Greeks, Arabians, and medieval Europeans continued to study musical sounds, their chief interest was the construction of musical scales for the composition, writing and reproduction of music. No really basic facts about the nature of musical sounds were uncovered in these civilizations. It was the seventeenth-century mathematicians and scientists, Galileo, his French pupil and colleague Mersenne, Gassendi, Halley, Huygens, Römer, and

Newton, who, with the characteristic vigor of that century, renewed the study of sound. One factor that enabled them to make progress was the pendulum clock: it made possible the measurement of small intervals of time. Throughout the eighteenth century the great mathematicians, the Bernoullis, Euler, D'Alembert, and Lagrange, all of whom we shall come to know better later, studied musical sounds and especially the vibrations of strings. But the really decisive step in the mathematical analysis of periodic phenomena was made by Joseph Fourier in the early nineteenth century, and it is through his work that the account we are about to relate became available.

To study any complex phenomenon we do best to follow the advice of Descartes: proceed from the simple to the complex. Hence let us start with simple sounds. But what is a simple sound? The ear gives us the answer. If we should listen to the sounds emitted by various instruments we would recognize that one in particular gives off what musicians call a pure tone and which, in fact, they use to standardize pitch, namely, the tuning fork. Hence it seems wise to begin with the sound of a tuning fork.

If either prong of a tuning fork is struck, both prongs will move inward and then outward very rapidly at a frequency determined by the material and dimensions of the fork. Let us follow the right

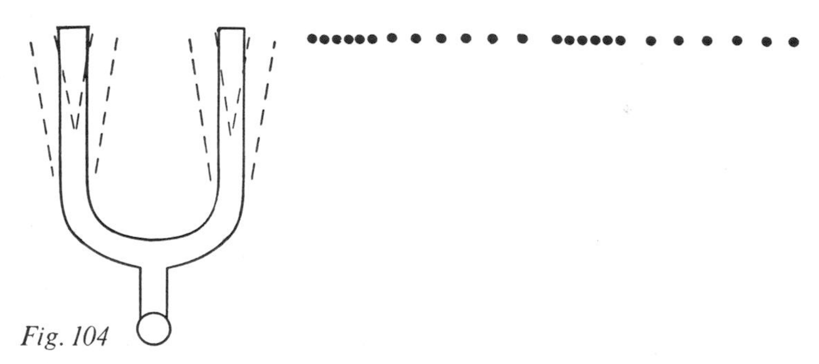

Fig. 104

prong (fig. 104). As the prong moves to the right for the first time its crowds together the air molecules alongside. This crowding is called a condensation. Because air pressure tends to equalize itself the crowded air particles move farther to the right where there is not so much crowding. The process is repeated and again the condensation moves to the right. While the condensations keep moving to the *right* the prong moves back to the left. This leaves a comparatively vacant region in the prong's former position. The more crowded air molecules situated to the right of this region will rush into the rarefied region, thereby creating another rarefied region in their former position. The molecules farther to the right will then move to the left into this new rarefied region, and so on. What is now happening is that a rarefaction moves to the *right,* away from the prong. Each movement of the prong to the right and to the left sends a condensation and rarefaction to the right. We have considered motions that take place to the right of the prong. Actually, condensations and rarefactions move off in all directions. It is the succession of condensations and rarefactions, or increases and reductions in the normal air pressure, moving outward from the fork that constitute the sound wave.

A sound wave, then, is something which travels out from the fork into space. But the behavior of the individual air molecules is quite different. First of all, the condensations and rarefactions are the consequences of the motion of billions of air molecules, and these molecules do not all do the same thing. But the average effect of these billions of molecules can be studied by considering a representative or typical molecule. Suppose this molecule is at the point *O* (fig. 105) before it is disturbed. A condensation causes it to be displaced to the right to *A*. The ensuing rarefaction then causes

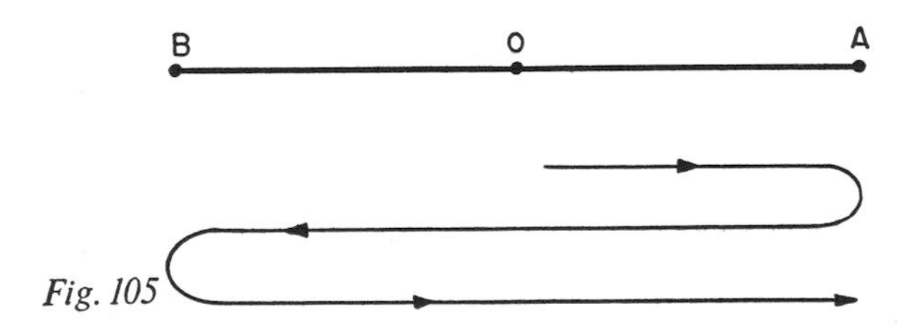

Fig. 105

it to move back past its original position to *B;* the next condensation causes it to move back to *O*. It has now made one complete vibration. Without stopping at *O,* however, the molecule goes through the whole set of motions again and again under the successive impulses arising from the tuning fork. Thus the displacement of the molecule from its original position varies periodically with the time that elapses from the instant it began to move.

We might begin to secure some mathematical grip on the phenomenon of sound if we could represent the motion of the typical air molecule. For this purpose one could consider the pressure waves that determine the motion of the typical air molecule. These pressure waves are created by a prong of the tuning fork, and they consist of condensations and rarefactions of the air. Air pressure always seeks to equalize itself, that is, to become uniform everywhere. In this respect it acts like a spring, for a spring that is extended or contracted always seeks to restore its natural length. Hence the air acts like an elastic medium, the condensations corresponding to contractions of the spring and the rarefactions to extensions. Hooke had noted this analogy and had extended his analysis of springs to, as he put it, "the spring of the air." Thus the air acts on the typical molecule as the spring does on a bob attached to it. We should expect, then, that the typical molecule executes sinusoidal motion; that is, the relation between the displacement and time of the typical air molecule is sinusoidal.

The fact that the relationship between displacement and time of the typical air molecule is representable by the sine function can be verified experimentally by using devices such as the phonodeik or the modern electronic engineer's oscilloscope. We shall rely upon experimental evidence to assert that for simple sounds the relationship between the displacement y of the typical air molecule and the time of motion t is

$$y = D \sin Ft, \tag{1}$$

where D, we may recall, is the amplitude or maximum displacement, F is the number of oscillations in 2π units of t, and t is the time. In the case of sound waves, t is generally measured in seconds and D in inches or centimeters. Since it is customary in the study of sound to speak of the number of vibrations per second we shall change formula (1) somewhat. F is the frequency in 2π seconds. If we denote by f the frequency in one second, then

$$f = \frac{F}{2\pi}.$$

Thus, from (1),

$$y = D \sin 2\pi ft. \tag{2}$$

To apply formula (2) to an actual sound created by a tuning

fork we must of course know D and f. An air molecule set into motion by a sound wave moves over a very short interval. A reasonable value of D is 0.001 inches. The frequency of the molecule's motion is the frequency of the tuning fork, and a typical value of f is 1000. Hence a typical formula for a simple sound is

$$y = .001 \sin 2\pi \cdot 1000\, t. \tag{3}$$

Formula (2) or (3) makes clearer just what we meant earlier when we said that intelligible sounds possess a pattern. The pattern in the case of simple sounds is the sine curve. If only part of the set or cycle of y-values that occur in one period reaches the ear, the ear cannot identify the sound because there is still no clearly established pattern. However, the repetition of several cycles serves to identify the sound to the ear, because by then the amplitude and frequency are established.

Thus far we have discussed the motion of a typical air molecule as it oscillates about its rest position. Each air molecule between the fork and, say, the ear of an observer goes through the same motion, but since time is required for the disturbance in pressure, or sound wave, to travel through space, these molecules do not begin their motion at the same time. We can understand this fact by means of an analogous situation. Suppose a person takes hold of one end of a long rope and shakes it up and down vigorously. We know from experience that a wave travels out along the rope. Each individual molecule moves *up and down* about some mean position just as the typical air molecule oscillates *back and forth* about its rest position. Also, the force transmitted by one molecule of the rope to another causes the successive molecules to go through the same motion. But the farther to the right a molecule is the later it will start its periodic motion because the transmission of the force along the rope takes time. Likewise, the condensations and rarefactions of the sound wave moving to the right take time to travel through the air and in this case, too, the farther a molecule is to the right the later it starts its sinusoidal motion.

Let us consider two molecules such as P and Q on the rope (fig. 106, p. 304). By the time that P is about to begin its downward motion, let us say for the third time, the molecule Q, because it has received later the force exerted by the hand, is about to execute its

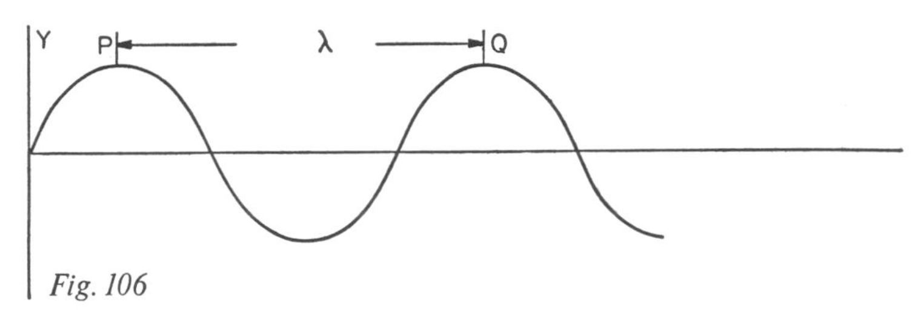

Fig. 106

downward motion for the second time. In *time,* Q is one period behind P. In *space* Q is some distance from P, generally denoted by λ (lambda) and called the *wave length.* The time then required for the wave motion to be transmitted from P to Q is the period T of the individual molecule's motion and the distance moved by the wave in that time is λ. Hence the speed v with which the wave moves to the right (which is not the speed of the individual molecule oscillating about its mean position) is λ/T. Moreover, since the period T of the molecule's motion is the reciprocal of the frequency f per second—formula (11) of chapter 18—we have the basic relation for wave motion that

$$v = \frac{\lambda}{T} = f\lambda. \tag{4}$$

Thus, since the velocity of sound waves is about 1100 feet per second, a sound of frequency 225 per second has a wave length of 4 feet.

Figure 106 suggests another fact about the wave motion, namely, that the relationship between displacement y and *distance* x is also sinusoidal. The maximum displacement in figure 106 is also the maximum displacement of the individual molecule, but the frequency in space, so to speak—that is, the number of cycles per foot of distance—is not the frequency of the individual molecule. In fact if the wave length or distance occupied by 1 space cycle is 4 feet, the number of cycles per feet of distance is ¼. Hence the number of cycles in 2π units of distance is $2\pi/4$ and the formula relating displacement and distance is

$$y = .001 \sin \frac{2\pi}{4} x. \tag{5}$$

More generally,

$$y = D \sin \frac{2\pi}{\lambda} x. \tag{6}$$

The full story of the mathematical representation of a simple sound is just a step away. Formulas (2) and (6) tell us that the displacement y is really a function of two variables, t and x. Fixing x means that we concentrate on a particular molecule x units away from the source, and then consider the displacement of that molecule as a function of time t. The functional relationship will not necessarily be formula (2) because (2) presupposes that the displacement is 0 when $t = 0$, and this will be true only of the first molecule and those a distance λ, 2λ, and so forth, away from it. Likewise, fixing t means that we "freeze" the motion of the molecules and consider the variation of displacement of all the molecules with distance at that instant. Again the functional relationship need not be (6) because (6) holds only at those times when the first molecule happens to be in its rest position. All of these facts can be conveniently represented by one formula, namely,

$$y = D \sin 2\pi \left(ft - \frac{x}{\lambda}\right). \tag{7}$$

A full discussion of formula (7) would take us farther into the analysis of wave motion than we need to go in order to understand the mathematics of musical sounds. We can concentrate on the relationship between displacement and time of the typical air molecule. Formula (2), which is of this type, expresses simple sounds mathematically. But very few actual sounds, vocal or instrumental, are simple. The flute, like the tuning fork, does produce some simple sounds but these instruments are exceptions. What does mathematics have to say about more complex sounds, and what value is there in the mathematical representation?

It is the ear that recognizes some tones as simple. Suppose, then, we ask the ear to listen to more complex sounds, such as the sound of a note on the cello, piano, or violin. Does the ear detect anything about the structure of these sounds that would help us to analyze them? Well, the ear is a remarkably sensitive organ. Its sensitivity to sounds differs with the frequency of the sound; it is most sensitive to a frequency of about 3500 per second. At this frequency it can detect displacements of air molecules as small as one ten-billionth of a centimeter. To put the matter more strikingly, if an insect whose weight were $1/10{,}000$ of the weight of a mosquito were to dance on a drumhead at the rate of 3500 times per second, the ear would hear the sound.

We should not be surprised, then, to learn that when the ear listens to a complex sound it is able to recognize in it a series of simple sounds. This is a fact of experience first enunciated by Georg S. Ohm, a teacher of mathematics and physics in Germany during the early nineteenth century. Ohm's law says that a complex sound is no more than a combination of simple sounds, each having its own frequency and amplitude. There are a number of experimental verifications of Ohm's law. For example, since the tuning fork gives off a simple sound, and a complex sound is presumably no more than a combination of simple sounds, it should be possible to duplicate a complex sound by striking simultaneously several tuning forks, each having the proper frequency and amplitude. This can be done. Another experimental verification is obtained if one first presses the pedal on a piano so as to lift the dampers from the strings and then strikes, say, middle C. A number of other strings whose basic frequencies are those of the simple sounds contained in middle C will also vibrate because each is excited by one of the simple sounds.

What does Ohm's law mean mathematically? How can a complex relationship of displacement versus time be a sum of sinusoidal functions? Just to make the mathematics simple for the moment, let us suppose that two simple tones having the mathematical representation $y = \sin t$ and $y = (1/2) \sin 2t$ are sounded simultaneously. The graphs of these individual simple sounds are shown in figure 107, a and b. If these two sounds are impressed simultaneously on the air molecules, then a typical molecule will move in accordance with the sum of the two displacements urged upon it. For example, at the instant $t = 3\pi/4$ the first sound displaces the molecule about 0.7 in the positive direction and the second displaces it about 0.5 in the negative direction. Then the molecule will move 0.2 units in the positive direction. This is the displacement in figure 107c at $t = 3\pi/4$.

Fig. 107

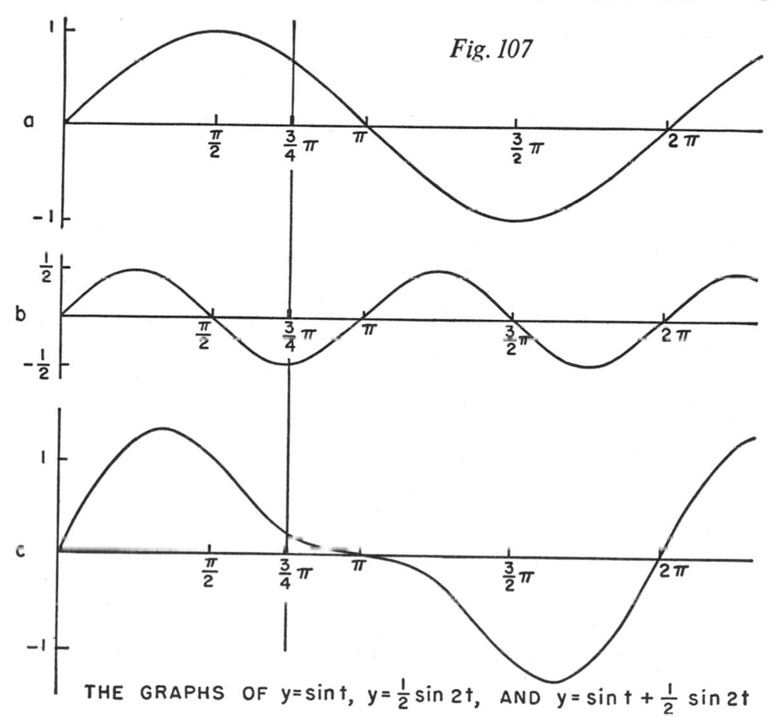

THE GRAPHS OF $y = \sin t$, $y = \frac{1}{2}\sin 2t$, AND $y = \sin t + \frac{1}{2}\sin 2t$

The entire graph in figure 107c is obtained in just this way. That is, the y-value at any value of t is the algebraic sum of the y-values of the two preceding graphs at the same value of t.

What will the full behavior of the third graph be? After each 2π units of time the first graph will repeat its behavior and so will the second graph, though the second one will at that instant be undertaking its third cycle. However, the combination of the two graphs will begin to repeat only after the slower one does. Hence the resulting sum of the two graphs will be periodic with a frequency that is the smaller of the two separate frequencies. Because the resulting graph is periodic, the corresponding sound will be a musical one. However, let us note that this graph is not sinusoidal. Indeed it can depart radically from a sinusoidal shape if the amplitudes and frequencies of the component graphs are properly chosen.

We see, therefore, how it is mathematically possible for two simple sounds to produce a complex musical sound. But Ohm's observation says far more than we have just pointed out. We showed that two specially chosen simple sounds will produce a complex musical sound, but Ohm says that every complex sound is a combination of simple sounds. His statement is not only more sweeping but a converse of what we showed. In mathematical language, his statement is that every periodic function representing a musical sound is the sum of sinusoidal functions representing simple sounds.

Ohm's declaration, based on what the ear recognizes, suggests an even broader mathematical assertion, for insofar as mathematics is concerned, a periodic function is no more than a periodic relation between two variables, whether or not these represent displacement and time, respectively. Hence the mathematical implication of Ohm's law is that every periodic function, no matter how complex the shape of the graph within the period, is a sum of sine functions. The dependence of the frequencies and amplitudes of these sine functions on the behavior of the complex function is not as yet known to us nor is the broad mathematical theorem proven, but what is clear thus far is that the ear has suggested a deep mathematical theorem. By one of the quirks of history a brilliant mathematician, Joseph Fourier, somewhat older but roughly a contemporary of Ohm, arrived independently at precisely the mathematical theorem that Ohm's physical studies suggested.

Though Fourier, born at Auxerre, France, in 1768, did exceedingly well in mathematics as a student, he had set his heart on becoming an artillery officer. However he showed such ability in mathematics that after graduation he was appointed a professor in the very military school he had attended. It was in 1807, after years of political-scientific service under Napoleon, that he presented to the French Academy the theorem, still named after him, which supplies the key to all oscillatory phenomena. Fourier was at the time actually working on the subject of the propagation of heat in solids—a simple example of this is furnished by the flow of heat along a metal rod one end of which is held in a flame—but the results of this major investigation proved not nearly so momentous as the broad mathematical theory he developed to obtain them. We shall apply it to the analysis of musical sounds.

Fourier's theorem, which is the heart of the vastly important subject termed Fourier analysis or harmonic analysis, does not at first sight evoke any shrieks of delight or cause one's hair to stand on end. Specifically, it says that *any periodic function* is a sum of simple sine functions of the form $a \sin bt$ and that the frequencies of these sine functions are all integral multiples of one lowest frequency. As applied to the study of musical sounds the theorem has the significant physical interpretation we have already indicated. Since a function of the form $a \sin bt$ corresponds to a simple sound—compare formula (1)—Fourier's theorem says first of all that *any* complex sound is a sum of simple sounds. The simple sound of lowest frequency is called the fundamental or first harmonic or first partial. The simple sound whose frequency is twice the fundamental is called the second harmonic or second partial, and so on. We may think of all of these simple sounds as imposing their motion simultaneously on the typical air molecule, and its resultant motion being the combined effect of these separate motions.

Just to fix our ideas, let us examine an actual complex sound. Figure 108 shows the note C sounded by a cello. The graph is obtained by an instrument that records the oscillation of a typical air molecule. The curve is of course periodic, but the shape of the curve within any one period is quite complex. However, Fourier's

Fig. 108

theorem tells us that the formula which represents this complex sound is a sum of simple sine functions. In the present case the formula that represents this sound is

$$(8) \quad y = .02 \sin 2\pi \cdot 60t + .04 \sin 2\pi \cdot 120t + \sin 2\pi \cdot 180t + .04 \sin 2\pi \cdot 240t.$$

This formula says that the complex sound may be regarded as a sum of simple sounds, the first having a frequency of 60 cycles per second, the second, 120 cycles per second, the third, 180 cycles per second, and the fourth, 240 cycles per second. Moreover, if the amplitude of the third harmonic, the 180-cycle sound, is arbitrarily chosen as 1, then the first harmonic has an amplitude of 0.02, the second, 0.04, and the fourth, 0.04. Theoretically, in order to represent a complex sound, it may be necessary to use an ending number of harmonics with higher and higher frequencies, all of which are integral multiples of the fundamental frequency. However, in actual instrumental and vocal sounds the amplitudes of the very high harmonies are so small that they have no effect or are not given off at all.

We have yet to indicate how formula (8) is obtained in practice. The graph of the displacement versus time of the typical air molecule is obtained directly by some electronic instrument such as an oscilloscope; then it is possible to obtain the corresponding formula, such as (8), by a mathematical machine called a harmonic analyzer. This machine does precisely what its name signifies, that is, it analyzes the complex sound and determines the frequencies and amplitudes of the harmonics which compose the sound. If one were to graph each of the separate terms on the right side of formula (8) and then add up the y-values of all these curves at each value of t, just as we added $y = \sin t$ and $y = (\frac{1}{2}) \sin 2t$ earlier, the resultant curve would have precisely the shape of the cello sound shown in figure 108.

Fourier's theorem in itself gives us a lot of interesting information about musical sounds. But it can also be put to further use. In the first place, it clarifies our understanding of the essential characteristics of musical sounds. We have spoken of the pitch of a sound. We can now interpret this physical effect mathematically. For simple sounds it was clear even before Fourier's work that the pitch varies directly with the frequency. Now complex sounds are also periodic; that is, the entire series of displacements which the typical air molecule undergoes repeats itself a number of times a second, and this frequency had seemed to determine the pitch of the sound. Fourier's theorem tells us that this must indeed be the case, for the higher harmonics all have frequencies that are multiples of the lowest one. Thus in the case of the cello's sound, the fundamental has a frequency of 60 per second; the second harmonic has a frequency of 120; the third harmonic, 180; and so on. The second harmonic will perform two complete oscillations while the first harmonic performs one, and the sum of these two harmonics will repeat itself when the first one does. A similar argument applies to the third and higher harmonics. Thus the graphs of all these separate harmonics execute an integral number of cycles precisely within the period of the first cycle of the lowest harmonic or fundamental, and each of the higher harmonics is ready to repeat its motion just when the first harmonic is about to undertake its second oscillation. Hence the entire graph repeats itself at the frequency of the fundamental tone and this frequency determines the frequency of the entire sound. It is this frequency, therefore, that must be associated with the pitch of the sound.

But two complex sounds of the same pitch do not necessarily sound the same. If a piano and a violin each produce a sound of the same pitch the ear will still recognize a difference in quality between them. The difference is due to the harmonics present in the sounds and to the amplitudes of these harmonics. Suppose both notes have a fundamental frequency of 500 and both contain second and third harmonics. The relative amplitudes of the three harmonics may be 1, ½, and ¼ in the first case and ½, 1, ⅓ in the second. As a consequence, the two notes will differ in quality. Graphically, the two notes will have the same frequency but will have different shapes.

Of course, two notes may differ in that one may have many more harmonics than the other in addition to having larger or smaller amplitudes in each of the harmonics present. The first six harmonics of notes on the piano or violin are generally strong compared to notes on other instruments. The vowels of the human voice differ from each other in the amplitudes of the upper harmonics. On the other hand, the tones of organ pipes are comparatively poor in upper harmonics. The tuning fork, as already observed, gives off only a simple sound, that is, one without upper harmonics. As a consequence, its note is soft and dull, never shrill or rough. The flute gives off an almost simple tone. The vowel "oo" in the word "too" is another almost simple sound. All of the qualities we describe by such adjectives as soft, piercing, braying, hollow, rich, dull, bright, and crisp are due to the harmonics present. For brightness and acuteness the upper harmonics are essential. However, if harmonics higher than the sixth or seventh are very distinctly present, the quality of the tone is piercing and rough. Generally the upper harmonics decrease in amplitude as their frequencies increase. However, if their amplitudes are comparatively too strong for the fundamental then the quality of poverty, as opposed to richness, pervades.

The quality of any instrument depends upon the device it embodies for generating the sound. The piano and violin use vibrating strings. The clarinet, oboe, and bassoon use vibrating reeds that act on air forced against them. The organ pipe contains a sharp-edged but rigid lip and air is forced past the edge. The action of the string, reed, or lip is modified by the sounding board of the piano, the hollow box of the violin, or the pipe of the wind instrument and organ. The quality depends further upon how the basic device is excited: whether the string is struck by a hammer or is bowed, and so on.

There is a third property of musical sounds that can be mathematically described, namely, their loudness. Strictly, loudness is the sensation experienced by the ear and brain. The intensity, on the other hand, is a physical property of sounds as they travel through the air; it is proportional to the square of the amplitude of the sound. Since the loudness depends upon the ear it does not follow that two equally intense sounds of different pitches will be heard equally loudly by the ear, and in fact this is generally not so. The ear is most sensitive to frequencies of about 3500 per second, and its sensitivity decreases as the frequency increases above that value or decreases below it. Above about 16,000 cycles the ear does not hear sounds at all, though the sound itself, as measured by some physical apparatus, may be intense. Loudness varies not only with the frequency for sounds of the same intensity, but varies with the wave shapes or quality at the same frequency. Moreover, of two sounds with the same wave shapes, but one twice as intense as the other, the more intense one will not be heard twice as loudly.

The entire analysis of the pitch, quality, and intensity of musical sounds shows how Fourier's theorem gives us understanding of the structure of these sounds. But the analysis can also be put to practical use. Though the design of musical instruments such as the violin is as at least as much a matter of experience as it is of science, man does make today a number of devices whose function is to record and reproduce musical sounds. The telephone, the phonograph, the radio, and the sound track of motion picture films are all scientifically designed. Since intelligible and in many cases faithful reproduction of sound is essential, it is necessary to know the effect of each component part in these instruments on the sound it is intended to reproduce. Faced with the infinite variety of sounds, the designer of such an instrument, even if he understands all the principles and technical knowledge involved, might well throw up his hands. But Fourier's theorem reduces the content of musical sounds to simple sounds of particular frequencies. Hence the designer need only consider how well his components will handle such sounds. Moreover, any one instrument does not have to handle all frequencies. Since the human ear can detect only frequencies lying between about 16 and 16,000 cycles per second, the problem of recording and reproducing is immediately and considerably simplified.

The designer may next limit himself to frequencies that the instrument's function requires. For example, motion picture sound tracks are generally not required to reproduce music faithfully. Hence the designer can ignore the very high audible frequencies, which generally appear only as very weak harmonics of sounds with much lower fundamental frequencies. He might decide, then, that he need concern himself with equipment that can "pass" frequencies between 16 and 8000 cycles per second and either not pass higher ones or pass them with loss of amplitude. Where intelligibility alone, rather than quality in the musical sense, is important, as in the telephone, dictaphone, and interoffice communication or loud speaker systems, a frequency range of from 100 to 1000 cycles is adequate. On the other hand, high fidelity musical instruments that stress quality of reproduction are designed to handle frequencies from 16 to 20,000 cycles per second with no loss of strength in any frequency of that range. Still other instruments, designed largely for industrial use, generate ultrasonic sounds, that is, sounds of frequencies too high for the human ear that can be directed into metals to test for flaws. In this last mentioned application we have the peculiar phenomenon of using the inaudible to test the invisible.

Of course most of these systems do not carry the sound as sound from input to output. The sound may be converted into electric current; the current is then passed through a number of electrical components and then converted into sound. But the functional relationship of the current strength and time must be as close as possible to the displacement and time relationship of the actual sound, and the Fourier analysis of the sound also applies to the electrical current.

Electrical currents are also used to perform other functions that have nothing to do with the reproduction of sound. For example, they are used to control machinery and to perform switching operations. Each of these applications requires special functions of current versus time, generally called pulses. The Fourier analysis of these pulses is the first step in the design of the appropriate electric circuits.

The full study of sound must include the action of strings, reeds, drumheads, and of devices such as hollow chambers, tubes, and pipes that accentuate the sounds created by the primary vibrating devices through the phenomenon known as resonance. Having acquired this knowledge one is prepared to study the action of the most versatile and most excellent of all musical instruments, the human voice, and the most complex and the most sensitive of all recording devices, the human ear. The vocal cords in the larynx act like reeds but are variable in shape and tension. They vibrate when air is forced against them and give off a great variety of sounds. The mouth is a resonant chamber which by the action of the lips and tongue can be formed into many shapes. Hence the human voice is capable of more qualities of tone than any man-made instrument.

The ear possesses a vibrating drumhead, hollow resonant chambers, and a series of about 30,000 nerve endings in what is called the cochlea. Any one frequency will excite a special group of these nerves; the number of possible groups permits the ear to distinguish a difference in pitch of one-half a cycle per second. A complex sound received by the ear excites just those endings that respond to the frequencies present in the complex sound. Hence the ear decomposes each complex sound into simple sounds and these are communicated to the brain.

We could pursue further the biology of the larynx and of the ear and see how the mathematical analysis of musical sounds and of instruments leads to sound inferences as to the function of the numerous parts of these organs. However, our purpose in referring to this subject is merely to point out that man's purely scientific study of an inanimate physical phenomenon leads back to man himself and to the science of biology. From biology it is a short step to the analysis and repair of biological disorders—that is, to medicine.

Fourier's theorem can also be put to nonscientific uses. Suppose one takes the outline of a human face and forms a periodic graph by repeating the profile (fig. 109). Fourier's theorem says that the formula of this periodic function is a sum of simple sine functions, each with a definite period and amplitude. Each of these sine functions can be matched by a simple sound and the combination of these sounds will yield a musical sound. Hence there is a musical sound determined by the repeated facial profile. The young man who tells his girl that her face reminds him of a melody is telling the truth—if he knows Fourier's theorem. Herein lies the real charm of Fourier's theorem.

Fig. 109

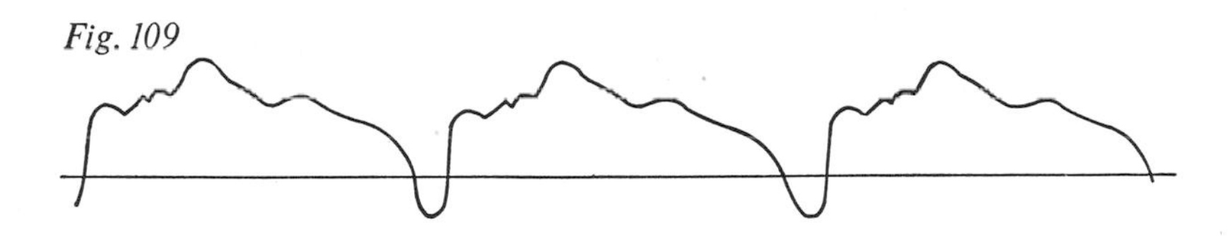

20: OLD FOES WITH NEW FACES

The magnet's name the observing Grecians drew
From the magnetic region where it grew.

—LUCRETIUS

THE world of nature is vaster than what the hand can touch, the eye see, or the ear hear. Beyond the senses lies a world that has been effectively explored only within the last hundred years or so. The mastery of this realm has altered our way of life and our understanding of nature far more rapidly and dramatically than even the spectacular and grand achievements in the field of motion. Of course the more man explores phenomena beyond the powers of his senses, the more he must rely upon reasoning to construct an account of such phenomena and to predict occurrences. The consequence has been an even greater stimulus to the creation and use of mathematics than the study of motion or of light provided.

The phenomena that defy sense perception but whose real effects can nevertheless not be denied are electricity and magnetism. The existence of these mysterious natural forces was known even to the Greeks. For example, Thales knew that iron ores containing lodestone, such as those found near Magnesia in Asia Minor, attract iron. He is also supposed to have known that after amber is rubbed it attracts light particles of matter. The medieval Europeans learned from the Chinese that if a piece of lodestone is allowed to rotate freely it will orient itself along the north-south line and hence can be used as a compass. In view of the immense practical value of this instrument one would expect to find that magnetism at least had been seriously studied in medieval times. But it was more to the taste of the Europeans of that period to allow their imaginations to run riot and to make one mystery the explanation of another. Thus, birds were supposed to navigate by magnetic power; love was a magnetic force; and magnets could reconcile estranged husbands to their wives.

The first purposefully directed and serious study of magnetism was made by William Gilbert, court physician to Queen Elizabeth. His *De Magnete*, published in 1600, is still a clearly readable account of simple experiments which proved, among other things, that the earth itself is a huge magnet. Gilbert also found that there are two kinds of magnetism, north-seeking and south-seeking, or more simply north and south, often labeled positive and negative, respectively. Two positive or two negative magnetic substances repel each other whereas oppositely endowed magnetic substances attract each other. These opposite types of polarities, as they are called, are found, for example, on the opposite ends of a bar magnet.

Gilbert also explored the second phenomenon observed by Thales, namely, the electrification of rubbed amber. He found that sealing wax, when rubbed with fur, and glass, when rubbed with silk, also acquire the property of attracting light particles of matter. These experiments suggested that there are two kinds of electricity. Glass rubbed with silk acquires what is called the positive kind, and wax rubbed with fur acquires the other or negative kind. Moreover, as in the case of magnetism, two objects possessing the same kind of electricity repel each other, and objects possessing different kinds attract each other. However, a long series of investigations whose details we need not follow showed that this account of electricity is not correct. We now know that there are not two kinds of electricity. Rather, physicists have discovered that there are tiny bits of matter, indeed the smallest pieces found in nature, which are called electrons. We cannot see electrons any more than we can see the larger bits of matter called atoms that contain the electrons, but the indirect evidence for their existence is quite strong. An object that is negatively charged, that is, exhibits the behavior of rubbed wax, contains an excess of electrons. On the other hand, objects that were formerly described as being positively charged, such as glass after being rubbed with silk, were recognized to be deficient in electrons. Apparently rubbing the glass with silk loosens a number of electrons from the glass and these attach themselves to the atoms of the silk. Hence the glass becomes positively charged and the silk negatively charged. Though we still speak of an object as being positively charged as though something were added to the object, actually the object has lost electrons. An object that contains the normal number of electrons is said to be electrically neutral.

With the proper experimental setup the action of electrically charged bodies and therefore of electrons can be studied. For example, if two little glass balls, positively charged, are hung on strings and placed alongside each other, the balls will repel each other because both are positively charged. Further, electrons can be isolated in large numbers. When metal is heated, electrons are ejected from the metal into the air. Of course the air hinders the motion of the electrons and the molecules of air absorb them. But when electrons are ejected into a vacuum, as they are constantly in the tubes of modern radio and television sets, they can travel long distances in space and during this travel their behavior and properties can be studied.

Since electrically charged objects will exert forces on each other, that is, they will attract or repel, and since magnetic poles will do likewise, it is evident that in the phenomena of electricity and magnetism we have forces at our disposal which can be studied to advantage. Let us investigate first the behavior of electricity.

The scientists of the late eighteenth century who became engrossed in the study of the forces exerted by electrically charged bodies had learned the lesson taught by Galileo and Newton, that is, to seek basic quantitative laws. The first law they discovered was indeed a surprise to them. Before examining it we should recognize explicitly what is almost obvious. An object can have more or less electricity in it. This means merely that if it is negatively charged it can have a large or small number of extra electrons, and the corresponding statement holds for positively charged bodies. Since the force exerted by one electrified object on another depends upon the quantity of electricity in each, it is necessary to have a measure of this electricity. Hence some quantity is chosen as a standard (just as some quantity of mass is chosen as a unit of mass) and the quantity of electricity in any object is then measured in terms of that standard. How these quantities are determined is a matter of experimental technique and of no consequence here.

It may, however, be of interest that one of the commonly used units of charge is called the coulomb, after the French physicist, Charles Augustin Coulomb (1736–1806), who discovered the law of force we are about to describe. Actually there are many different units of charge in use, just as there are many different units of length, the yard, the foot, the meter, and so on. One can then measure the quantity q of electrical charge, positive or negative, that is contained in a body. If one has two quantities of charge, q_1 and q_2, these will attract or repel each other depending upon whether they are unlike or like. Coulomb found the remarkable law that the force F of attraction or repulsion is given by the formula

$$F = k\frac{q_1 q_2}{r^2}, \qquad (1)$$

where r is the distance between the two collections q_1 and q_2, and k is a constant which is the same no matter what q_1, q_2, and r are. The value of k depends upon the units used to measure charge, distance, and force.

The remarkable fact about formula (1) is its identity with the form of the law of gravitation. The charges q_1 and q_2 act like masses, and the force between them varies with distance in exactly the same way as the force of gravitation between two masses. Of course we must remember that the electrical force can be attractive or repulsive whereas the gravitational force is always attractive.

The formal identity of the two laws of force immediately suggested to the mathematicians that the immense knowledge already acquired about gravitation could be applied at once to the behavior of electricity. Thus they could predict that a positively charged body could be made to move in an ellipse around a negatively charged body under the attractive electrical force. However, to describe mathematically what must happen and to make objects actually behave in that way are totally different matters. In the late eighteenth and early nineteenth centuries the techniques of isolating, measuring, and maintaining electrical charges were not well developed; many decades of ingenious experimental work had to pass before electrical phenomena were physically realized.

The chief practical difficulty which the early workers in this field encountered is that they generated electricity by friction, and it is hard to measure and control the electricity so obtained. Fortunately, a series of discoveries made by the Italian biologist Galvani and the Italian physicist Volta produced the first stable generator of electricity, the battery. The battery can take many forms, for example, the wet battery now used in automobiles and the dry battery used in flashlights. What is important about a battery is that it provides us with positively and negatively charged poles. That is, at one terminal, the negative pole, there is an excess of electrons and at the other terminal, the positive pole, a deficiency of electrons. Of course the electrons are not all heaped up at the negative pole nor are all the positively charged atoms heaped up at the positive pole, but for the purposes of utilizing the action of batteries it is just as well to think in these terms and to ignore the chemical and physical phenomena within the battery.

Batteries have different "strengths" created by the internal chemical action; that is, some batteries seek to force the electrons at the negative terminal to join the positively charged atoms at the positive terminal through some external medium more strongly than others do. The strength is measured in volts, in honor of Volta. When there is no external connection between the terminals, the electrons cannot flow, but the pressure is always there.

If one were to connect the positive and negative terminals of a battery by a piece of metal, through which electrons flow freely, the negative electrons heaped up on the negative terminal, being attracted to the positively charged atoms on the positive terminal, would rush through the connection until all the atoms in the battery became neutral or uncharged. Obviously to permit this flow of electrons, or this current, to take place freely would destroy the usefulness of the battery. The force acting between the electrons at the negative terminal and the positive atoms at the other should be utilized to make it do work for us. For example, in a flashlight, the battery terminals are joined by the thin wire in the bulb, and as the electrons from the negative terminal rush through this very thin wire to reach the positive terminal they heat up the wire. The heated wire glows and furnishes light, at least until the battery is discharged, that is, until all the excess electrons at the negative terminal have forced their way to the positive terminal. A battery, then, provides us with negatively and positively charged bodies, and we can use it to make electrons do some of the things we should like to have them do.

We have already pointed out that the force acting between electrical charges obeys the same law as the force of gravitation. Let us now explore this mathematical fact further to see if it suggests what might be done with electrical charges. Formula (1) can be written as

$$F = q_1 \frac{kq_2}{r^2} \,. \tag{2}$$

Speaking in terms of the gravitational analogy, we may regard q_1 as a mass being attracted by the mass q_2 in accordance with the inverse square law of gravitation. In this analogy Newton's second law of motion, $F = ma$, suggests that the quantity kq_2/r^2 is the acceleration with which q_1 would move. But this acceleration varies with r. We may remember, however, that the simpler gravitational phenomena, the ones investigated by Galileo, dealt with an acceleration which was everywhere constant in amount, namely, 32 ft/sec^2, and constant in direction, namely, straight down, perpendicular to a flat earth. As we have seen, the use of a flat earth and of a gravitational force everywhere perpendicular to it is merely an approximation of the exact situation, but it suffices when we deal with objects moving over restricted regions of the earth's surface. To obtain the analogous simpler phenomena of electricity we should look for some situation having a force or acceleration, constant in direction and magnitude. Is there such an analogous situation in electrical actions? Yes, and it is a most useful one.

Suppose we place two large plates of metal parallel to each other (fig. 110) and connect the upper plate by a wire to the negative terminal of a battery and the lower plate to the positive terminal.

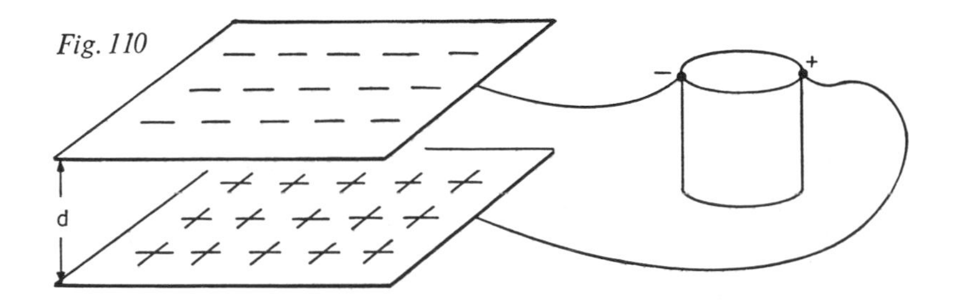

Fig. 110

The electrons crowded together at the negative terminal repel each other and hence will spread out over the upper plate. Moreover, if there should be a greater concentration of electrons at any one point on the plate than at any other, the greater repulsive forces acting at that point will cause the electrons to disperse until they are spread out evenly over the plate. Likewise, the positively charged atoms at the positive terminal repel each other, and they too will distribute themselves evenly over the lower plate. (Strictly speaking, the positively charged atoms at the positive terminal will attract electrons belonging to neutral atoms of the lower plate, and as a consequence the lower plate will become positively charged.)

We have mentioned that it is possible to isolate electrons, for example, by ejecting them from heated metal. Let us suppose, without examining the experimental technique, that it is possible to place a single isolated electron e between the two plates. What force will the positively and negatively charged plates exert on the electron? Of course the electrons on the upper plate will repel the electron e because all electrons bear negative charges; at the same time the positively charged atoms on the lower plate will attract e. There is, then, some kind of downward force exerted on the electron. We must, however, be more specific about this force if we are to work with it mathematically.

Imagine the electron e to be somewhere in the middle of the space between the plates. According to Coulomb's law, the force acting between any two charges either pushes them directly apart or attracts them directly toward each other. Hence electrons on the upper plate but somewhat to the right of e will push e downward and toward the left. Electrons on the upper plate and to the left of e will push e downward and toward the right. But as we pointed out earlier, the electrons on the upper plate are uniformly distributed. Consequently, as many will push e to the left as toward the right and so no sideways motion of e will take place. Therefore, insofar as the force exerted by the upper plate on e is concerned, its direction is straight down. If we analyze similarly the pull exerted by the positive charges of the lower plate on e, we find that the net force is also straight down.

What we have determined thus far is that the force acting between the upper and lower plates is straight down and this force acts everywhere between the plates (except at the edges where the direction is no longer straight down). It is helpful to imagine vertical lines (fig. 111) extending from the negative charges on the upper plate to the positive charges. These lines show the direction of the force which the plates exert on a negative charge placed

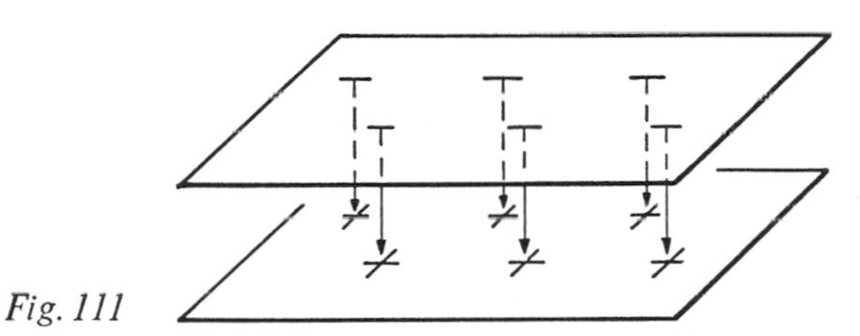

Fig. 111

between them. There are of course no actual lines extending between the plates, but thinking about the action of the plates in terms of the lines of force to which they give rise is very helpful. Now insofar as the action of the charged plates on the electron *e* is concerned, we could forget about the plates and just keep in mind the lines of force that show how the plates act. This collection of lines of force is called an *electric field.*

Just to become better acquainted with the notion of a field of lines of force let us note that we can also use this concept in connection with the action of gravitation. Let us go back to the action of gravity as Galileo conceived it when he studied the motion of bodies on or near the surface of the earth and regarded the earth as flat. Under this admittedly limited and only approximate view of the action of gravity one can imagine a collection of lines (fig. 112) extending straight down and describing how gravity will pull

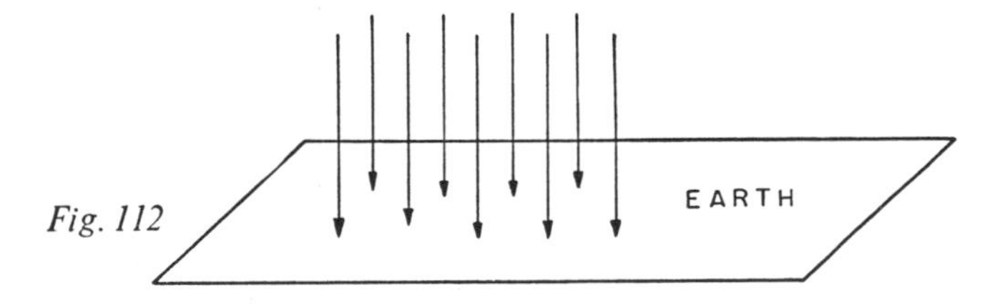

Fig. 112

an object. In this situation one would speak of a gravitational field of lines of force, as opposed to the electrical field of lines of force created by the charged plates.

The analogy between electrical and gravitational fields suggests the next question we should raise about the force that the plates exert on the electron between them. The significant fact about the gravitational field insofar as local motions over a "flat earth" are concerned is that the acceleration which the field exerts on any mass m is always 32 ft/sec^2 and the force of gravity acting on m is

$$F = 32m. \tag{3}$$

It is reasonable to ask, now that we know the direction of the electric field between two plates, What is the magnitude of the force this field exerts on the electron e?

When e is closer to the upper plate it will be repelled more strongly, since Coulomb's law says that the force is greater when the separation of the charges is smaller. At the same time the attractive force exerted by the positively charged atoms on the lower plate will be weaker because e is farther away. As e moves down, the repulsive force will become weaker and the attractive force stronger. One might hope that the combined forces acting on e would be the same as they were at the previous positions of e. Nature is often obliging, especially to the mathematician, and she has so ordained. The strength E of the electrical field everywhere between the plates is constant, and the force F that acts on the electron e is given by the formula

$$F = Ee. \tag{4}$$

We have a physical argument that formula (4) might hold. It is a mathematical consequence of formula (1); however, we shall not take the time to prove it. We shall be more concerned with the use of formula (4). Before applying it we should note the analogy between formulas (3) and (4). The quantity E in (4) acts like the quantity 32 in (3). The number 32, though technically called an acceleration, could be described as the gravitational field strength which, acting on a mass m, produces the gravitational force of $32m$. Likewise, E is an electrical field strength, which, acting on a charge e, produces the electrical force Ee.

How large is E? The field strength created by the charged plates depends upon how much charge is forced upon the plates by the battery. That is, a stronger battery will force more electrons to crowd on a plate of given area than a weaker battery because the stronger battery better overcomes the mutual repulsion of the electrons. The same is true of the positively charged atoms on the lower plate. Hence the voltage V of the battery to which the plates are connected is the first factor that determines E. Next, if the plates are separated widely and we imagine an electron halfway between them, then formula (1) tells us that the electrons on the upper plate will repel with less force and the positive charges on the lower plate will attract with less force than if the plates were closer to each other. Thus E decreases as the distance between the plates increases. These considerations suggest (but do not prove) that if V is the voltage of the battery and d the distance between the plates then

$$E = \frac{V}{d}. \tag{5}$$

The importance of formula (5) is that it tells us how we can alter E. For example, to increase E we must use a stronger battery or decrease the separation of the plates. Formula (4) then tells us that the force F on the electron e will be stronger.

Now that we have learned something about the force acting on an electron between parallel plates, and we see that such an electron will be acted upon by a constant force once V and d are fixed, we can investigate how the electron will move. Let us suppose that the electron is placed between the plates with no initial velocity. Then the electrical force F given by formula (4) will act on it and cause it to move downward. The electron has a mass m and will also be pulled downward by the gravitational force. However, the mass of the electron is so small that the gravitational force is for present purposes negligible. (For ordinary values of V, the electrical force is about 10^{14} greater than the gravitational force.) However, Newton's second law of motion, which says that a force F acting on a mass m gives it an acceleration a, still applies because the electrical force is considerable. If we apply Newton's law to the electrical force F, we may say, in view of formula (4), that

$$ma = Ee$$

and therefore that

$$a = \frac{Ee}{m}. \tag{6}$$

Now the charge e is constant, as is the mass m of the electron. Moreover E is constant everywhere between the plates, as we discovered above. Hence the electron placed in the region between the two charged plates is acted upon by a constant acceleration that is directed straight down. We know how masses behave when acted upon by a constant acceleration. All of our work with the gravitational acceleration of 32 ft/sec^2 applies here with a taking the place of 32. Hence the electron will fall straight down with a velocity v at time t given by

$$v = at,$$

and it will fall distance d in time t given by

$$d = \frac{1}{2}at^2. \tag{7}$$

We see from all this that electric fields can be used to direct and control the motion of electrons. Of course in place of the electron we might very well have put a positive ion, that is, a positively charged particle. The same electric field would cause the ion to rise instead of fall. Moreover, in place of e in formula (6) we should have the amount q of positive electrical charge on the ion.

Since we can control the motion of charged particles let us see if we can put this fact to use. When metal is heated it ejects electrons into space. This process, as already mentioned, is used in radio tubes. Let us suppose that we shoot an electron horizontally (fig. 113) into the space between two parallel plates. The

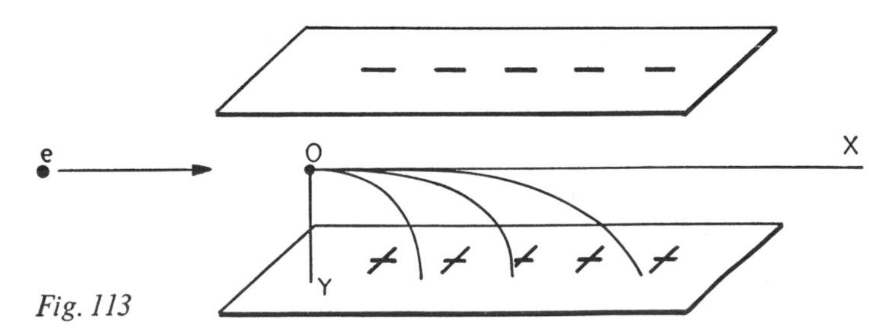

Fig. 113

electron will be pulled downward as it continues to move horizontally. However, we know from our study of projectile motion that the resulting curvilinear motion can be analyzed on the basis of Galileo's principle that the horizontal and vertical motions do not interfere with each other. Let x and y be the horizontal and vertical distances the electron moves, measured from the point O at which the field begins to act on the electron. Then the horizontal distance traveled is governed by the formula

$$x = vt, \tag{8}$$

where v is the original horizontal velocity with which the electron is shot between the plates and t is the time that elapses from the instant the electron is at the origin of the coordinate system. The downward motion is governed by formula (7). Hence

$$y = \frac{1}{2} at^2, \tag{9}$$

where a is given by formula (6). The resultant path is obtained in exactly the same way as in the case of projectile motion. The value of t in formula (8) is substituted in formula (9), and the equation of the electron's path is

$$y = \frac{1}{2} a \frac{x^2}{v^2}. \tag{10}$$

By substituting the value of a from formula (6) into (10) we have

$$y = \frac{1}{2} \frac{Ee}{m} \frac{x^2}{v^2}. \tag{11}$$

One major fact about equation (11) is readily recognized. The quantities E, e, m, and v are all constant. Then the relation between y and x is of the form $y = cx^2$, and the path is a parabola. Hence we can make electrons follow parabolic paths.

In the case of a projectile dropped from a plane, say, and falling under gravitational attraction, we know that we can change the parabolic path by changing the horizontal velocity of the projectile. To do the same thing for the path of an electron we have much more freedom. The width of the parabola represented by equation (11) is determined by the value of c, namely,

$$c = \frac{1}{2} \frac{Ee}{mv^2}. \tag{12}$$

This constant depends upon several factors, for example, the strength of the electric field E. But E can be varied by varying the voltage V of the battery applied to the plates. We can therefore make the electrons take different parabolic paths by varying V.

This power to vary the electron's path may seem to be of academic interest, but it is the essence of a phenomenon many people witness daily. The major component of a television receiver is the picture tube or cathode-ray tube (fig. 114). In this tube there is a gun, as it is called, from which electrons are shot out hori-

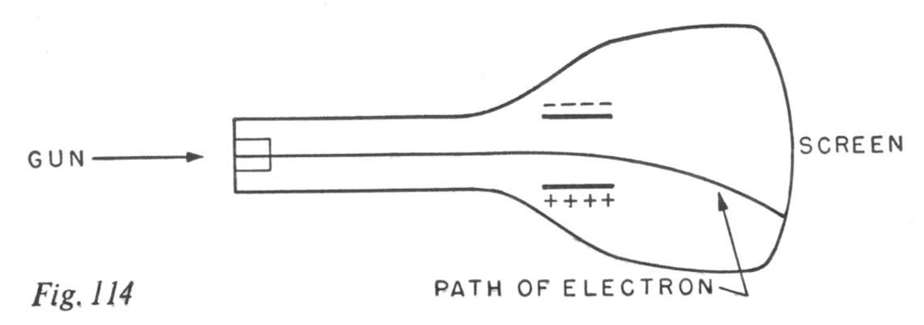

Fig. 114

zontally. The electrons pass between two charged parallel plates and then pursue a parabolic path. By varying the charges on the plates the paths of the electrons are varied and they therefore strike different spots on the inside face of the tube. This inside face is a flourescent screen; when struck by an electron, it gives off light. Because the electrons can be made to strike different parts of the screen, the picture we see, which consists of a bunch of flashes of light, can be varied. This account of cathode-ray tube action in a television set is of course a considerably simplified one, but it may make clear how basic principles of electron motion, seemingly of little significance, find their application in modern life.

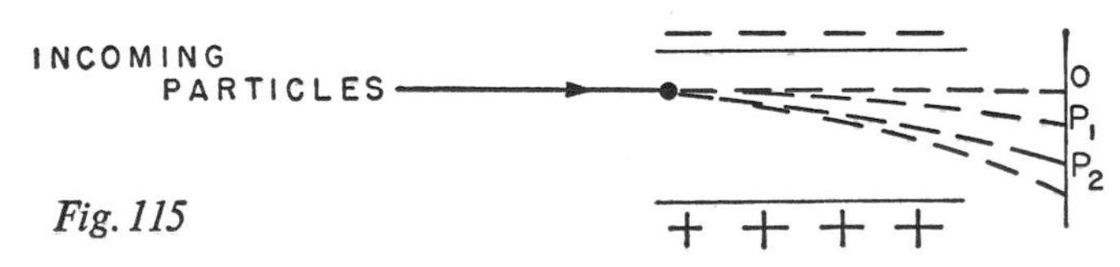

Fig. 115

There is another fact about formula (11) that can be utilized. The path of the parabola represented by formula (11) depends, as we have seen, upon the value of c given by formula (12). Now c depends upon the value of m, the mass of the moving particles. Suppose that particles with different masses but all carrying the same charge e are shot into a region between parallel plates (fig. 115). The heavier the particle, the larger its mass, and therefore the smaller the value of c. Hence the heavier particles will traverse a shallower parabola and strike the plate OP_1P_2 closer to O. If OP_1P_2 is a photographic plate, then the spots developed on the plate can be used to analyze the varieties of the particles in the original beam.

If the plate OP_1P_2 contains many holes, each leading to a box of some sort, then the various particles that are thus separated are collected in separate boxes. We may recall that any element such as oxygen really consists of several types called isotopes, which differ in weight. Hence, if a quantity of oxygen is electrified and then passed between plates, the isotopes of different weights will be separated. As in the case of the cathode-ray tube, the description of the actual process of separating isotopes is considerably simplified but the principle involved is evident.

Thus far in our consideration of the motion of charged particles we have neglected the gravitational force on the mass of the particle. The reason was simply that the electrical forces in normal electric fields are very much greater. However, it is possible to take the gravitational force into account and deduce some very valuable information. Let us suppose that an electron is placed between two charged parallel plates but this time let us charge the upper plate positively and the lower plate negatively. Then the electron will rise under the action of the electrical force Ee. However, because the electron has mass it will be pulled downward by the gravitational force of magnitude $32m$. If the charge on the plates is adjusted so the electrical force just offsets the gravitational force,

$$Ee = 32m,$$

or

$$\frac{e}{m} = \frac{32}{E}. \tag{13}$$

Now the value of E can readily be determined because the voltage V of the battery is readily measured, as is the distance d between the plates. Formula (5) then tells us what E is. By using (13) we can calculate the ratio of the charge to the mass of an electron. It is possible to go further and calculate m by means of a theory we have not presented and hence to obtain e. The results are of interest. The mass of an electron is $9.1 \cdot 10^{-31}$ kilograms. The charge

e on an electron is $1.602 \cdot 10^{-19}$ coulombs, the coulomb being the unit of charge. To gain some idea of the minuteness of this charge we might note that one coulomb contains $6 \cdot 10^{18}$ electrons, and that the coulomb is the amount of charge which would flow during one second past any point in a wire carrying one ampere. A current of about this magnitude flows in a 100-watt bulb.

The motion of charged particles in the space between parallel plates might be called the "flat earth" theory of electrical motions. However, when one charge attracts another and both are concentrated essentially at points, then we must use Coulomb's law, formula (1), directly. We have already remarked that this law is quantitatively exactly the same as the law of gravitation except that the force between charges can be repulsive as well as attractive. In the case of gravitation, when one of the bodies is very heavy, as is the sun, any other body moving in a closed orbit and attracted to the first one, must, in accordance with the inverse square law, follow an elliptical path. Is there any analogous situation in the behavior of electrical charges?

There is. The most useful current model of the atom is that it contains a central mass called the nucleus and this mass is positively charged. The amount of charge depends upon which element, that is, hydrogen, oxygen, gold, silver, and so forth, one considers. Revolving around any nucleus are a number of electrons, each of which is, of course, negatively charged. Hence each electron is attracted to the nucleus in accordance with Coulomb's law. What, then, is the path of each electron? We need not redo Newton's work. The mathematics he produced to show that each planet moves in an elliptical orbit applies at once to the atomic situation since the mathematical formulas involved are the same. Here again, as in so many other situations, we must appreciate that the mathematical formulas do not deal with physical objects but with numerical quantities. Though Newton, in using the law

$$F = G\,\frac{mM}{r^2}$$

was concerned with masses such as the sun, the earth, and the distance between them, in our application we have but to think of m as the amount of one charge, M the amount of the other, and r

as the distance between them. F now stands for the electrical force of attraction instead of the gravitational force. We prefer to write q_1 for m, q_2 for M, and k for G, merely to remind ourselves that we are dealing with an electrical problem and quantities of charge. We may conclude at once, then, that each electron revolves around the nucleus of an atom in an elliptic orbit, and that the numerous electrons and nucleus all form in any atom a miniature solar system. The number of "planets" in these various "solar systems" may be as small as 1, which is the case in the hydrogen atom, and as large as 92, which is the number of electrons surrounding the nucleus in the uranium atom.

In the case of atomic problems we have reason to raise a question that does not arise in planetary motion. What happens when one charged particle moves in the neighborhood of another and is repelled by the second one in accordance with the inverse-square law? Specifically, suppose a charge q_1 at P (fig. 116) is moving in the direction PP' but is repelled by the charge q_2. How will q_1 move? The answer is that q_1 will move along the hyperbola shown in the figure. The proof of this assertion is a bit too difficult to undertake here, but we can at least note what both cases of Coulomb's law lead to mathematically.

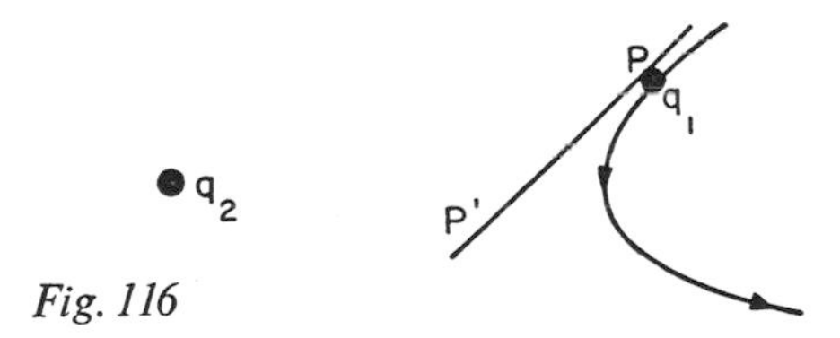

Fig. 116

As a matter of history, this second case of a repulsive force proved to be the key to one of the great discoveries in modern atomic theory. The distinguished British physicist Lord Ernest Rutherford had set one of his students to the problem of observing what happens when positively charged particles are shot into an atom. At the time (1910), the prevailing theory of the atom was that the very light negatively charged electrons and the heavy positively charged particles contained in the atom were distributed homogeneously throughout. If a positive charge of large mass were shot into the atom, it was thought that it would not be deflected appreciably by the very light electrons. However, atomic physicists, including Lord Rutherford, expected that the heavier positively charged particles in the atom would, so to speak, kick the incoming positive charge around somewhat and let it emerge slightly deflected from its original course. But Rutherford's student observed the positive particle to be deflected along the hyperbolic path shown in figure 116. This path, it will be noted, shows that the course of the particle is almost reversed. Rutherford attempted to account for this phenomenon, and concluded that there must be a large concentration of positive charge at the center of each atom. Thereby was conceived the notion that the atom consists of a highly concentrated nucleus surrounded only by electrons.

We see, therefore, from a few examples, how a little mathematical reasoning about the action of charges on each other gives us much information about the sizes of these charges and atomic structure, and guides us in putting these charges to work for us. But the action of electric charges alone is but a small part of the world of electricity and magnetism, and the use we have made of mathematics thus far is trifling. Let us wade in a little further.

We remarked at the outset that the phenomenon of magnetism had also been observed in ancient times and was studied systematically from the time of William Gilbert onward. The peculiar property of magnets that caused people to notice them is their power to attract ordinary or unmagnetized iron or steel. Magnets, as most of us know, differ in the attracting force they exert, a stronger magnet being able to pull a heavier mass of iron to itself. Thus the strength of a magnet is like the force that a quantity of electric charge can exert on other charges. There is, however, an essential difference between magnets and electric charges. We can rub glass with silk and cause the glass to be positively charged and the silk negatively charged. We can then separate the glass and silk and have at our command a positive charge on the glass that is completely independent of the negative charge on the silk. While there are likewise two kinds of magnetism, positive and negative, or north and south, and while, as in the case of electric charges, the opposite types attract each other and like types repel each other, the two types cannot be separated. As every child learns from playing with magnets, a magnetic rod will possess one type of magnetic pole, as it is called, at one end, and the other type at the other end.

A horseshoe magnet will likewise possess different types at each end of the shoe.

It is more convenient and more useful to think of the combined action of these two poles at once. It is possible to shape a magnet so that the two poles face each other (fig. 117). The two faces are like the parallel plates we considered in connection with the electrical charges. Thus, if a north pole is placed between the two poles of the curved magnet—this north pole can be one end of a bar magnet sufficiently long so that the distant south pole can be ignored—it will be repelled by the north pole of the curved magnet and attracted by the south pole. It is helpful in this case, too, to

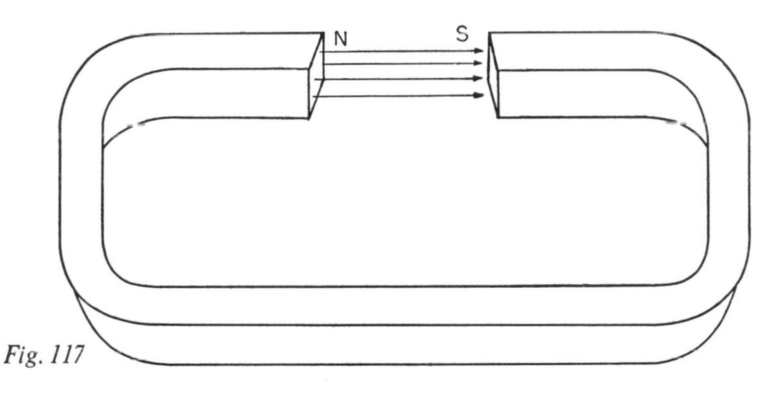

Fig. 117

think of a field of force extending between the poles of the horseshoe magnet and directed from the north pole to the south pole. The strength of this magnetic field is denoted by H, which is entirely analogous to the electric field strength E we discussed earlier.

The action of magnetic poles on each other, or of the magnetic field between two poles on another pole, can be studied and utilized, but the most interesting and most useful phenomena of magnetism concern the action of magnetic fields on electric charges. And here we meet with some rather peculiar phenomena. If an electric charge q is at rest in a magnetic field nothing happens. However, if the charge is in motion, then the magnetic field (or, what amounts to the same thing, the combined effect of the two magnetic poles) exerts a force on the charge. To utilize this force effectively we must learn precisely how it acts. For this purpose mathematics must be used.

F O H v

Fig. 118

Suppose that H (fig. 118) denotes the magnitude of a magnetic field that has the direction shown. The magnitude of H will of course depend upon the strength of the magnetic poles that give rise to it. Incidentally, H is often measured in gausses, the unit having been named after the great nineteenth-century mathematician who contributed many laws to the sciences of electricity and magnetism and was one of the inventors of the electric telegraph. Let us suppose further that q units of positive charge are moving past the point O with a velocity v in a direction perpendicular to the magnetic field and out of the paper. Then the force F that the magnetic field exerts on the charge q will be perpendicular to the directions of the magnetic field and the velocity of the charge. The direction of F, as shown in the figure, is upward. If the charges were negative, the direction of F would be downward instead of upward; if the charges were negative but moving into the paper, then the direction of F would again be upward. The magnitude of the force F, determined experimentally, is

$$F = qHv. \tag{14}$$

How will the charges move? To start with, they have the velocity v in the given direction. They are acted upon by a force that imparts an upward velocity. Hence the charges will move in some direction between that of F and v but in the plane determined by F and v. As the charges move in this new direction, the force exerted by the magnetic field continues to act. But since the direction of motion of the charges has changed, the direction of F will change accordingly. The direction of motion of the charges and of the force F acting on them will therefore keep changing and the particles will move in some curved path. Let us note one of the very practical uses of what we have just learned.

Suppose that a rectangular loop of wire is placed in a horizontal magnetic field as shown in figure 119. The loop is rigidly attached to a horizontal rod OO' which is free to rotate. Now suppose that a battery is somehow connected at P and Q so that electrons can flow through the wire loop in the direction from P to A to B and so on around the loop to Q. This flow of electrons is a current. We now have electrons moving in a magnetic field, and in view of what we said above there will be forces acting on them.

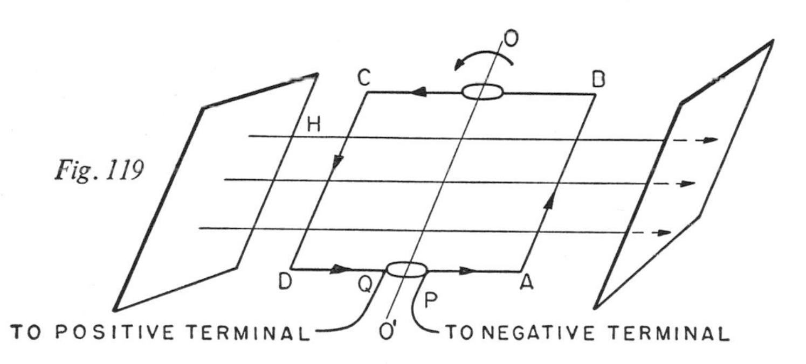

Fig. 119

Consider first what happens when the loop is horizontal. The electrons in the side AB will be moving from A to B in the horizontal magnetic field that extends to the right. (Electron motion from A to B is usually described as positive charge moving from B to A.) Then, according to the principle by which magnetic fields act on moving charges the electrons will be acted upon by a force F that will be directed upward. Since these electrons are in the wire, the force will be exerted on the side AB of the wire, and this side will be forced upward. The electrons in the side CD will be moving from C to D, and this direction is opposite to that from A to B. Hence the force on the electrons in CD and therefore on the wire segment CD will be downward, and this side will be forced downward. The action of the magnetic field H on the electrons in sides DA and BC of the loop can be ignored since the forces on these two sides will not be effective in causing the loop to rotate. What we have observed so far is that the loop will rotate with the side AB moving upward while the side CD moves downward. Since the entire loop is attached to the rod OO', the rod will rotate and the mechanical rotation can be used to make some other rod or wheel rotate. Thus we have the possibility, at least, of obtaining mechanical forces and energy from the behavior of electrons in magnetic fields. We have, in other words, the principle of an electric motor. Moreover, we can say a good deal about the strength of the mechanical force. Formula (14) tells us that if we increase q or H or both we shall increase F. We can increase q by increasing the voltage of the battery that forces the charges to flow in the loop, and we can increase H by increasing the strength of the magnet that creates the magnetic field. Hence we can increase F and make a more powerful motor.

The simple scheme of placing a rectangular piece of wire in a magnetic field will not provide a very practical motor, for many reasons. But the mathematician and scientist can now turn to the engineer and say, "We have given you a brand new idea, the principle of an electric motor. We have given you also the key mathematical formula that tells how the motor should respond. The problem of designing a practical motor is yours."

The conversion of electrical energy to mechanical energy has made an enormous difference in our civilization. Everyone knows a thousand uses of electric motors, to run machinery, to start an automobile, to run electric fans, and so forth. But these uses all depend upon having electrical energy to start with. The batteries we have referred to thus far are a partial answer to this need and are used, for example, in the automobile. But batteries operate by chemical action, and it is not practical to build powerful or high-voltage batteries by such means. Hence there is wisdom in seeking new ways of obtaining electricity. Nature provides man with excellent forms of energy. Coal, oil, and waterfalls are sources that can be used to create mechanical energy; for example, coal and oil are burned to drive steam engines. The problem then becomes whether the mechanical energy can be used to create electrical energy.

Long before the various uses of electrical energy that we know so well today were ever envisioned two distinguished scientists, the Englishman Michael Faraday (1791–1867) and the American Joseph Henry (1799–1878), found the answer. Faraday, the son of a blacksmith, received a poor education and at the age of thirteen started work as an errand boy and apprentice to a bookbinder. He studied chemistry and physics by himself and attended scientific lectures given by the famous scientists of his day, among them the noted chemist Sir Humphrey Davy. Faraday wrote to Davy to ask for a job and enclosed notes he had taken of Davy's lectures. Davy was impressed and employed Faraday as an assistant. Faraday, thus launched on his scientific career, rose rapidly to become the greatest nineteenth-century experimenter in electricity and magnetism. Henry started life from as lowly a position as Faraday. At fifteen he became a watchmaker's apprentice with the ambition of becoming an actor and writer. Like Faraday, he ran across books on science and acquired a love for the subject. He entered the Albany Academy in New York for further education and soon became professor of mathematics there, and later professor of natural philosophy [science] at what is now Princeton.

In the work of Faraday and Henry we find another of those curious incidents in the history of science wherein two men, working independently and in this case without knowledge of each other's work, discovered the same fundamental principle. The nature of this principle—electromagnetic induction—will be taken up in the next chapter. Here we shall discuss its more limited application to the generation of electricity.

We have seen how the force exerted by a magnetic field upon electrons moving in a wire causes the wire to move and thus produces mechanical energy. Now, however, let us not feed current into the wire loop of figure 119 but let us make the loop turn by applying mechanical force to it. As before, the magnetic field supplied by a magnet is to be present. The pertinent question is, Would the electrons that are normally present in the metal wire loop be forced to move along the wire?

Let us consider the behavior of any typical electron, say in the length AB of the wire. When this segment is horizontal and is moving upward the velocity v of the electrons in it is then upward. The magnetic field is horizontal and is directed from left to right. Hence each electron should be acted upon by a force F which is given by formula (14) and which is perpendicular to the magnetic field and the direction of motion of the electron. It is therefore in the direction from B to A. At the same time the electrons in the segment CD are moving downward and the resulting force on them is from D to C. The forces acting on the electrons in DA and BC will be perpendicular to the wire and hence cannot make electrons move *along* the wire. We shall therefore neglect these two segments.

We now have a force causing the electrons in the loop to move (in the direction opposite to that shown in figure 119). However, we must consider a new element in this picture, one we did not pursue when considering the electric motor. Suppose that the loop rotates at a constant speed and is in a position that makes an angle L with the vertical (fig. 120). Each electron in side AB will move in a circle at a constant speed v, but the direction of this velocity is tangential to the circle. The velocity of the electron *perpendicular to the magnetic field* is no longer v but $v \sin L$.* The

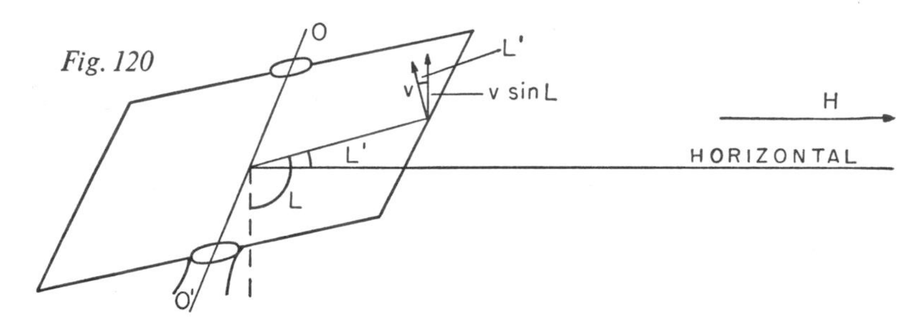

Fig. 120

force on each electron is still given by formula (14), provided that we replace v there by $v \sin L$. Hence

$$(15) \qquad F = qHv \sin L,$$

wherein we may regard q as standing for the charge of each electron in the wire.

Formula (15) contains a new feature. The force on the electrons depends upon the angle that the loop makes with the vertical and varies with the sine of this angle. This force on the electrons will still cause them to move *along* the wire, and a current, which we shall denote by I, will still flow in the wire. However the current will vary as the force does. Since q, H, and v are constant, the force F varies sinusoidally. Hence the current will also have the functional form

$$(16) \qquad I = D \sin L,$$

where D is the maximum amplitude of the current. The quantity D depends upon the nature of the material of the wire and other factors. We now have the essence of Faraday's and Henry's principle of the generation of electricity. The force F which their

* The velocity perpendicular to the magnetic is $v \cos L'$ where L' is the angle shown in the figure. But $L = 90° + L'$ and by an elementary result of trigonometry, which we have not presented, it follows that $\cos L' = \sin L$.

scheme produces and which drives electrons around the loop is no different in character from the voltage supplied by batteries.

In the theory of generators of electricity it is convenient to have the force on the electrons and the current expressed as functions of time. This is simply done. Suppose the wire loop makes f rotations per second; than it requires $1/f$ seconds to make one rotation. If t is the time that elapses during the rotation of the loop through angle L, then this amount of time is to the time of one rotation as L is to 2π. Thus

$$\frac{t}{1/f} = \frac{L}{2\pi}$$

or

$$L = 2\pi ft.$$

Hence formula (16) for the current becomes

$$I = D \sin 2\pi ft. \tag{17}$$

Formula (17) is both illuminating and useful for illumination. In the first place it tells us much about the nature of the current generated. This current will increase to a maximum, decrease to zero, decrease to a maximum in the opposite direction, and then increase to zero again, in accordance with the usual sinusoidal behavior. The current is called alternating because it changes its direction every half-period. Moreover, the number of these complete oscillations or cycles in one second will be f, precisely the number of complete rotations the loop makes in one second. Hence, if one wishes to generate alternating current of a given frequency, one must make the loop rotate at that frequency. Alternating current is used to light electric bulbs and to run most home appliances and machinery in factories. For home use a frequency of 60 cycles per second has been adopted in this country. Thus the current that passes through the filament of an electric light bulb goes through 60 complete cycles in one second, and it is actually zero 120 times a second. We do not see these variations in the current strength not only because they are too rapid for the eye to follow but because the filament becomes white-hot when the current is strong and does not lose enough heat while the current decreases to zero and increases again. This alternating current, once generated, can be converted to direct current and used to run motors in the manner we discussed above or used directly as alternating current to produce a steadily running motor.

The modern reader does not have to be told about the widespread uses of electricity and of the effect of this type of power on our civilization, but it may be necessary to emphasize that the principles of generating electricity by mechanical means and of converting electrical power to mechanical power were investigated long before such applications were even dreamed of. While Faraday was doing his early experiments in electricity he was asked by a visitor what use his principle of inducing electricity in wires might be and Faraday replied, "What is the use of a child?—it grows to be a man." On another occasion Faraday was visited by Mr. Gladstone when he was Chancellor of the Exchequer. Gladstone asked the same question and this time Faraday answered, "Why, sir, presently you will be able to tax it."

We could continue to show how the use of magnetic and electric fields to make electrons or positive charges move according to our will produces fundamental scientific knowledge and almost miraculous devices. Some of the sensational and most effective atom smashers, such as the cyclotron and betatron, are designed on the basis of the few principles we have already examined. Of course the design of such expensive pieces of equipment is thoroughly worked out mathematically before they are built.

The motion of electric charges is of concern to geophysicists and astronomers. There is excellent scientific evidence for the belief that billions of positively and negatively charged particles which emanate from the sun and stars are constantly entering the atmosphere of our earth. Near the earth these particles interact with the magnetic field of the earth and are compelled to move to the extreme northern and southern latitudes where they cause the auroras, the spectacular displays of light. Of course it is impossible to observe the motion of particles so far from the earth, but mathematics can be used to predict what these incoming beams of electrons or positive ions will do when they reach the earth's magnetic field, and the mathematical predictions can then be checked against observations.

21: MATHEMATICAL OSCILLATIONS OF THE ETHER

For many parts of nature can neither be invented with sufficient subtlety, nor demonstrated with sufficient perspicuity, nor accommodated unto use with sufficient dexterity without the aid and intervention of mathematics.

—FRANCIS BACON

THE developments in electricity and magnetism that we have examined so far might be said to constitute the childhood of a new science. But a child, as Faraday remarked, can grow to be a man, provided that it is properly nourished. This nourishment, if history is any guide, must include larger and larger portions of mathematics. While Faraday's own work set the youthful sciences of electricity and magnetism on the road to manhood, it was his notable successor and devoted admirer, James Clerk Maxwell, who supplied the mathematical sustenance.

Let us resume our account of Faraday's work at the point where we left it in the preceding chapter. Faraday had found that when a wire loop is rotated in a magnetic field the electrons in the wire are acted upon by a force which causes the electrons to flow along the wire and thus creates a current. In this method of generating electrical current the wire moves in a stationary magnetic field, namely, the field between the poles of a fixed magnet. It then occurred to Faraday to ask the following question. Suppose the wire remained stationary and the magnetic field moved past the wire: Would not the same effect take place, that is, would not the electrons in the wire still experience a force which would set them moving? It seemed likely that only the relative motion of the field and wire should matter, rather than which one was stationary.

To test this conjecture by some experimental arrangement that would produce more than a momentary effect and therefore permit convenient study, Faraday resorted to a somewhat elaborate setup in which he utilized a notable discovery (1820) of the Danish physicist Hans Christian Oersted and the Frenchman André Marie Ampère. These men had found a fundamental link between electricity and magnetism, namely, that when current flows in a wire a magnetic field surrounds the wire. Moreover, if the current varies in strength, the magnetic field also varies. This occurrence of a magnetic field is of course somewhat surprising and not predictable from the phenomena we have so far examined. It offered a totally new way of producing magnetic fields, and is the principle now used in electromagnets.

Oersted's and Ampère's discovery provided the moving magnetic field that Faraday wished to employ. He placed two coils of wire near each other as shown in figure 121. His plan was to make a current flow in the left-hand coil and have this current set up a magnetic field whose direction is shown in the figure. This magnetic field would extend past the wire in the second coil. However, Faraday wanted a moving magnetic field. Hence he connected the terminals *A* and *B* of the first coil to a source of alternating current.

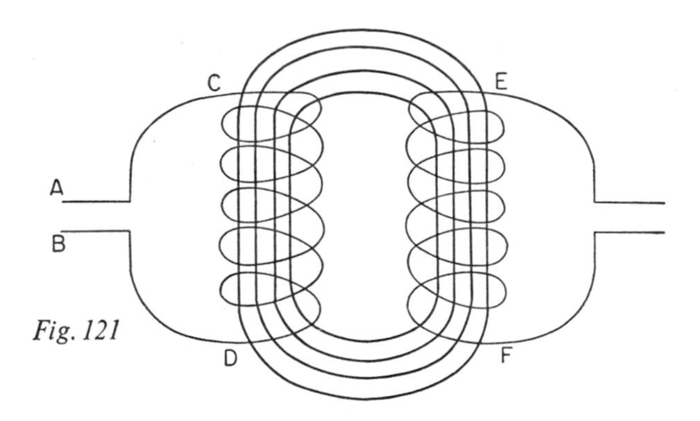

Fig. 121

This alternating current passes through the coil and sets up a varying magnetic field about it in accordance with Oersted's principle. Thus, as the alternating current increases in strength, a stronger magnetic field appears around the coil *CD*, and as it decreases in strength the magnetic field decreases. Insofar as the coil *EF* is concerned, since it is alongside the coil *CD*, a magnetic field surges out to cross it and then recedes. Thus Faraday had a magnetic field moving past a wire, the coil *EF*. If a magnetic field moving past a fixed wire produces a force in the wire, this magnetic field should set up a force or voltage and therefore a current in the coil *EF*. Moreover, since the magnetic field not only moves past the coil *EF* but increases and decreases in strength, the current induced in the coil *EF* should do likewise. That is, the current in the coil *EF* should be alternating. Faraday further expected that this current should persist as long as the alternating current was maintained in the first coil and hence that he would be able to study the induced current at length.

Faraday did find that an alternating current appeared in the coil *EF* and moreover, as he had expected, the frequency of this current was precisely the frequency of the current applied at the terminals *A* and *B* of the first coil. This discovery of Faraday's, that a changing magnetic field will induce a force on the electrons in a wire placed in that magnetic field and therefore cause a current to flow, is called electromagnetic induction. One apparent use of this principle is to transfer electric current from one coil to another, even though the second one is not connected to the first one and may in fact be moving independently. A more important use is made in what are now called voltage transformers, but we shall not pursue this idea because it would take us too far afield from mathematics.

With Faraday's discovery of the important principle of electromagnetic induction, and thereby of a new link between magnetism and electricity, the science of electromagnetism—the word used to denote the interaction of electricity and magnetism—now had a number of significant advances to its credit. But the phenomena were becoming more involved, and Faraday was beginning to have difficulty in thinking about them. In the cases of simple electric and magnetic fields it had been rather easy to construct some physical pictures and to obtain, either by measurement or simple reasoning, the appropriate mathematical laws. In the case of electromagnetic induction the problem of determining the electric force and current induced in the second coil, knowing the current in the first one, was already too complex to be easily analyzed. This phenomenon involved, first of all, calculating the strength of the magnetic field that accompanied the current in the first coil and then calculating the voltage and current in the second coil. Moreover, because he had discovered what seemed to be a promising physical process, Faraday wished to know just how he could increase its effectiveness. Would more current in the first coil or a longer coil or a wider coil produce more current in the second one? How should the coils be placed relative to each other?

There is no doubt that continued experimentation on an extensive scale might have answered many of Faraday's questions. But the time and expense would have been enormous. Moreover, experimentation by itself gives a set of disconnected results from which the complete and accurate picture may never emerge. Let us just imagine the experimentation that would be needed in the simpler problem of finding the dependence of the range of a projectile on the angle of fire. One would have to fire at least dozens of projectiles at different angles, making certain that other factors such as the velocity with which the projectiles were fired, the shape of the projectiles, and the state of the atmosphere were constant, and one would have to measure the range and angle accurately each time. With all these precautions taken and the information secured the experimenter might obtain some limited information. He might, for example, learn that as the angle of fire increases in one-degree steps from 1° to 40° the range increases steadily, but he might miss the all-important fact that above 45° the range decreases. The dependence of range upon velocity would still be entirely unknown and require further experimentation. But we have seen how with some simple mathematics costing only pencil and paper Galileo answered all these questions and many more.

Faraday was well aware of what mathematics can accomplish but his genius was confined to experimentation and physical thinking. In the case of complicated electromagnetic phenomena, unlike the motion of projectiles, physical thinking is at a considerable disadvantage. It is easy to visualize the motion of a projectile, its angle of fire, and its range. Electric and magnetic fields, on the other hand, are not visible and their configurations not readily obtained. Despite his past successes in concocting physical pictures Faraday realized that physical thinking was not going to carry him much further. Faraday was at that stage where the physics becomes too difficult for the physicists and requires the services of a mathematician.

Fortunately, the great nineteenth-century mathematical physicist James Clerk Maxwell was assiduously preparing himself for the task. As a youth Maxwell already showed promise of being able to make first-class contributions. A paper he wrote at the age of fifteen on mechanical methods of generating some curves was published in the Proceedings of the Royal Society of Edinburgh. During his attendance at the universities of Edinburgh and Cambridge his professors and fellow students recognized his brilliance and originality. In 1856 he was chosen to be professor of physics at Marischal College in Aberdeen. A few years later he transferred to King's College in London and in 1871 moved to Cambridge University.

Like all scientists Maxwell took up the challenging problems of his day, and he began his work in electromagnetism by reading Faraday's *Experimental Researches*. In 1855, at the age of twenty-three, he published his first paper on the subject, "On Faraday's Lines of Force." In this and later papers Maxwell undertook the task of translating Faraday's physical researches into mathematical form. We shall not present Maxwell's mathematical formulations. These involve complicated differential equations, and though we shall have something to say on this mathematical subject in a later chapter, we shall not be able to encompass Maxwell's work even in that place. However, we can describe the nature of his mathematical work and be more specific subsequently about the mathematical and physical consequences.

Maxwell's first step was to write down in precise mathematical form four key laws of electromagnetism. The first of these is not much more than a restatement of Coulomb's law, formula (1) of the preceding chapter. Instead, however, of describing the force exerted by one electric charge on another, Maxwell gave a quantitative statement of the field that any electric charge creates in the space around itself. Maxwell's second law states the analagous fact about magnetic poles. Next, Maxwell tackled the relationship between electric currents and magnetic fields. Oersted and Ampère had shown that a magnetic field accompanies an electric current in a wire and surrounds the wire. In its original form Maxwell's third law states the precise quantitative relationship between the magnetic field surrounding the wire and the current in it. His fourth law describes the mathematical behavior of electromagnetic induction. In this phenomenon a changing magnetic field induces an electric current in a wire present in that field (we refer to the current induced in the coil *EF*). In pursuance of Faraday's own line of physical thinking Maxwell viewed this phenomenon in a slightly different manner. The current induced in the wire consists of electrons that move along the wire. These electrons move because some force acts on them. But the force acting on electrons that were stationary to start with must be an electric field. This conclusion is recommended to us by the experience of seeing electrons between charged parallel plates move under the force exerted by the field between the plates. Though there are no charged plates present in the phenomenon of electromagnetic induction it seems fair to conclude that the changing magnetic field sets up or creates an electric field that drives the electrons in the wire. Maxwell's fourth law states the quantitative relationship between the changing magnetic field and the electric field to which the magnetic field gives rise. As we saw earlier, this electric field changes with time if, for example, the magnetic field varies sinusoidally.

Up to this point Maxwell had merely formulated in mathematical language some well-established physical principles. He was therefore prepared to treat quantitatively a number of electromagnetic phenomena and to answer by mathematical reasoning questions such as Faraday raised about electromagnetic induction. But Maxwell's labor in formulating the mathematical laws of electromagnetism paid an unexpected dividend. On mathematical grounds Maxwell found that the four equations he had thus far formulated were inconsistent with a fifth equation, known as the equation of continuity in physics, which expresses the fact that electric charge cannot be created or destroyed. The inconsistency Maxwell noted is of the same nature as if from the set of four equations he had deduced that $x = 2$ and by using the fifth equation had found that $x = 3$. Evidently something was missing in the four equations that should be added to 2 so that x would amount to 3.

Maxwell pondered on this inconsistency and decided that to eliminate it he must add a new term to his formulation of Ampère's law. The nature of this term requires some explanation. Following Faraday's line of physical thinking, Maxwell regarded electric and magnetic fields as existing in the space around magnets and around wires carrying currents. Ampère's law itself deals with a current flowing in a wire. But when this current alternates (for example, suppose it varies sinusoidally in time), the electrons in the wire move rapidly back and forth. Now even when the electrons are stationary the attraction or repulsion which they exert on other charges can be explained by supposing that the electrons set up an electric field around them. If these electrons should be moving back and forth, then the electric field they set up will also move, and at any point in space outside the wire the strength of this electric field will vary with time. Hence an alternating current in a wire can be thought to have accompanying it a varying electric field in the space surrounding the wire. Maxwell not only accepted the reality of this varying electric field but observed that it had the mathematical properties of a current, even though regarded by itself, that is, apart from the wire which gave rise to it, it did not consist of the motion of electrons. He called this varying electric field a displacement current, because it amounted to the displacement or variation of an electric field. Maxwell's own words in his *Treatise on Electricity and Magnetism* clearly describe this conclusion: "One of the chief peculiarities of this treatise is the doctrine which it asserts, that the true electric current, that on which the electromagnetic phenomena depend, is not the same thing as the current of conduction [the current in the wire], but that the time variation of the electric displacement must be taken into account in estimating the total movement of electricity. . . ."

Maxwell pursued the implication of the existence of displacement current. Ampère's law said that a current in a wire is accompanied by a magnetic field. But since Maxwell had added the displacement current to the conduction current, or current in the wire, he concluded that the displacement current also gave rise to a magnetic field, and this magnetic field was part of the one formerly regarded as produced by the conduction current alone. In other words, the magnetic field surrounding the wire must be due to both currents, conduction and displacement.

To recapitulate, the first bold step Maxwell made was to introduce the concept of a displacement current and to suppose that this current, which existed in space rather than in a wire, also gave rise to a magnetic field. Thus he revised Ampère's law to relate the total current, that is, conduction and displacement, to the magnetic field around the wire. A most important part of this law is that a changing electric field itself creates a magnetic field. If we now recall that Faraday's law, as formulated by Maxwell, states that a changing magnetic field creates an electric field, we see that Maxwell introduced the reciprocal relationship.

Maxwell's next step was to study the implications of the revised laws of electromagnetism as to the variation of electric and magnetic fields in space and time. He succeeded in making further capital discoveries. To follow his thinking let us return to the two coils of Faraday, and let us suppose now that the coil *EF* is far removed from coil *CD*. As in the earlier situation let us suppose that an alternating current is fed into the coil *CD*. This alternating current is accompanied by a changing electric field in the space around it, the displacement current, and both currents create a sinusoidally varying magnetic field in the space about the coil *CD*. Now this sinusoidally varying magnetic field sets up a sinusoidally varying electric field, in accordance with Faraday's law. According to Maxwell's revision of Ampère's law, this varying electric field is accompanied by a magnetic field. This magnetic field will also vary with time, because the electric field happens to vary sinusoidally. Hence the magnetic field in turn will create a new electric field. Thus a succession of changing magnetic and electric fields are created, one inducing the other.

We are in a position now to foresee by a physical argument what Maxwell predicted from mathematical reasoning. If the end of a stick is put into a large body of water and moved back and forth rapidly, a series of waves is created in the water. As new waves are created they force the existing waves to spread out and travel a great distance from the stick. The stick is comparable to the sinusoidal current in the coil *CD* and the varying electric field in the space around it. These create the varying magnetic field. But the magnetic field, we have seen, creates a varying electric field and this, in turn, a varying magnetic field. What will these fields do under the constant "pressure" exerted by the current in the coil *CD?* The answer is almost obvious. They will spread out into space and reach points far removed from the coil *CD*. They might even reach the distant coil *EF*. There the changing electric field will cause a current to flow in the wire, and this current can be used for any purposes currents serve. Thus Maxwell discovered that an electromagnetic field—that is, a combination of a changing electric and magnetic field—will travel far out into space. Incidentally, Faraday had already suspected that this might be possible when he considered what would happen if the coil *EF* were somewhat removed from the coil *CD*. But what Faraday conjectured on physical grounds, without fully comprehending the mechanism or recognizing the existence of displacement currents, Maxwell established on mathematical grounds.

In 1887, about twenty-five years after Maxwell had predicted the existence and propagation of electromagnetic waves, Henrich Hertz, using simple coils, produced such waves and received them some distance away. These electromagnetic waves were none other than the radio waves used in modern radio and television.

Waves, as we noted in our study of sound, have a wave length and a frequency of variation per second. In the case of electromagnetic waves the wave length is determined (though this may not be immediately obvious) by the size of the coil used. (As a matter of fact, coils are not the most efficient sources of electromagnetic waves.) To keep the coils, or whatever wires are used to send electromagnetic waves into space, reasonably small, it is necessary that the wave length be small. But we also know that the wave length λ and frequency f are related by the formula—see (4) of chapter 19—

$$\lambda f = c, \tag{1}$$

where c is the velocity of the wave motion. Now c, in the case of electromagnetic waves, is enormously large. It is about 186,000 miles per second. Hence, if the sources of the radiation, or the antennas as they are called, are to be kept small, say of the order of 10 feet or so, then λ will be about this size, and the frequency f, according to formula (1), must be 98,208,000 per second. As a matter of fact, frequencies of about this magnitude are used in modern television. In ordinary radio broadcasting larger antennas and lower frequencies, of the order of 1,000,000 per second, are used.

From his equations Maxwell could predict the precise mathematical form of the electromagnetic field traveling in space. If the current in the coil or antenna acting as a source is sinusoidal and of frequency f, then the wave length is given by (1). The traveling electric field E has the form—compare formula (7) of chapter 19—

$$E = D \sin 2\pi \left(ft - \frac{x}{\lambda}\right) \tag{2}$$

and the traveling magnetic field will have the same form. Of course equation (2) does not tell the whole story. We know from earlier work that electric and magnetic fields have directions attached to them, and equation (2) says nothing about the direction of the electric field. But the essential behavior of the field, its sinusoidal behavior in time at a fixed place (fixed x-value) and its sinusoidal behavior in space at a fixed time (fixed t-value), are contained in it. Figure 122 shows the variation of E with space or time, depending upon whether we interpret the variation along the horizontal axis as a variation in distance at a given time or as a variation in time at a given distance from the source. The figure also shows that the directions of the electric field E and the magnetic field H are at right angles to each other.

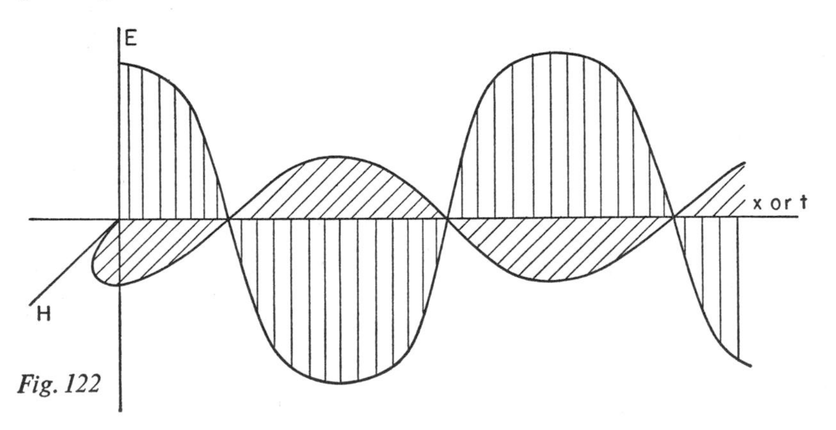

Fig. 122

Maxwell showed that electromagnetic waves could be transmitted over great distances. This idea in itself, when applied to a practical apparatus, makes wireless telegraphy possible. In telegraphy all one needs to do is to send out long and short bursts of radio waves. These are picked up at one or several receiving antennas and the message decoded. The wireless transmission of vocal and musical sounds and of television images presents greater difficulties. However, the mathematical study of musical sounds had shown that these consist of sinusoidal waves of frequencies between about 16 and 16,000 cycles per second. Moreover, work on the telephone had shown that these sounds could be converted into electrical currents and the currents then reconverted into sound waves. Hence the problem of transmitting musical sounds became that of somehow imposing the corresponding electrical currents onto the high-frequency current that is propagated into space as a radio wave. Several schemes for doing this were developed.

Let us utilize formula (2) but ignore the variation in space and concentrate on the variation of the electric field with changing time at some fixed point in space. What happens at one point will happen at another, though the sinusoidal behavior may begin a little sooner or later. We can then study, as we did in the case of musical sounds, the formula

$$E = D \sin 2\pi ft. \tag{3}$$

If what is sent out is merely this sinusoidally varying field, then the same signal will be received. Would there be any way of varying this signal so as to make it convey a musical sound? Well, the amplitude D of this signal is constant. Suppose the amplitude were made to vary in accordance with the varying strength of an electric current that represents a musical sound. Then this variation, if properly detected at the receiving station, would convey the musical sound. Since all musical sounds are reducible to a sum of sinusoidal sounds, let us consider the problem of transmitting a simple sound of, say, 400 cycles per second on a high-frequency electromagnetic wave of, say, 1,000,000 cycles per second. The suggestion of varying the amplitude D in equation (3) amounts, then, to varying D sinusoidally at the rate of 400 cycles per second. What we wish to send out, then, is a 1,000,000-cycle radio wave, imposed on which is a sinusoidal variation in amplitude at the rate of 400 cycles per second. Mathematically, what we wish to send out is represented as

$$E = D(1 + .3 \sin 2\pi \cdot 400t) \sin 2\pi \cdot 1{,}000{,}000t, \tag{4}$$

wherein the quantity 0.3 is merely a typical value. Formula (4) says that we are dealing with a 1,000,000-cycle wave whose amplitude is the quantity

$$D(1 + .3 \sin 2\pi \cdot 400t). \tag{5}$$

But this amplitude is itself a varying quantity that will change in value from D to $D + .3D$ to D to $D - .3D$ to D at the rate of 400 times a second. The shape of the graph of formula (4) is shown in figure 123. Formula (4) represents what is called an amplitude-modulated radio wave. It conveys a message, namely, a 400-cycle

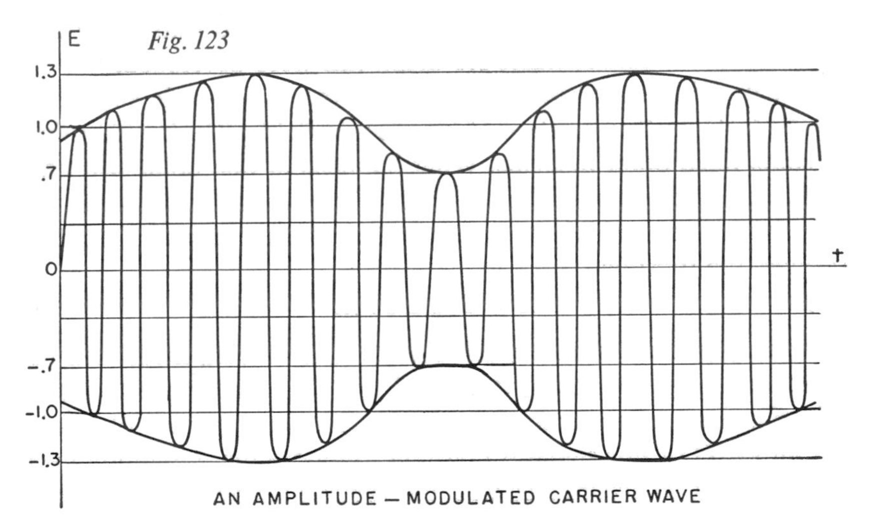

Fig. 123

AN AMPLITUDE—MODULATED CARRIER WAVE

musical note. The quantity 0.3 used above represents the amplitude of the sound itself. That is, the louder the note the larger this quantity is, but, of course, it cannot be larger than 1 since it can only add or subtract from the amplitude of the 1,000,000-cycle wave.

The technique for imposing this amplitude modulation or variation on the 1,000,000-cycle carrier, as it is called, and the technique for removing this amplitude variation and converting it into a 400-cycle electric current which is then converted into a sound wave are matters which electronic engineers have to handle and do handle successfully. We shall add only that whereas we have considered the problem of sending out just a simple sound, the same theory applies to more complex sounds. In place of the amplitude variation given by expression (5) above there would be the amplitude variation belonging to the complex sound.

Amplitude modulation, then, is one of the means of conveying voice or music on sinusoidal electromagnetic waves, and this is the process used by the AM stations that we tune in on our radios. Of course we tune in on the carrier frequency of any one of these stations; the radio set not only picks up this carrier but demodulates it, that is, converts the amplitude variation into ordinary electric current, and the loud speaker converts the current variations into sound waves.

There is another commonly used system for the transmission of vocal and musical sounds by radio. It is called frequency modulation. The suggestion for this system comes likewise from studying formula (3). As long as a radio wave of constant frequency f is transmitted, no information beyond the frequency itself is contained in the received signal. However, if the frequency f is varied, then there is the possibility that this variation can be used to carry a message. To make the variation in f detectable, it must be large. One therefore starts with a carrier frequency that is large, say 100,000,000 cycles per second. Suppose that a 400-cycle note is to be imposed on the carrier. Then the scheme is to vary the 100,000,000 frequency at the rate of 400 times a second. That is, to convey the frequency of the musical note the frequency of the carrier is varied sinusoidally from 100,000,000 to some value above 100,000,000, then back to 100,000,000, then to some value below 100,000,000, and finally back to 100,000,000, at the rate of 400 times a second. The amount of the periodic change in the basic frequency depends upon the intensity or loudness of the musical note.

The mathematical function representing the electric field propagated into space is then

$$(6) \qquad E = D \sin (2\pi \cdot 100{,}000{,}000t + a \sin 2\pi \cdot 400t),$$

where the quantity a depends upon the loudness of the musical note. The function (6) that represents the electric field of the frequency-modulated wave is no longer the simple type of sinusoidal function of the form $\sin 2\pi\, ft$ that we encountered earlier in several places. Hence it is not immediately obvious that the frequency of the function (6) is 100,000,000 cycles per second, increased and decreased sinusoidally by a fixed amount at the rate of 400 times per second. The graph of a frequency-modulated electric field is shown in figure 124.

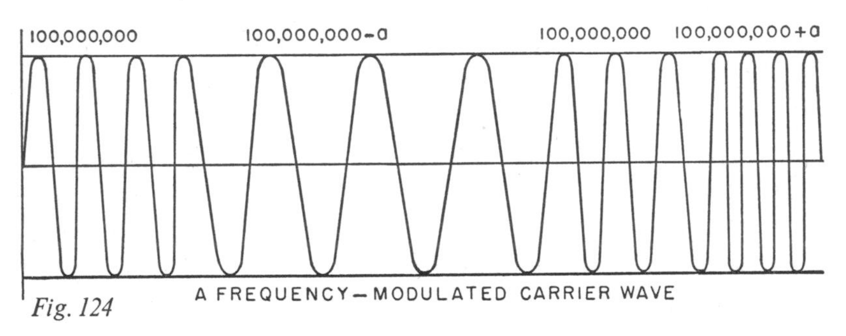

Fig. 124

As in the case of amplitude modulation the technique of imposing frequency modulation on the carrier and of abstracting this modulation from the carrier must be supplied by the electronic engineer.

We have digressed from Maxwell's work to give some inkling of how his basic discovery of electromagnetic waves is now employed in the science of radio. It is perhaps unnecessary to point out to anyone living in this second half of the twentieth century how this application of Maxwell's work has led to profound changes in our way of life. But we have yet to present Maxwell's second sensational discovery, which has had equally significant impress on science and daily life. We merely mentioned above that electromagnetic waves travel out into space at a velocity of 186,000 miles per second. Maxwell obtained this figure by mathematical reasoning and was immediately struck by the fact that may already have occurred to the reader. This velocity is also the velocity of light. Now evidence had been accumulating during the early part of the nineteenth century, notably through the work of Thomas Young and Augustin Fresnel, that light is a wave motion with frequencies of the order of 500 billion billion ($5 \cdot 10^{14}$) per second. The identity of the velocities and the fact that electromagnetic radiation and light were both wave motions caused Maxwell to draw the patent inference. He declared unhesitatingly that light is an electromagnetic wave.

This new theory of light called for reinterpretation of all of the numerous experiments performed and facts discovered in this subject since the early seventeenth century. The reinterpretation not only proved to be possible but provided a better understanding and unified diverse phenomena. In particular, it became clear that light waves of various colors are electromagnetic waves differing in frequency, the frequencies ranging from $4 \cdot 10^{14}$ for red light to $7 \cdot 10^{14}$ for violet light. Light waves of a single frequency are thus analogous to sound waves of a given frequency. Combinations of light waves of different frequencies produce color effects in the eye that are quite different from the individual frequencies. For example, the combination of all frequencies from red to violet produces white light, as Newton showed. Such combinations are analogous to combinations of sounds, such as chords. Just as some chords are harmonious to the ear, some light combinations are pleasing to the eye. Though the eye can detect to some extent the basic colors in any given combination it is not as sensitive in this respect as the trained ear, which can not only recognize the component frequencies of a complex sound and the relative strengths of these frequencies, but can also distinguish the separate tones of a chord. The implications of Maxwell's new view of light are still being explored.

The assertion that light is an electromagnetic wave raises the question of whether what we had to say about light in earlier chapters is correct. There we spoke about light as some form of energy traveling along rays and obeying such laws as the law of reflection and the law of refraction. Nothing was said about wave motion or about electric and magnetic fields. The two mathematical treatments of light can be reconciled. As we have just indicated, the frequencies of light waves are so large and the wave lengths accordingly so small that the variation in strength of light waves with time or space is not only not noticeable but has very little effect in most common phenomena. Hence it is usually all right to treat light as some unanalyzed quantity traveling along rays and reflected or refracted according to the laws we have hitherto used.

However, the older version of light, called geometrical optics, is only an approximate theory. In phenomena requiring a careful analysis of the behavior of light it is the electromagnetic theory that must be used.

If the knowledge that light is an electromagnetic wave grants us new information about light, the reverse is also true. The basic phenomena of light such as reflection, refraction, and the concentration of light by mirrors and lenses should be applicable to electromagnetic waves. This indeed is the case. Hertz showed experimentally that electromagnetic waves are reflected from large polished metal surfaces; these act as mirrors do on light waves. Radar sets utilize this reflection of electromagnetic waves from metallic surfaces. Moreover, just as light from a source placed at the focus of a paraboloid will be concentrated into a beam issuing parallel to the axis of the paraboloid, so radio waves issuing from a small source (called a dipole) placed at the focus of a large paraboloid will be sent out essentially parallel to the axis of the paraboloid (fig. 125). The source and paraboloid are used in the field of radio as an antenna system, and paraboloidal antenna systems are very commonly used to concentrate radio waves for locating airplanes.

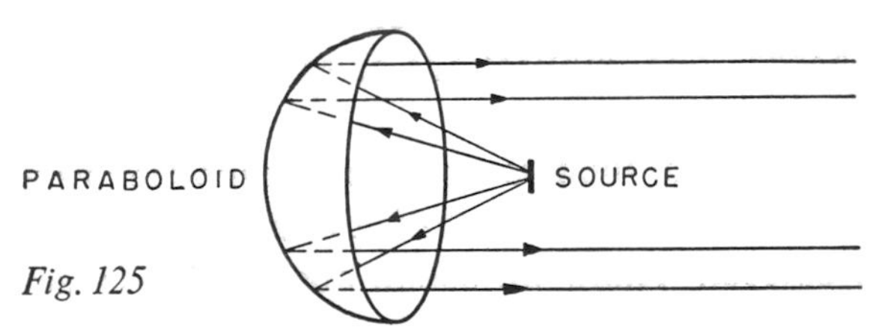

Fig. 125

It would seem that Maxwell's mathematical work had already produced as much return as any one man's efforts ever had, but additional returns now become available almost for the asking. Radio waves with frequencies from a few hundred thousand cycles per second to frequencies of about a hundred million cycles per second were produced experimentally and put to use rather early in this century. But Maxwell's work had shown that light, too, is an electromagnetic wave with frequencies of the order of 10^{14} per second. The obvious question was, Are there waves with frequencies larger than 100 million and in fact up to 10^{14} per second and possibly even beyond? Undoubtedly there were, and the scientists had to either detect those that existed in nature or create them. Since the behavior of some radio frequencies and the visible frequencies was known, many properties of these unknown frequencies could also be predicted. And it was clear that the unknown frequencies might be profitably employed in applications where the existing radio waves and light waves were not the most advantageous. For example, the smaller the wave length of an electromagnetic wave, the more it might be expected to behave like a light wave, and hence the more readily could it be directed in a powerful beam. Research was therefore devoted to obtaining generators of electromagnetic waves at frequencies higher than 100 million. Today radio waves of frequencies up to 30 billion ($3 \cdot 10^{10}$) per second are being employed, and still higher radio frequencies are commonly produced in laboratories.

But between $3 \cdot 10^{10}$ cycles per second and 10^{14} cycles per second and beyond 10^{14} there are vast ranges of numbers at least. Are there any electromagnetic phenomena behaving at frequencies in these ranges? In Maxwell's time physicists had already become somewhat informed as to the existence and properties of ultraviolet waves, which, though invisible to the eye, make their presence known by blackening photographic film. Also, infrared rays, likewise invisible, convey heat readily registered by a thermometer. Both of these rays are present in the sun's radiation. They can also be generated by passing electric current through special filaments in the same manner that visible light is created by passing current through tungsten. The conjecture that infrared rays and ultraviolet rays are electromagnetic waves was readily established experimentally, and infrared rays were found to have frequencies a little below those of visible light whereas ultraviolet rays proved to have frequencies a little above.

What about frequencies above the ultraviolet range? In 1895, Roentgen discovered X rays, and it was immediately suspected that these rays were electromagnetic and of extremely high frequency. In 1913, Von Laue and his students measured the frequencies of X rays and found them to lie in the range 10^{16} to 10^{18} per second. At about the same time work on radioactive elements led to the discovery of gamma rays, which are emitted by the nuclei of atoms and have proved to have a frequency of about 10^{20} per second. Indeed, the conscious search for and accidental discovery of new types of electromagnetic waves has led through devious channels into a wave theory for all matter. The exploration of this theme would involve us in too much of modern physics to pursue it clearly here. It may suffice to mention that even electrons behave as waves and are used in the electron microscope in the same way that light waves are used in ordinary microscopes.

The basic identity of the many forms of electromagnetic waves is now utilized so commonly that we lose sight of this remarkable feature. Let us consider, for example, the process of television. The variations in brightness of a scene being televised are converted into electric current; this is converted into radio waves that travel through space; the waves induce electric current in a receiving antenna; from here the current passes through an electric circuit; finally, by means of a cathode-ray tube the current is transformed into light waves.

The various classes of electromagnetic waves carry energy. Hence it is possible to convert other forms of energy into electromagnetic energy, utilize the advantages of this form, and then retransform the electromagnetic energy into a more serviceable form. Ordinary radio broadcasting utilizes this idea in converting sound waves into radio waves and then returning to sound waves.

Though we are unable to present the full details, Maxwell's work is an unexcelled example of how mathematical reasoning has produced new knowledge of the physical world and, in view of the numerous inventions based on it and now being put to use, mastery of the physical world. The example is so impressive and has so many implications for the role of mathematics that it warrants further discussion. First of all, one mathematical scheme embraces a number of major and seemingly diverse physical phenomena: electrical currents such as are used for light and power, radio waves, infrared rays, light, ultraviolet rays, X rays, gamma rays, and still other forms of radiation. The mathematics covers known waves whose wave lengths range from 0.000,000,000,000,01 cm. to 1,000,000,000 cm., or, expressed in terms of frequencies, frequencies as low as 30 or less per second and as high as $1 \cdot 10^{24}$. The same mathematical laws therefore describe the physical properties of these varieties such as reflection, refraction, propagation in space, and interference. Hence it is possible to transfer knowledge from one physical domain to another. One of the values of mathematics is of course to provide reason and order in the welter of sensations and effects that nature proffers. No other plan has ever organized such a wealth of phenomena.

Mathematics permits us to know nature through the analysis and synthesis it provides. But in the case of electromagnetic theory there is a deeper significance to the knowledge of nature that we acquire thereby. It is possible that the reader has felt uneasy and dissatisfied with all of the exposition of the preceding and present chapters because it deals so much with fields. We have spoken of fixed electric fields and magnetic fields, changing fields, fields moving through space, and fields interacting with electrons. We have spoken of changing magnetic fields creating electric fields, and the converse. But our pictures of these fields were sketchy, and we gave no physical explanation of what these fields consist of, or how one field creates another. It is time to face these problems.

The notion of fields originated with and became significant with the work of Faraday, and he constructed an ingenious account of the action of electric and magnetic fields. In the case of a magnet that pulls a piece of steel toward itself Faraday believed that lines extended from one pole to the other and covered the entire region of space about the magnet. These lines were elastic and, like extended rubber bands, wished to contract. Moreover, lines lying alongside each other repelled each other and hence became curved. These lines pushed or pulled on objects placed in the space they covered. He therefore called them lines of force. He postulated that electric fields also consist of lines of force. When he discovered the more complicated phenomenon of electromagnetic induction, Faraday tried to extend the action of lines under tension to account for these new happenings. Maxwell devoted many long research papers to improving Faraday's physical reasoning about lines of force stretching out from electric charges and magnetic poles, and then attempted a complicated mechanical model of the action of these lines in space, in conductors and in insulating media. But the evidence for the physical existence of such lines was never present. Moreover, even if the physical description were regarded as fiction or pure hypothesis it was not adequate to explain the multitude of electromagnetic phenomena. Neither Maxwell nor anyone else has ever succeeded in giving a satisfactory physical picture.

An example of the difficulty Maxwell faced was the problem of explaining what medium carried electromagnetic waves. When he proved that electromagnetic waves travel with the velocity of light, Maxwell concluded that these waves travel in ether, because since Newton's days ether had been accepted as the medium in which light moved. But since electromagnetic waves travel through all substances this means that ether must pervade all substances, including empty space. Moreover, since the waves move with enormous velocity the ether has to be highly rigid, for the more rigid a body the faster waves travel through it. On the other hand if ether pervades space it must be completely transparent and the planets must move through it with no friction. These conditions imposed on ether are contradictory. Moreover, ether cannot be touched, smelled, or isolated from other substances. Such a medium is physically incredible. We must conclude that it is a fiction, a mere word satisfying only those minds that do not look behind words. Further, the entire account in terms of fields is a crutch that helps the human mind to propel itself forward but must not be taken literally or seriously.

To sum up, then, we not only do not have any physical account of the action of electric and magnetic fields but we have no physical knowledge of the electromagnetic waves as waves. Only when we introduce conductors such as radio antennas in electromagnetic fields do we obtain any evidence that these fields exist. Yet we send radio waves bearing complex messages thousands of miles. Just what substance travels through space we do not know. The precise and comprehensive account of electromagnetism is the mathematical account. Hence electromagnetic theory is entirely a mathematical theory illustrated by a few crude physical pictures. These pictures are no more than the clothes that dress up the body of mathematics and make it appear presentable in the society of sciences. This fact may disturb or elate the mathematical physicist, depending upon whether the mathematician or the physicist in him is dominant.

No one appreciated the thoroughly mathematical character of electromagnetic theory more than Maxwell. Though he had tried almost desperately to build a physical account of electromagnetic phenomena, in his classic *Treatise on Electricity and Magnetism* (1873) he omitted most of this material and emphasized the highly polished, complex mathematical theory. He himself had once advised a preacher who seemed to be speaking over the heads of his congregation, "Why don't you give it to them thinner?" But his own attempt to thin down the mathematics of electromagnetic fields with an intuitively understandable explanation was unsuccessful. Radio waves and light waves operate in a physical darkness illuminated only for those who would carry the torch of mathematics. And whereas it is possible in some branches of physics to fit mathematical theory to physical facts, about the best one can do in electromagnetic theory is to fit inadequate physical theories to mathematical facts.

The physical mystery presented by electromagnetic waves reminds one at once of a similar problem presented by the force of gravitation. By mathematical laws, the laws of motion and gravitation in the latter case and Maxwell's equations in the former, the two most magnificent and most perfected branches of physical science have been organized and developed. Yet both fields are marked and marred by physical ignorance. The two fields are linked in other striking ways. The laws of attraction and repulsion of electric charges and magnetic poles are mathematically the same as the law of gravitation. Further, in the electric generator mechanical power such as is furnished by the fall of water under gravity is converted to electric power and the latter used to generate electromagnetic waves through electronic devices; conversely, the energy derived from electromagnetic waves may be converted to electrical current and then mechanical power. Hence there seem to be positive reasons for seeking some common link between these two fields. That a physical link may be discovered seems at present remote. Of mathematical links we shall say more in a later chapter.

The success of electromagnetic theory rests squarely upon the method of investigating nature initiated by Galileo and so remarkably developed by Newton. This method, we recall, is to seek basic mathematical formulas relating quantitative or measurable properties of the physical objects and processes and then use these formulas to deduce new knowledge by mathematical reasoning. Of course in concentrating on quantitative relationships one abstracts the seemingly incidental properties of the true phenomenon under study and leaves the phenomenon behind. To those uninitiated into the working of this process it would seem that abstract relationships must be sterile. But we have now several times seen its power. It not only leads to significant deductions about the phenomenon in question, as Maxwell's laws of electromagnetism led to the prediction of radio waves, but the abstract formulations often reveal unsuspected relationships because the quantitative laws turn out to be the same for apparently unrelated phenomena. This statement is nowhere better illustrated than by Maxwell's mathematical discovery that light and electromagnetic waves possess the same physical properties.

Alfred North Whitehead, the late mathematician and philosopher, emphasized the power of abstraction: "Nothing is more impressive than the fact that as mathematics withdrew increasingly into the upper regions of ever greater extremes of abstract thought, it returned back to earth with a corresponding growth of importance for the analysis of concrete fact. . . . The paradox is now fully established that the utmost abstractions are the true weapons with which to control our thought of concrete fact." Those who, admitting the paradox, still deplore that to achieve success the physical sciences have to pay the price of mathematical abstractness must reconsider what it is they would look for in our ultimate scientific exposition of the nature of the physical world. The answer of physicist Arthur Stanley Eddington is that a knowledge of mathematical relations is all that the science of physics can give us. And his colleague, Sir James Jeans, said that the mathematical description of the universe *is* the ultimate reality. The pictures and models we use to assist us in the understanding are, to him, a step away from reality. They are like "graven images of a spirit." We go beyond the mathematical formula at our own risk.

22: THE DIFFERENTIAL CALCULUS

It's enough if you understand the Propositions with some of the Demonstrations which are easier than the rest.

—Newton's advice to prospective readers of his *Mathematical Principles*

THROUGH the study of the motion of projectiles, planets, pendulums, sound, and light the scientific world of the seventeenth century became conscious of the pervasiveness of change. It also had become aware of the extraordinary usefulness of the concept of a function to represent relationships between variables and to deduce new scientific laws. But the continued exploration of changing quantities soon forced the scientists to realize that they required a profounder mathematical tool, and they hastened to create it. This tool is now called the differential and integral calculus. Next to the creation of Euclidean geometry the calculus has proved to be the most original and most fruitful concept in all of mathematics.

It is not difficult to grasp the problem which the exploration of change thrust upon the scientists. In dealing with variables, that is, quantities which change continually, it soon becomes necessary to distinguish between change and *rate of change*. The blood in our veins and arteries changes continually. However, if it were to flow too slowly we would die of suffocation and hunger; on the other hand, if it were to flow too rapidly our hearts would have to work so hard that they might fail under the strain. As one travels, his distance from the starting point continually changes, as does the time that elapses; but what matters in travel is the speed, that is, the rate of change of distance compared to time. At one mile per hour one might never attempt a cross-country trip; at sixty miles per hour, and at the risk only of his neck, a person can make the trip in a few days. As we rise above or descend below the ground the air pressure changes; but if we rise or descend rapidly the sudden change in pressure may cause discomfort or even a burst eardrum. A person traveling in a train at ninety miles per hour may hardly be conscious of the speed, but a sudden increase or decrease in the speed of the train will throw him out of his seat. Thus, though change is a fundamental process, a moment's reflection makes us aware that the rate of change may be far more significant than the fact of change itself.

Let us pursue this notion of rate of change a little further. When a person makes a trip, by automobile, say, his speed is of course important if he is to arrive at his destination in a reasonable time. But in such a situation what matters is average speed. That is, if he plans to travel a hundred miles and if he hopes to make the trip in four hours, then it suffices for him to know that he must travel at an average speed of twenty-five miles per hour. Again, the rate of increase of the population of the United States is of interest to real estate dealers, builders, boards of education, and others. But in this case, too, what matters is the average rate of change. The rate of change per year meets the needs of the interested parties. On the other hand, when an automobile strikes a tree what matters is the speed, or rate of change of distance compared to time, *at the instant the automobile strikes.* Likewise, as a bullet travels through the air, its average speed may be a thousand miles per hour, but what counts when it strikes a person is the speed at the instant of striking. If at that instant the speed is one mile per hour, the bullet will drop to the ground; if it is one thousand miles per hour, the person will drop. In other words, we must distinguish between the average rate of change and the rate of change at a given instant or the instantaneous rate of change, and we must recognize that the latter rate is often the more significant one.

Immediately several difficulties present themselves. What is an instant? It may be difficult to give a good physical definition of an instant, because the very notion of time itself is an elusive one. Nevertheless, the notion of an instant does have some physical meaning. When two objects collide we think of this happening at an instant. A lightning flash is practically instantaneous. We speak of an event happening at six o'clock and refer thereby to an instant. Thus even in common situations we think of and utilize the notion of an instant. Mathematically we have less trouble with the concept of instant. The mathematician thinks of time as a measurable quantity measured, say, in seconds. Then the passage of time is recorded by a number of seconds measured from some event that is represented as happening at zero time. The mathematician has of course idealized physical experience. But having done so he speaks of an instant as something that happens at a value of the variable t which represents time. Thus $t = 2$ is an instant two seconds after the event that the mathematician has selected as happening at zero time. Actually we have used this notion of instant right along, but it is important now to be clear as to its meaning.

If the notion of instant is clear, then the notion of instantaneous speed is not. Here the difficulty is not one of physical reality. The reader who accepts the physical notion of an instant must also and probably does accept the notion of instantaneous speed. For example, a person traveling in an automobile has a speed at each instant during the time he is traveling. Collision with a tree would convince the doubter that he is moving at the instant of the collision. But there is difficulty in stating just what we mean by instantaneous speed, and if we do not know precisely what it means then we certainly shall have trouble in calculating it.

To see the difficulty more clearly let us reconsider what average speed is and how this is calculated. Since speed is the rate of change of distance compared to time, the average speed, which applies over an interval of time rather than at an instant, is the distance traveled during that interval divided by the time. If a person travels ninety miles in three hours, his average speed over the three hours is thirty miles per hour. The concept of average speed, then, is distance divided by time, and this definition permits us to calculate average speed. We are tempted to define and calculate instantaneous speed in the same way. But at an instant zero distance is traveled and zero time elapses. Hence, to define instantaneous speed as distance divided by time leads to the expression 0/0, which is meaningless. Here, then, is the problem. Physically we have every reason to believe that there is such a thing as an instantaneous speed; yet we cannot define it and calculate it mathematically.

But why do scientists bother with instantaneous speeds? Is it to determine the speeds with which automobiles are daily striking trees? The answer is of course that many scientific problems require the use of instantaneous speeds. For example, since an object near the surface of the earth falls with varying speed, to know its speed at any time means of course to know an instantaneous speed. When an object is far from the earth and falls toward it under gravitational attraction, then not only its velocity but its acceleration varies from instant to instant. Kepler's second law of motion says that the planets move with varying speed in their elliptical paths around the sun. A bob on a spring and on a pendulum moves with constantly varying speed and acceleration. A deep investigation of all these motions requires understanding of and calculation of instantaneous speed and instantaneous acceleration. Moreover, the problem which scientists since the seventeenth century have faced is not just that of treating instantaneous speed and acceleration but also instantaneous rates of change of forces, intensities of light and sound, energies, and hundreds of other instantaneous rates of change.

The reader may have noticed, now that his attention is directed to instantaneous speeds, that we worked with such speeds earlier. This is indeed the case, and we may as well confess that we had no right to do so. We relied upon the physical meaningfulness of these rates to speak about them at all, and fortunately we were able to calculate those we needed by relying upon physical arguments. Had we attempted to calculate any of these instantaneous speeds by dividing distance by time we should have run into the meaningless expression 0/0. But the very progress we seek to achieve with the calculus, to say nothing of remedying the deficiencies of the past, will no longer permit the use of such substitutes for mathematical reasoning.

The problem of defining and calculating instantaneous rates, of which speed and acceleration were the most pressing, attracted almost all of the mathematicians of the seventeenth century, and the roster of those who contributed to the subject and achieved limited successes is extensive. The decisive steps were taken by Newton and Gottfried Wilhelm Leibniz. Since the latter is a newcomer to our account let us make our acquaintance with him.

This man of universal gifts and interests, son and grandson of professors, was born in Leipzig in 1646. His early formal education was but a fraction of what he learned by reading for himself. At the age of fifteen he entered the University of Leipzig with the announced intention of studying law and with the unannounced intention of studying everything. There he received the bachelor's and master's degrees in law. His brilliance so excited the jealousy and the envy of his teachers that they never granted him his doctor's degree. This he received in 1666 at the University of Altdorf. Fortunately, an essay on the teaching of law, written soon after he left Leipzig, attracted the Elector of Mainz and he employed Leibniz as a diplomat. During the years in which he acquired his legal training Leibniz was also busily studying mathematics and physics.

Brief sojourns in London and Paris brought Leibniz into contact with great mathematicians and scientists, notably Huygens, who fired him with passionate interest in their problems. Unfortunately, his time was limited, for the necessity of earning a living forced him to continue to serve as a diplomat. In 1676 he was appointed councilor and librarian to the Elector of Hannover, and this position allowed him some leisure. Thereafter he was able to make extensive contributions to mathematics and physics along with profound contributions to law, theology, politics, economics, history, philosophy, philology, and logic. This man, who became a member of several distinguished scientific academies, was a "whole academy in himself." Unlike Newton, Leibniz died neglected.

Let us consider how Newton and Leibniz approached the problem of defining and calculating instantaneous rates. Though there were differences in their approaches we shall ignore them and examine the subject in the form in which it has been standardized in recent years. To start with, let us take the formula that relates the distance an object falls to the time it falls, and let us calculate the instantaneous speed of a ball exactly three seconds after it is dropped. According to Galileo, the relation between the distance the ball falls when measured in feet and the time of fall measured in seconds is

$$d = 16t^2. \tag{1}$$

At the end of three seconds the ball will have fallen a distance d_3, which is obtained by substituting 3 for t in formula (1). That is,

$$d_3 = 144. \tag{2}$$

We shall calculate next the distance the ball falls in the time $3 + h$ seconds, where h represents some additional time interval after the 3 seconds. We therefore substitute the value $3 + h$ for t in (1). Before calculating the new distance let us note that it will be more than d_3. Let us call it $d_3 + k$ where k is the additional distance the ball falls during the additional h seconds. Then

$$d_3 + k = 16(3 + h)^2.$$

Multiplication of $3 + h$ by itself yields

$$d_3 + k = 16(9 + 6h + h^2),$$

and a further multiplication gives the equation

$$d_3 + k = 144 + 96h + 16h^2. \tag{3}$$

The next step is to obtain the average speed in the time interval h. Since k is the distance traveled in the h seconds, we shall find k and then divide by h. To get k we subtract equation (2) from equation (3). This gives

$$k = 96h + 16h^2. \tag{4}$$

Since the average speed during the h seconds is k/h, we divide both sides of formula (4) by h and obtain

$$\frac{k}{h} = \frac{96h + 16h^2}{h}. \tag{5}$$

We note in formula (5) that the average speed, k/h, over the interval of h seconds, is a function of h.

Now for any value of h *other than* 0 we can divide the numerator and denominator of the right side of (5) by h and obtain the simpler expression

$$\frac{k}{h} = 96 + 16h. \tag{6}$$

Thus far Newton and Leibniz had calculated the average speed of the falling body in the interval of time h seconds after the third second of fall. Moreover, since h can be very small and the above algebra still hold, they had obtained the average speed over any small interval of time just after the third second. But the problem they set out to solve was to calculate the speed at the end of the third second, that is, when $h = 0$. One is tempted to let $h = 0$ in (6) and obtain the answer 96. The answer happens to be correct but the reasoning is incorrect. When $h = 0$, (6) does not yield the correct expression for the value of k/h because the derivation of (6) from (5) is not valid in this case. To determine the value of k/h when $h = 0$ we should use (5). But if we substitute $h = 0$ in (5) we see that we obtain $0/0$, the very difficulty in obtaining instantaneous speed that we mentioned at the outset.

The situation is exasperating. The answer we seek is obviously at hand in formula (6) but we cannot use formula (6). Why must these mathematicians be so punctilious? One is tempted to cheat a little and use formula (6). But cheating will not do, not merely on moral grounds but because it will not produce results later. The new idea that Newton and Leibniz contributed comes in at this point. Let us examine (6) when h is not 0 and let us see what happens as h gets closer and closer to 0 in value. For such values of h, formula (6) is valid, and we see from this formula that as h gets closer and closer to 0 the right side of (6) approaches 96. We therefore take 96 to be the speed when $h = 0$, that is, the speed at the end of the third second.

In other words, what we do is consider what happens as h gets smaller and smaller, and see what number the expression $96 + 16h$ approaches as h approaches 0. This number, which is called the *limit* of $96 + 16h$ as h approaches 0, is taken to be the value of the speed in question. Now it is true that 96 is also the value of the expression $96 + 16h$ when h is 0. That the two results are the same is due only to the fact that we used a very simple function to start with. The verbal expression of the general fact about speed at an instant that we have just illustrated is this: The speed at an instant is the limit approached by the average speed k/h as h approaches 0. Physically, this means that we take as the speed at the end of the third second the speed approached by the average speeds over smaller and smaller intervals of time commencing at the end of the third second.

The person who meets this concept of instantaneous speed for the first time gets the impression that despite the avowed renunciation of an attempt to cheat, cheating has been done nevertheless.

It appears as if, after stressing that we could not use the expression (6) to obtain the value of k/h when h is 0, we inserted some mumbo jumbo about limits and then used expression (6) anyway. We did use the expression (6) but the *manner* in which we used it is all important. If we are to think correctly in the calculus and avoid serious errors when the algebraic expressions that represent k/h are not so simple as in (5), we must understand and respect the distinction between the limit of a function of h as h gets smaller and smaller and the value of the function when h is 0. The limit concept is of course a complex and subtle one, and the way to get to know such an idea is like the way in which one gets to know a person: one must live with it.

To convince ourselves that the limit concept is not much ado about nothing, let us consider a more difficult example. We know from our discussion of the motion of the bob on a spring that the function which relates displacement and time, when the amplitude is 1 and the period is 2π seconds, is

$$y = \sin t. \tag{7}$$

Let us calculate the velocity of this bob at the instant $t = 0$. The value $t = 0$ need not, of course, be the instant at which the bob is set into motion. Rather, it may be pulled down, released, and allowed to execute its sinusoidal motion. If, however, we begin to measure time from the instant the bob passes through the rest position and is moving upward, then (7) is the correct formula, and the velocity we seek at $t = 0$ is the velocity of the bob at that instant.

We shall follow the procedure used in steps (2) to (5); this time, however, the function is given by (7) rather than (1). The first step is to calculate y at $t = 0$. For the function (7), $y = 0$ at this instant. We introduce an interval h of time beginning at the instant 0, and we calculate y at the instant $t = 0 + h$. This new value of y will be the old value, 0, plus some increment k. Then

$$0 + k = \sin (0 + h) = \sin h. \tag{8}$$

The analogue of step (4) above is to subtract from equation (8) the equation

$$0 = \sin 0.$$

In this case the operation is trivial, the result being

$$k = \sin h. \tag{9}$$

We now divide both sides of this equation by h to obtain the average speed in the h seconds. Then

$$\frac{k}{h} = \frac{\sin h}{h}. \tag{10}$$

To obtain the speed at $t = 0$, that is when $h = 0$, we might try to cheat and let $h = 0$ on the right side of formula (10). But when $h = 0$, $\sin h = 0$, and the right side yields $0/0$. Unfortunately one cannot repeat the trick here that allowed us to go from formula (5) to formula (6), that is, we cannot divide the numerator and denominator of the right side of (10) by h to obtain a simpler expression. We are forced, then, to consider the limit or number approached by the right side as h approaches 0.

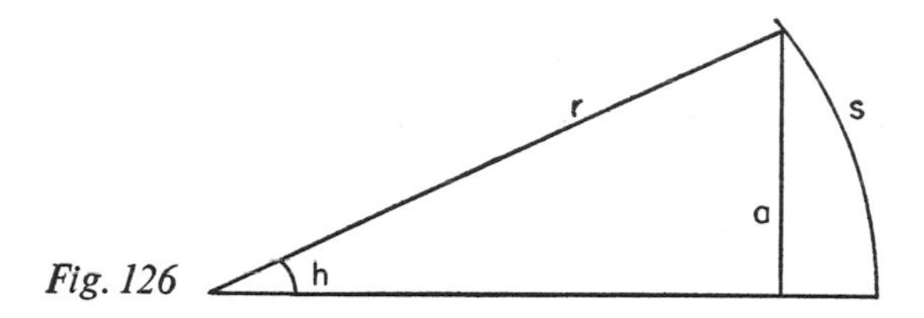

Fig. 126

A minor argument will show us what this limit is. We can think of the quantity h as the size of an angle (fig. 126). Then $\sin h = a/r$. However, the angle h, if we measure it in radians, is s/r. Then

$$\frac{\sin h}{h} = \frac{\frac{a}{r}}{\frac{s}{r}} = \frac{a}{s}.$$

Let us rely upon visualization to see that as h gets smaller and smaller, the lengths a and s differ less and less. Hence the limit of $\sin h/h$ as h approaches 0 is 1. Of course the argument given here to obtain the limit is not mathematical, since we relied upon what our eyesight suggests. But let us not lose sight of the main point for the sake of a rigorous argument. Our objective was to determine the limit of the right side of formula (10) as h approaches 0. We see now that this limit is 1, and we may conclude therefore that the velocity of the bob on the spring at the instant $t = 0$ is 1.

The problem of determining the instantaneous velocity by observing the trend of average velocities over intervals neighboring the instant in question may be compared, if we allow ourselves to use a very loose analogy, with trying to determine where a falling bomb will land. We have two choices. We can study the trajectory described by the bomb as it falls, and try to predict, on the basis of the path pursued by the bomb while it is still in the air, where it will land. This choice is analogous to calculating average velocities for smaller and smaller h and then trying to predict the limit. Alternatively, we can try to observe where the bomb lands. But to do this effectively one should be at the spot, and this procedure, which is analogous to substituting 0 for h in the expression for k/h, leads only to disaster for the investigator. The experienced mathematician prefers to calculate limits rather than be blown to bits.

To appreciate the full generality of the process of computing instantaneous rates we must go a step further. Let us consider the function $y = 16x^2$, where y and x are any variables so related, and let us ask for the instantaneous rate of change of y with respect to x at the value x_1 of x. If x is time t, if y is the distance d, and if x_1 is 3, then our present problem reduces to the earlier one. To find the instantaneous rate of change of y with respect to x at $x = x_1$ we first calculate y at $x = x_1$. This value of y, denoted by y_1, is of course

$$y_1 = 16x_1^2. \tag{11}$$

We now increase the value of x by an arbitrary amount h and calculate the value of y for the value $x_1 + h$ of x. This new value of y will be the old one plus some quantity k. Then

$$y_1 + k = 16(x_1 + h)^2.$$

By carrying out the indicated multiplications, one obtains the equation

$$y_1 + k = 16x_1^2 + 32x_1h + 16h^2. \tag{12}$$

The change k in y, which is due to the change h in x, is now obtained by subtracting (11) from (12). Thus

$$k = 32x_1h + 16h^2. \tag{13}$$

To get the average rate of change of y with respect to x in the interval h we divide both sides of (13) by h and obtain

$$\frac{k}{h} = \frac{32x_1h + 16h^2}{h}. \tag{14}$$

Fortunately, we can divide numerator and denominator on the right side of (14) by h, and obtain

$$\frac{k}{h} = 32x_1 + 16h; \tag{15}$$

and now the crucial step is to see what happens on the right side of (15) when h approaches 0. In this case the answer is obvious. As h approaches 0 in value, so does $16h$, and the limit is $32x_1$. Thus the instantaneous rate of change of y with respect to x at $x = x_1$ is $32x_1$.

Let us introduce some notation. To indicate the instantaneous rate of change of y with respect to x, Newton used $\dot{y}$. Leibniz used the suggestive but also misleading notation dy/dx. This suggests that the instantaneous rate comes from considering an average rate which is indeed a quotient, the quotient k/h in our notation. In Leibniz' notation dy is our k and dx is our h. However, the Leibnizian notation is misleading in that it represents the instantaneous rate in the form of a quotient, whereas the instantaneous rate is not a quotient but the limit approached by a quotient. We shall use the Newtonian notation and write

$$\dot{y} = 32x_1. \tag{16}$$

Formula (16) has several valuable things to tell us. The quantity x_1 was any value of x. Hence in the steps from (11) to (16) we obtained the instantaneous rate of change of y with respect to x at any value of x. We may emphasize this fact by dropping the subscript and writing

$$\dot{y} = 32x. \tag{17}$$

Thus we have calculated the rate of change of y with respect to x for an infinite number of values of x in one operation. In fact (17) is a new formula that can be combined with other formulas to produce new conclusions. It is called the *derivative* of $y = 16x^2$. As a check on (17) let us note that for $x = 3$, $\dot{y} = 96$, and this result agrees with the conclusion derived from formula (6).

The second valuable implication of (16) or (17) is that the result holds regardless of the physical meaning of y or x. As usual, the mathematics proper treats only pure numbers or pure spatial relationships. Hence we can apply the result to thousands of physical situations in which the original function $y = 16x^2$ applies.

Moreover, the process we used to calculate (17) can be applied to any function: *we can calculate the rate of change of one variable with respect to the other at a value of the second one by the same mathematical procedure* we used to calculate the rate of change of y with respect to x when $y = 16x^2$. For example, if v represents speed and t time, we can calculate the rate of change of speed compared to time at an instant; this instantaneous rate of change of speed is instantaneous acceleration. Thus from the formula $v = 32t$ we can derive the formula $\dot{v} = 32$, and $\dot{v}$ is the acceleration. As another example, the pressure of the atmosphere varies with height above the surface of the earth. Given the formula that relates pressure and height, we can calculate the rate of change of pressure compared to height *at any given height*. Or, if the variable P represents the price level of a commodity and t represents time, we can compute the rate of change of price compared to time *at any instant*. Thus our method enables us to define and calculate thousands of significant and useful rates of change of one variable with respect to another *at a value of the second variable*. All such rates are referred to as instantaneous rates, despite the fact that time may not be one of the variables involved, because the original calculus problems of speed and acceleration did involve time and were concerned with rates at an instant of time.

A rather interesting application of the process of finding derivatives or rates of change, which at the same time may shed more light on the concept, is to find the instantaneous rate of change of the area of a circle with respect to the radius. The formula that relates area and radius is, of course,

(18) $$A = \pi r^2.$$

We should now carry out the process previously pursued in steps (11) to (17) for the function $y = 16x^2$. There is, however, no need to repeat all the details. The two functions are practically alike, the only difference being that π occurs in formula (18) where the number 16 occurs in $y = 16x^2$. Of course h in the present case is an increase in the length of the radius (fig. 127) and k is the increase in area that results from the increase h in r. Instead of ob-

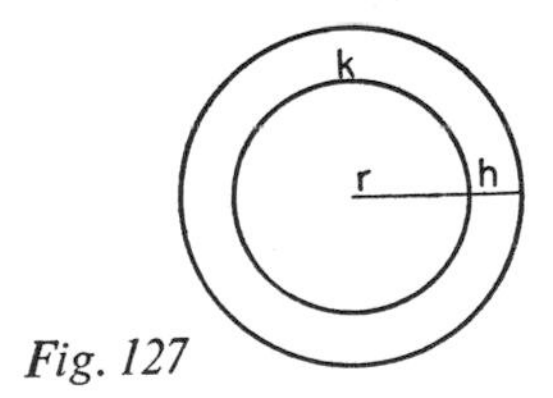

Fig. 127

taining $\dot{y} = 32x$ for the derivative we now obtain

(19) $$\dot{A} = 2\pi r.$$

The result (19) is of interest in that it says that the rate of change of area with respect to radius is the circumference of the circle. This result is intuitively clear, for as the radius increases, one might say that "successive" circumferences are added to the area.

With the preceding result at hand one might guess that the rate of change of volume of a sphere with respect to the radius is the area of the surface of the sphere. And it is the case for the formula

$$V = \frac{4}{3}\pi r^3$$

that

$$\dot{V} = 4\pi r^2,$$

which is the formula for the area of the surface of a sphere.

The importance of the derivative lies to a large extent in its uses in more advanced branches of mathematics, some of which we shall examine in later chapters. However, a direct major application of the concept is found in a class of problems called maxima and minima. We have had a few occasions to consider such problems. For example, we found the rectangle with maximum area among all rectangles with a given perimeter, the maximum range of a projectile, and the maximum height reached by a ball thrown up into the air. In these instances we were able to obtain the solution without requiring the calculus. But usually the calculus must be employed.

To see the role of the derivative in such problems, let us reconsider the problem of finding the maximum height reached by a ball that is thrown straight up with an initial velocity of 128 feet per second. We may recall (chapter 12) that the formula for the height d at any time t, is

$$d = 128t - 16t^2. \qquad (20)$$

(We use d here for the height instead of h because in this chapter h has been used to represent an increase in the independent variable.) We should note now that we can derive the formula for the velocity from (20) merely by finding the instantaneous rate of change of d with respect to t. We shall not bother to carry out the details. As a matter of fact, we derived the result by a physical argument in chapter 12—see formula (10) there. The formula for the velocity is

$$\dot{d} = v = 128 - 32t. \qquad (21)$$

The important point insofar as maxima and minima are concerned is that the velocity is 0 at the highest point on the ball's path; otherwise the ball would continue to rise. We then set $v = 0$, find the value t_1 of t at which v is 0, and then substitute this value of t in formula (20). Thus we are able to obtain the maximum height of the ball.

The argument that the velocity must be zero at the highest point is a physical rather than a mathematical one. Moreover, this argument certainly does not apply to maximum and minimum problems in which velocity is not involved. But physical thinking has given us a most important lead. The formula for velocity is the derivative of the formula for height, and the velocity is zero when the height is a maximum. One might therefore guess that the derivative of any function is zero when the function is a maximum, and possibly also when the function is a minimum. This guess can be established by a rigorous mathematical argument. Hence we now have a purely mathematical method for finding maximum and minimum values of any function.

To illustrate the use of the method of maxima and minima in finding important physical laws, let us return to the behavior of light. When light travels from one medium to another the key to its behavior is the law of refraction, namely,

$$\frac{\sin i}{\sin r} = \frac{v_1}{v_2}, \tag{22}$$

where v_1 and v_2 are the velocities in the respective media (chapter 7). As we remarked in another connection, the law of refraction eluded hundreds of top-notch investigators until its discovery by Snell and Descartes, in the seventeenth century. It is of course understandable that the discovery of this law was so long in coming, because it presupposes knowledge about the velocity of light in different media, and even with two fixed media, the relationship between the angles i and r is not readily conjectured, let alone established.

But with the calculus the derivation of the law of refraction is almost immediate. Of course this law is a physical one, and new physical facts are not derivable from mathematics alone. One must also have physical facts to start with. The physical fact we shall use here is Fermat's principle of least time. The Greek mathematician and engineer Heron had already observed that light takes the shortest path in going from one point to another after reflection in a mirror. Since this motion takes place in one medium and the velocity of light is constant in any one (homogeneous) medium, the shortest path requires the least time. On the basis of this evidence and other evidence Fermat asserted that light, in going from one point to another, always takes the path requiring least time. Let us use this physical principle to derive the law of refraction.

Suppose that light travels from point P in the first medium to point Q in the second one (fig. 128, p. 378). We know that the light will travel along a straight line to some point O on the boundary between the two media, enter the second medium, and then travel in a straight line from O to Q. If we could determine the location of the point O, we would also know the angles i and r. The light travels the distance PO with velocity v_1 and OQ with velocity v_2. Hence the time the light takes to travel from P to Q is

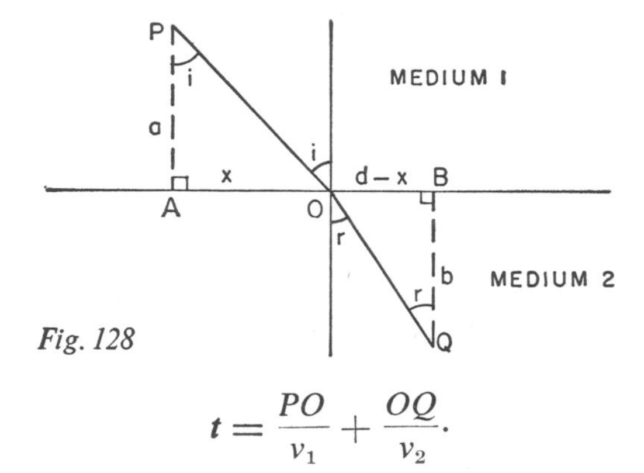

Fig. 128

$$t = \frac{PO}{v_1} + \frac{OQ}{v_2}. \tag{23}$$

Since we do not know the location of O let us denote by x its distance from A, the foot of the perpendicular from P to the boundary between the two media. The positions of Q and P are fixed. If, therefore, we drop a perpendicular from Q to the boundary, the distance AB is fixed. We call this distance d, and the distance OB is therefore $d - x$.

Let us label the fixed distances PA and QB, a and b respectively. Then by the Pythagorean theorem

$$PO = \sqrt{a^2 + x^2}$$

and

$$OQ = \sqrt{b^2 + (d - x)^2}.$$

If we substitute these values in (23) we obtain

$$t = \frac{\sqrt{a^2 + x^2}}{v_1} + \frac{\sqrt{b^2 + (d - x)^2}}{v_2}. \tag{24}$$

We now have the time as a function of x. If the time of travel from P to Q is to be least, then the rate of change of t with respect to x must be zero for that value of x which fixes the path requiring least time.

The function (24) is more complicated than ones we treated earlier; hence we cannot find the rate of change of t with respect to x in one or two steps. To pursue the methods of the calculus effectively one must learn techniques for handling complicated functions. The acquisition of technical skills is unfortunately a necessity in all branches of mathematics, and for that matter in any science, art, or trade. However, students of the calculus learn to differentiate, as the process of finding rates of change is called, almost mechanically. A trained student can write down at once that

$$\dot{t} = \frac{x}{v_1 \sqrt{a^2 + x^2}} - \frac{d - x}{v_2 \sqrt{b^2 + (d - x)^2}}.$$

We learned earlier that t is a minimum when $\dot{t} = 0$. Hence let us seek the implications of the equation

$$\frac{x}{v_1 \sqrt{a^2 + x^2}} - \frac{d - x}{v_2 \sqrt{b^2 + (d - x)^2}} = 0.$$

A glance at figure 128 shows at once that

$$\sin i = \frac{x}{\sqrt{a^2 + x^2}} \text{ and } \sin r = \frac{d - x}{\sqrt{b^2 + (d - x)^2}}.$$

Thus we have

$$\frac{\sin i}{v_1} - \frac{\sin r}{v_2} = 0,$$

which gives at once the law of refraction as stated in (22). In a matter of minutes we have derived a law that was sought for almost two thousand years.

In the applications of the concept of instantaneous rate of change thus far described we started with the formula relating two variables and then found the rate of change. From the standpoint of mathematics proper we could also consider the question, Given the rate of change of one variable with respect to another, what is the formula relating the two variables? Is this question worth considering? In view of the fact that the most weighty laws of nature are expressed as formulas, the answer is obviously yes. The value of the process of finding the formula from a knowledge of the rate of change of the variables involved depends, then, upon knowing some important rates of change. Fortunately this information can be readily obtained in many physical phenomena. From it we proceed to the formula and to the solution of many problems. The significant uses of this procedure will be taken up later; here let us just become a little clearer as to what is involved.

Suppose we were interested in finding the formula that relates the distance an object falls and the time it falls. If the object is near the earth, its acceleration, as Galileo discovered by simple experiments, is constant. That is, the rate of change of speed compared to time is 32 ft/sec^2 at each instant of time. In symbols, if a stands for acceleration,

$$a = 32. \tag{25}$$

Since a is the instantaneous rate of change of speed compared to time, that is, $a = \dot{v}$, we can think of it as coming from a formula relating speed v and time t. If we could find this formula we would have the expression for the speed in terms of the time. But the mathematician knows, as a purely mathematical fact, that the derivative of $v = 32t$ is $\dot{v} = 32$. Hence he knows that if $\dot{v} = 32$ then

$$v = 32t. \tag{26}$$

Formula (26) gives us the speed at each instant the body falls in terms of the time it has been falling. This is desirable information, but we are seeking the relation between distance and time. However, the speed is the instantaneous rate of change of distance compared to time, that is, $v = \dot{d}$. Let us therefore regard formula (26) as the rate of change derived from the formula relating distance and time, and let us consider once more the process of reversing the rate of change. Again experience with finding rates of change tells us that (26) is the rate of change of $d = 16t^2$. Hence the formula relating the distance the body falls to t, the number of seconds, is

$$d = 16t^2. \tag{27}$$

Thus by starting with the physical information contained in (25), and by twice reversing the process of finding rates, we have found the formula relating the distance and time of a falling body.

Though we have used very familiar facts to illustrate the process of reversing rates of change, we should not presume that we see at once the implications of what we have done. We started with a formula for acceleration and by *purely mathematical processes* derived the formula for velocity and then for distance. When, earlier, we treated falling bodies we had to supply a physical argument to derive the velocity from the acceleration and another physical argument to derive the distance from the velocity. Given the calculus, we can dispense with these two physical arguments. Mathematical deduction takes over the role of physical reasoning. Thus we see how with the enlargement of mathematical ideas and techniques the power of science to deduce physical knowledge is strengthened. It was the aim of Galileo and Newton to deduce all the phenomena of nature from two or three basic physical principles by the application of mathematical reasoning. The calculus is the major tool in the realization of this aim.

To be able to apply freely the process of reversing rates of change one must learn a great deal of technique about finding rates and about the inverse process of finding the original formulas from known rates. This technical knowledge is gained in the usual courses in the calculus, where, unfortunately, it is often presented as the substance of the calculus rather than what it truly is, the incidental mechanics of larger and more significant processes and applications. As in algebra courses, technique has been overemphasized to the exclusion of ideas and their uses.

The technique of the calculus is straightforward and requires rather little thinking for its mastery, but the opposite is true of the limit concept on which the calculus rests. We have spoken rather glibly of the limit as the number approached by a function. For example, to obtain instantaneous speed we found the limit approached by the average speed, which is a function of h, as h approaches zero. In using the word *approach* we relied upon its intuitive meaning. But far more must be said if the notion is to have the precision on which mathematics insists. Let us suppose that the average speed takes on the values 1, $3/4$, $5/8$, $9/16$, $17/32$, . . . , as h approaches zero. What number are these average speeds approaching? They certainly are approaching zero because they are clearly getting smaller, and yet do not go below zero in value. Would zero then be the limit? The answer is no, because these speeds are always larger than $1/2$ and it is clear from the rule of formation that they will remain so. Hence the notion of approach that we apparently want in connection with limits is that the function values —average speeds in our application—should come as close as one may wish to the limit. This degree of closeness must be specified accurately, for the word *close* is in itself rather useless. A meteor is close if it comes within a few miles of the earth. A bullet is close if it passes within one inch of one's head. In the case of instantaneous speed, a more accurate statement of what we mean by the average speeds approaching $\frac{1}{2}$ is that if we take the intervals of time sufficiently small, we can make the average speed over those intervals differ from $\frac{1}{2}$ by as little as we please. Even this formulation, however, is not sufficiently precise. We shall not present a rigorous definition of the limit concept, but it is desirable to know that mathematics has refined it so as to eliminate all vagueness and ambiguity.

The difficulty in attempting to present a rigorous as well as clear statement of the theory of limits is inherent in the subject. The concept is subtle and elusive. If the reader has found some difficulty in grasping it he may be less discouraged when he is told that it eluded even Newton and Leibniz.

No one can read the details of their writings on the calculus without being amazed by the number of times they changed their explanation of the limit concept and still failed to get it right. Some of these explanations contained outright contradictions of earlier ones. It is fair to say that though both men had their hands on a sound idea they could not grasp it securely. Newton came close to the limit concept, somewhat as we have presented it, but, because he did not understand it very well, could not formulate it accurately. Leibniz actually thought of $\dot{y}$ as a quotient of dy and dx, which quantities he considered to be the values of k and h when h was "sufficiently small." He called dy and dx infinitesimals and regarded them as neither zero nor numbers of actual size such as $\frac{1}{2}$. He believed that there was a realm of quantity intermediate between zero and definite nonzero numbers. Carlyle tells in his *Frederick the Great* that Leibniz used to give this explanation of the infinitesimal or infinitely little quantity to Queen Sophia Charlotte of Prussia. Once she replied that she needed no further instruction on the subject because the behavior of her courtiers had made her thoroughly familiar with it. For about one hundred years after Leibniz' work on the calculus, the mathematicians who adopted his views referred to the calculus as the Infinitesimal Calculus.

Because of the difficulties in the very foundation of the calculus, conflicts and debates on the soundness of the whole subject were prolonged. Many contemporaries of Newton, among them Michel Rolle, who contributed a now famous theorem, taught that the calculus was a collection of ingenious fallacies. Colin Maclaurin, after whom another famous theorem was named, decided that he would found the calculus properly, and he published a book on the subject in 1742. The book was undoubtedly profound but also unintelligible. One hundred years after the time of Newton and Leibniz, Joseph Louis Lagrange, one of the greatest mathematicians of all times, still believed that the calculus was unsound and gave correct results only because errors were offsetting each other. He, too, formulated his own foundation for the calculus, but it was incorrect. Near the end of the eighteenth century D'Alembert had to advise students of the calculus to keep on with their study: faith would eventually come to them. This is not bad advice to beginners, who cannot be expected to see a difficult idea at once, but it is no substitute for rigor and proof.

Attacks came from all quarters. Because mathematics was the principal tool with which the new science of mechanics was invading the heavens, the privileged domain of religion, some of the strongest criticisms came from religious leaders. Of these the most famous is the highly original philosopher Bishop George Berkeley.

He proceeded to demolish mathematics and was shrewd enough to attack the heel of the Achilles. Since the fundamental concept of the calculus was not clearly understood and therefore not well presented by either Newton or Leibniz, Berkeley was able to enter the fray with justification and conviction. In *The Analyst* (1734), addressed to an infidel mathematician, he did not mince words. Instantaneous rates of change of functions he condemned as "neither finite quantities, nor quantities infinitely small, nor yet nothing." These rates of change were but "the ghosts of departed quantities. Certainly . . . he who can digest a second or third fluxion [Newton's technical name for instantaneous rate of change] . . . need not, methinks, be squeamish about any point in Divinity." To account for the fact that the calculus gave correct results, Berkeley, like Lagrange, argued that somewhere errors were compensating for each other. Berekley's attack was based on the *tu quoque* principle. If we theologians speak nonsense then so do you mathematicians.

About a century and a half after the creation of the calculus, the mathematician Augustin Louis Cauchy, a foremost nineteenth-century French mathematician, finally gave a definitive formulation of the limit concept that removed doubts as to the soundness of the subject. His formulation has withstood the test of time.

The history of the controversies about the calculus is most illuminating as to the growth of mathematics. Here was a subject whose soundness was doubted by the greatest mathematicians of the eighteenth century, yet it was not only applied freely, but some of the greatest developments in mathematics, differential equations, the calculus of variations, differential geometry, potential theory, and a host of other subjects comprising what is now called analysis were developed and explored by means of the calculus. Why did the mathematicians adopt this illegitimate child? The answer is that it proved so immensely useful in the exploration of nature that their hearts were touched even though their minds remained critical. They had an idea that made physical sense, and, since mathematics and physical science were closely intertwined and even identified in the seventeenth and eighteenth centuries, they were not greatly concerned about the lack of mathematical rigor. One might say that in their minds the end justified the means.

Whether or not this is a sound ethical principle we should be grateful that these men acted as they did. The calculus is the greatest single idea of modern mathematics; yet it might have been lost to us forever had the mathematicians of that age been too concerned with rigor. We know now that even in mathematics intuition and physical thinking produce the big ideas and that the logical perfection of these ideas must come afterward. We also see more clearly today that the pursuit of absolute rigor in mathematics is an unending endeavor, calling for patience. The understanding and mastery of nature must be sought with the best tools available.

23: THE INTEGRAL CALCULUS

As God calculates, so the world is made.

—LEIBNIZ

ONE of the fascinating facts about the development of mathematics is that an idea that is created to solve one type of problem often solves another which on superficial examination seems to be totally unrelated. This turn of events occurred in the history of the calculus. The concept of the derivative, or the instantaneous rate of change of one variable with respect to another, was introduced to treat velocity and acceleration. But the very same concept solves another class of problems that had baffled mathematicians for two thousand years; namely, the problems of finding the areas of figures bounded by curves and the volumes of figures bounded by surfaces.

The few results on areas and volumes of curved figures that were available to seventeenth-century Europe had been obtained by Euclidean geometry. Though this subject is one of the most impressive intellectual creations of mankind, from the standpoint of mathematical power it is a very limited achievement. The axioms and therefore the theorems of Euclidean geometry state properties primarily about straight lines. As a consequence, even such problems as finding the area and circumference of a circle are difficult to solve with the techniques of that subject. To obtain the theorem

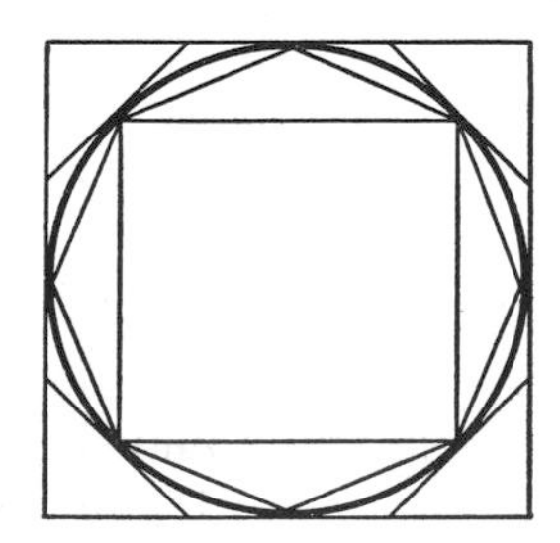

Fig. 129

that the area of a circle is a constant times the square of the radius, one must first approximate the circle by inscribed and circumscribed regular polygons of more and more sides (fig. 129). Then, after proving some relevant facts about these polygons and using the fact that the circle lies between any inscribed polygon and any circumscribed polygon, one obtains the desired theorem about the circle.

The mediation of polygons to treat the circle is necessitated by the fact that polygons are figures bounded by line segments. It is easier to obtain theorems about such figures, after which one may take the next step of deriving rigorous results about the circle. The entire method, which is due to Eudoxus, the first of the great theorists in astronomy and one of the founders of Euclidean geometry, is often referred to as the method of exhaustion, not only because it is sufficiently complex to exhaust the reader but because the successive inscribed polygons gradually exhaust the circle's area.

The proof that the area of a circle is a constant times the square of the radius does not really complete the task of finding the area of a circle, because one still does not know the value of that constant. To calculate it one must resort again to the inscribed and circumscribed polygons. In view of the fact that the Greeks had only a clumsy arithmetic and practically no algebra at their disposal, the determination of π, the constant in question, required the genius of Archimedes. His famous result is that π lies between $3\frac{1}{7}$ and $3\frac{10}{71}$. Incidentally, this result is so remarkable in view of the ill-devised arithmetic available at the time that historians are still uncertain as to how Archimedes obtained it.

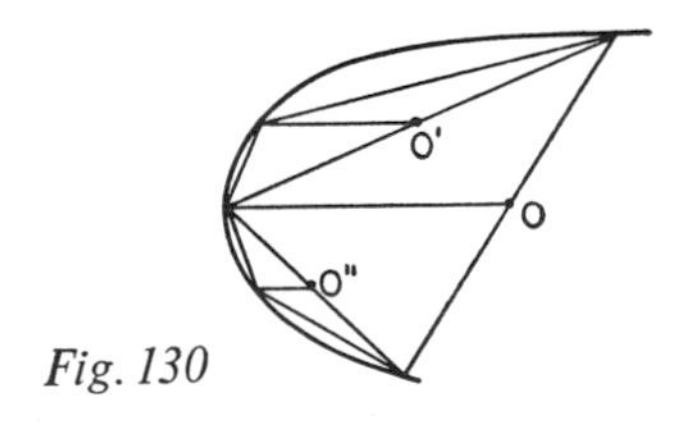

Fig. 130

Euclidean geometry suffers further, as Descrates pointed out, from lack of method. When Archimedes sought the area bounded by an arc of a parabola and a chord (fig. 130), he could no longer use inscribed or circumscribed regular polygons. He was obliged in this case to use triangles and approximate the area more and more closely by adding smaller and smaller triangles that gradually exhausted the area in question.

By the seventeenth century the questions raised about even such familiar curves as the conic sections and the trigonometric curves, to say nothing of the newly discovered curves such as the cycloid and others we shall not have occasion to introduce, called for far more effective methods of calculating areas, volumes, lengths of curves, and other quantities we shall mention later. As in the case of the problem of finding instantaneous velocities and accelerations, numerous limited methods were invented by the seventeenth-century mathematicians, but again the decisive steps were made by Newton and Leibniz, more especially the latter in this present problem.

The Newton-Leibniz method, like the Euclidean, begins by approximating a desired area by rectilinear, that is, straight-line, figures, and then proceeds to obtain it by using the notion of limit already introduced in the differential calculus. The curvilinear area proves to be the limit approached by the areas of rectilinear figures much as the circle is approached by the inscribed and circumscribed regular polygons and the parabolic area is approached by inscribed triangles. In fact the Greek method of exhaustion is often described as a limit process. Actually, this description is inaccurate, though with hindsight it is possible to see the germ of the limit concept in the work of Eudoxus and Archimedes. In addition to introducing the limit concept explicitly, Newton and Leibniz realized that polygons and triangles as approximating figures would not work for all problems of area, and they resorted to rectangles. Finally, Newton and Leibniz took advantage of algebra, coordinate geometry, and the differential calculus, all created in their own century, and arrived at a general method of finding areas.

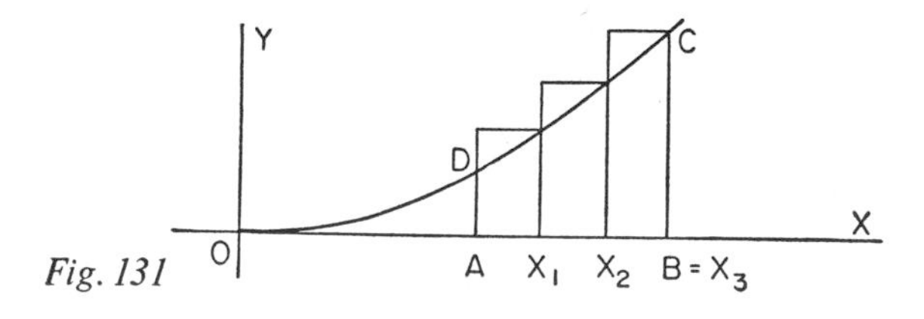

Fig. 131

To follow their thinking let us choose the problem of finding the area *ABCD* of figure 131. This area is bounded by a curve only on top; however, the solution of the problem will not only illustrate the Newton-Leibniz method but provide the basic method that is applicable to more complicated areas.

One begins by subdividing the length *AB*, which we shall suppose extends from $x = 3$ to $x = 5$, into a number of equal parts (we have used three for simplicity of figure) and constructs the rectangles shown in the figure. The sum of these three rectangles is larger than the area we seek but it is an approximation to that area.

One can improve on the approximation by making the bases of these rectangles a little smaller. Figure 132 shows what happens when the middle rectangle of figure 131 is subdivided into two. Following the construction of using the y-value of the point of subdivision as the height of a rectangle, one now has eliminated the shaded area from the sum of the rectangles. The sum of the four

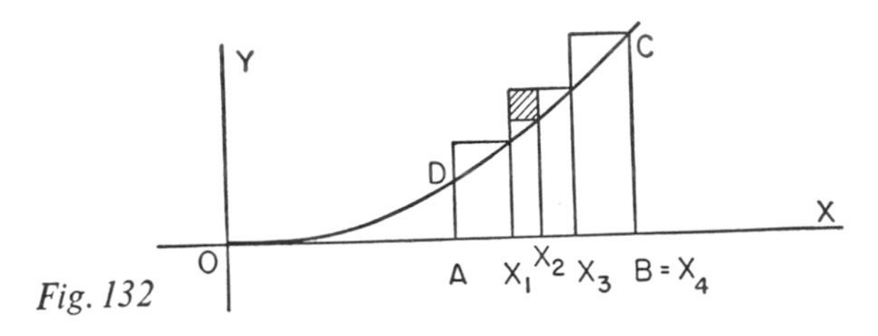

Fig. 132

rectangles is, then, a better approximation to the area *ABCD* than the sum of the original three. The important point to note is that one can continually improve the approximation by using more and more rectangles with, of course, shorter bases. Hence it is intuitively clear that the area we seek is the limit approached by the sum of the rectangles as the number becomes larger and larger.

Thus far the Newton-Leibniz method is very much like that of the Greeks; the only difference is that it uses rectangles whereas the Greeks used regular polygons, triangles, or other simple figures to suit the area being sought. At the moment the decision to use rectangles seems to possess no merit.

The next step is to translate the problem into algebraic terms. To be general, let us label the points that subdivide the interval *AB*, $x_1, x_2, \ldots, x_{n-1}$, and let x_n be the x-value of the point *B*. Thus in figure 132, the points of subdivision are labeled x_1, x_2, and x_3, and the point *B* is denoted by x_4. However, if we should introduce 100 rectangles, then the points of subdivision would be x_1, x_2, . . . , x_{99}, and B would be x_{100}. The value of n in this case would be 100. We denote the height of the first rectangle by y_1, of the second by y_2, and so on. Let the length of the base of each rectangle be denoted by h. We can now state that the sum of the areas of all the rectangles is

$$y_1h + y_2h + \cdots + y_nh. \tag{1}$$

The mathematician, with his characteristic physical laziness, prefers to write this expression as

$$\sum_{i=1}^{n} y_ih, \tag{2}$$

which means no more than (1), the $\sum$ being the Greek capital letter sigma, here standing for sum. The notation $i = 1$ and the superscript n indicate that the sum in question is what one would obtain by letting i be 1, 2, . . . , n in y_ih, and then adding. Of course mathematicians could have used the Latin letter S instead of the $\sum$, but then the symbolism would look less impressive, and there is no point to learning mathematics if one cannot impress his fellow man.

Let us keep in mind that the sum (1) or (2) is only an approximation to the area we are seeking. What we really want is the limit approached by (1) or (2) as h approaches 0 in size and, consequently, as the number of rectangles increases. The notation that indicates the desired limit is

$$\lim_{h \to 0} \sum_{i=1}^{n} y_ih, \tag{3}$$

wherein lim is an abbreviation for limit and the symbols $h \to 0$ indicate that we seek the limit as h approaches 0. The expression (3) is usually written in the somewhat briefer notation

$$\int_{2}^{5} y\,dx \tag{4}$$

wherein $\int$, called the integral sign, is actually an elongated capital S, here denoting the *limit of a sum* and the $y\,dx$ is *symbolic* of the rectangles whose bases are small intervals of the x-axis, denoted by dx, and whose heights are y-values. The 3 and 5 alongside the integral sign are the x-values of A and B, and they indicate the range of x over which the area is to be taken.

So far we have done no more than make a simple concept difficult. We have come to view the simple area *ABCD* as a limit approached by a set of rectangles that in themselves only approximate the area. What is gained by this more complicated view? Here we are reminded of a famous remark about mathematics to the effect that it distorts a problem beyond all recognition and then solves it. Now that we have distorted our usual understanding of area we can show how Newton and Leibniz obtained it.

The Newton-Leibniz prescription for finding the exact area *ABCD* is the following. The area is bounded above by an arc of a curve. We are supposed to know the equation of that curve, a not unreasonable expectation in view of what Descartes and Fermat have taught us. Suppose, then, that the equation is $y = x^2$. *Find the function whose derivative, or rate of change, is* $y = x^2$*; substitute 5 for x in this new function; substitute 3; subtract the second result from the first. The number obtained is the area* ABCD. Without pretending to understand in the slightest the rationale of this procedure, let us carry it out.

The function whose derivative or rate of change is $y = x^2$ is $y = x^3/3$. Here, as we pointed out in the preceding chapter, we must call upon the knowledge of functions and their derivatives that the professional mathematician builds up as his stock in trade. (Of course the reader could find the derivative of $y = x^3/3$ by the process we examined in the preceding chapter and check that it is $y = x^2$.) To follow the Newton-Leibniz procedure we now substitute 5 for x in $y = x^3/3$ and obtain $^{125}/_3$. We substitute 3 and obtain $^{27}/_3$. The difference is $^{98}/_3$ and this is the area *ABCD*.

The procedure for calculating the area is simple enough, but how did Newton and Leibniz arrive at it? Mathematicians do not pull rabbits out of hats even though the presentations in textbooks often seem to suggest just this type of conjuring. We may be sure that underlying every mathematical theorem or procedure there is a sound, readily understandable, though not necessarily readily discoverable, intuition. And there is in this case, too, though the geniuses of Newton and Leibniz were required to ferret it out. Let us consider the problem from a slightly different point of view. We may imagine the area to be generated by a vertical line that starts from the position $x = 3$ and moves to the right (fig. 133). Suppose it reaches the position P; then the area thus far swept out is a function of x, the x-value of P. Now suppose that the line moves a

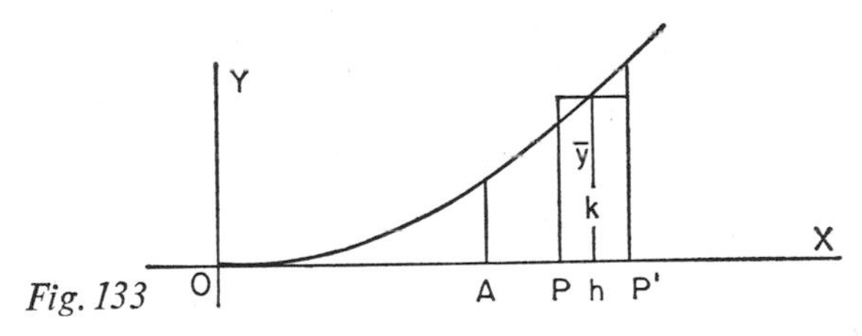

Fig. 133

little farther to the right to P' so that the x-value of P′ is $x + h$. Then the area swept out increases by an amount k. Let ȳ be some value of y in the interval PP' such that the area of the rectangle $\bar{y}h$ is k, the increase in area which results when x is increased by h. We may therefore write that

$$k = \bar{y}h,$$

and hence that

$$\frac{k}{h} = \bar{y}. \tag{5}$$

This ratio k/h is the average rate of increase in the actual area under the curve due to the increase h in x.

The discussion thus far should remind us of a similar one in the preceding chapter when we considered the instantaneous rate of change of a function of x. There we first obtained the average rate of change, k/h, and then obtained the instantaneous rate as the limit of the average rate as h approached 0. But equation (5) deals with precisely the same situation. What then does the limit of the ratio k/h in (5) yield as h approaches 0? It is the instantaneous rate of change of area A with respect to x at the point P. The limiting value of k/h is obtained from the right side of (5). As h approaches 0, $\bar{y}$ approaches the y-value at the point P. What we have shown then is that

$$\dot{A} = y, \tag{6}$$

where $\dot{A}$ is the Newtonian notation for the derivative of A with respect to x. Of course y is the function of x determined by the curve CD itself, and in our example above $y = x^2$.

Equation (6) tells us that the instantaneous rate of change of area with respect to x at any value of x is the y-value of the curve at that point x. Hence, to find A, we must find the function whose derivative is x^2. This function is $x^3/3$. Hence

$$A = \frac{x^3}{3}. \tag{7}$$

But when $x = 5$, $A = 125/3$, and when $x = 3$, $A = 27/3$. Hence the area between $x = 3$ and $x = 5$ is $125/3 - 27/3$. Thus we see that the area originally approached as the limit of a certain sum of rectangles, expression (3), can be obtained essentially by the simple process of reversing differentiation. But this is precisely what the Newton-Leibniz procedure calls for.

It is difficult for the modern student to appreciate how much simplification is introduced into the problem of finding areas by the procedures of the calculus. The calculus replaces the lengthy, complicated, and specialized summations that were formerly required to find areas by the rapid, general process of reversing differentiation as illustrated and intuitively rationalized above. Thus the new method solves "in an instant" problems that took months of original thought on the part of the best Greek and Renaissance masters.

After learning any mathematical process that is created to solve a particular problem, we should expect from our experiences in this book that the process may be capable of generalization, and thus applicable to a much larger class of problems than the one which gave rise to it. The problem of finding the area bounded by a curve was reformulated above so as to call for the limit of a special type of sum, namely, the limit in (3) or the integral (4). This limit was then evaluated by reversing differentiation as in (7), followed by a couple of incidental steps. The essential mathematical point in this process is that a limit of a sum (of the kind in question) can be obtained by reversing differentiation. The fact that the limit represented area is extraneous. Hence we might expect that any limit of a sum of the form (3), regardless of its origin or physical meaning, could be evaluated by reversing differentiation. This is indeed the case, and this general relationship is the fundamental theorem of the calculus. The fundamental theorem enables us to solve any number of physical problems. Let us see how it can be used in a totally different situation from that of finding areas.

In Newton's time and for a century afterward the gravitational attraction exerted by one body on another was a key problem in working out the implications of the law of gravitation. If physical objects were concentrated at points there would be far less difficulty in applying the law, because the distance between two objects would be the distance between two points. But real objects are extended in space. Hence the question arises, What is the distance between them? Suppose, for example, the two objects are spheres. Is the distance between them the smallest distance between a point on one and a point on the other, or is it the farthest distance between points on each, or is it the distance between the centers? This problem troubled Newton when he tried to verify the law of gravitation. The diameter of the earth is about 7900 miles and the diameter of the sun is about 865,390 miles. Obviously there is room for considerable error in deciding upon what the distance between the sun and the earth should be. One of Newton's great achievements was to prove that the gravitational attraction exerted by a sphere acts as though the entire mass of the sphere were concentrated at the center. Indeed, it was the lack of this proof that delayed his publication of the law of gravitation for twenty years.

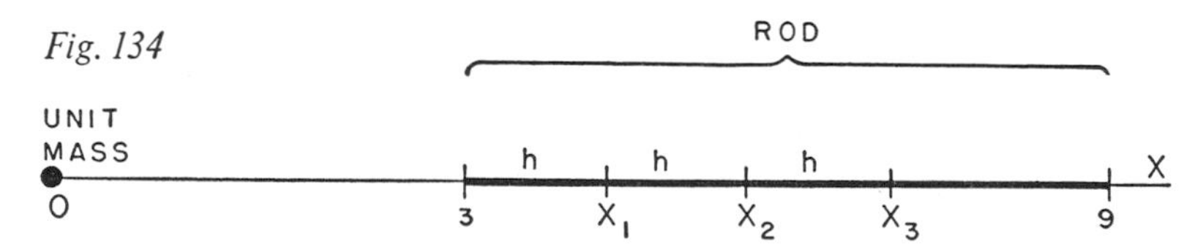

Fig. 134

To see how the integral calculus enables us to solve such problems let us consider a somewhat simpler one of the same nature. Suppose a rod 6 units long exerts a gravitational force on a unit mass lying in the line of the rod (fig. 134) but three units away from one end. What gravitational force does the rod exert on the unit mass? Of course to apply the law of gravitation we must know the mass of the rod, and so we shall suppose that its mass is 12 units. Moreover, a rod is not a line but has thickness. However, the thickness is small compared to the length and so we shall idealize our problem and assume that mass is distributed along a straight line. But we are still faced with the difficulty that the distance from the unit mass to the rod varies from 3 to 9 units. What distance shall we choose? Intuition suggests an average value, namely 6 units, and that we compute the force of attraction with this value of the distance. Now let us see if this value is correct.

Let us divide the rod into n little pieces each h units long. Then $nh = 6$. Since the mass of 12 units is divided into n pieces, each piece has a mass of $12/n$. But $n = 6/h$. Hence each piece has a mass of $12(h/6)$ or $2h$ units of mass. Now let us suppose that the first piece, which extends from 3 to $3 + h$ units of distance from the unit mass, is concentrated at the distance x_1, which is $3 + h$ units from the unit mass. Of course this supposition is only an approximation, analogous to replacing a piece of area under a curve by a rectangle. The gravitational force F_1 that this piece exerts on the unit mass is, according to the law of gravitation,

$$F_1 = G\frac{1 \cdot 2h}{x_1^2},$$

where G denotes the gravitational constant. Likewise, let us suppose that the second piece is concentrated at the distance x_2, which is $3 + 2h$ units from the unit mass. The gravitational force this piece exerts on the unit mass is

$$F_2 = G\,\frac{1 \cdot 2h}{x_2^2}.$$

We can thus form F_3, F_4, and so forth. Since there are n pieces, the last or nth piece will be regarded as concentrated at the distance x_n, which is $3 + nh$ units from the unit mass, and the force that this piece exerts is

$$F_n = G\frac{1 \cdot 2h}{x_n^2}.$$

To obtain the total force exerted by the entire rod we can add these forces, since they are all in the same direction. However, this sum is only an approximation because we assumed that the mass of each piece was concentrated at the right-hand end of that piece. If we take h smaller and smaller, our approximation improves, because the smaller the pieces the more accurate it is to suppose that the mass of each piece is concentrated at one point. The correct force F exerted by the entire rod is actually the limit approached by the sum $F_1 + F_2 + \ldots + F_n$ as h approaches 0. In the notation already employed in expression (3) above, the correct force F is

$$F = \lim_{h \to 0} \sum_{i=1}^{n} G \frac{1 \cdot 2h}{x_i^2}.$$

We may as well humor mathematicians by writing this limit in the standard form as

$$(8) \qquad F = \int_3^9 \frac{2G dx}{x^2},$$

wherein we recognize—see expression (4)—that dx is symbolic for the small length h, the integral sign $\int$ indicates that we are dealing with the limit of a sum, and the numbers 3 and 9 remind us that we were summing the effects of little pieces of mass lying between $x = 3$ and $x = 9$. Further to relate (8) with (4), we should also note that y in (4) stands for some function of x, the function representing the equation of the curve CD. In the present case the function of x is $2G/x^2$, which we could denote by y if we wished to.

We have arrived at the integral we wish to evaluate. Now the fundamental theorem of the calculus tells us that we can obtain that value by reversing differentiation. The procedure requires that we first find the function whose derivative is $2G/x^2$. Here we must call upon the purely technical knowledge that one easily acquires in the study of the calculus to state that the function in question is $-2G/x$. The next step in the Newton-Leibniz procedure is to substitute 9 for x, which yields $-2G/9$. We then substitute 3 for x, which yields $-2G/3$, and subtract the second result from the first. Hence

$$F = -\frac{2G}{9} - \frac{(-2G)}{3}$$

or

$$(9) \qquad F = \frac{2G}{3} - \frac{2G}{9}.$$

Now the value of F is clearly $4G/9$, but let us deliberately write the result (9) as

$$F = \frac{18G}{27} - \frac{6G}{27} = \frac{12G}{27}, \tag{10}$$

because the physical interpretation of the result is more readily made. The number 12 in the numerator is the mass of the rod. One might fear that this 12 is only coincidentally the same number 12 as the mass, and that we have no right to interpret this number as representing the mass. However, if we had carried through the entire process using M for the mass instead of 12, so that our procedure would be general, we would see that M appears in the numerator where 12 now appears. Here we see, incidentally, how the use of letters helps us in keeping distinct in the answer those quantities that remain constant throughout the mathematical work.

What, now, is the significance of the 27 in the denominator? Well, we were seeking the effect of the entire rod on the unit mass. If this rod were concentrated at one point, at a distance r from the unit mass, the force it would exert is

$$F = G\frac{12 \cdot 1}{r^2}. \tag{11}$$

If we equate the right sides of (10) and (11) we see that $r^2 = 27$ and $r = \sqrt{27}$. Thus the rod acts as though its mass were concentrated at a point a little more than 5 units away from the unit mass. We should compare this result with the figure of 6 units arrived at by intuition. Apparently intuition is not good enough to determine the gravitational effect of a simple rod.

When the mathematician obtains a result from theory and calculations that seem to conflict with his intuitive expectations, his first move generally is to check not his intuition but his mathematics. He trusts his intuition more because intuition is at least in part based on experience, and if mathematical conclusions do not agree with our experiences in the physical world then something is wrong with the mathematics. If, having checked his mathematics carefully, the mathematician becomes convinced that it is correct, he then asks himself what may be wrong about his intuition. This is the question we should now ask about our expectation that the mass of the rod could be regarded as concentrated at its mid-point.

It is not hard to find our mistake. The choice of the mid-point is based on the belief that two small pieces of the rod, one to the left of center and the other as far to the right of center, are equivalent to the same two pieces placed at the center. But consider a small piece of mass m lying around $x = 4$. This piece will attract the unit mass with a force $Gm/4^2$. The symmetrically placed piece is one unit from the other end, that is, at $x = 8$. This piece will attract the unit mass with a force $Gm/8^2$. The sum of these two forces, $Gm/4^2 + Gm/8^2$, is not $2Gm/6^2$, as a little arithmetic would show. Put otherwise, by regarding the mass at $x = 4$ to be acting as though it were at $x = 6$, we sacrifice more gravitational force than we gain by regarding the mass at $x = 8$ to be at $x = 6$. The reason is that the gravitational force depends upon $1/x^2$. Hence one should expect, if the entire mass of the rod could be concentrated at one point, that the point would have to be closer to the unit mass than the geometrical mid point of the rod. This is what the mathematical result shows; the mass of the rod must be regarded as concentrated at $x = \sqrt{27}$ and not at $x = 6$.

The mathematician who works with a variety of functions and geometrical figures gradually improves and refines his intuitive understanding; as a consequence the intuitive thinking that guides him in difficult physical and mathematical problems is far keener than that of the layman. When one speaks, therefore, of the intuition used by mathematicians and scientists to guide them in their anticipation of theorems and proofs, he means this trained intuition that enables them to foresee a conclusion where the inexperienced person would be unable to detect a glimmer of light.

The problem of the gravitational attraction of the rod perhaps points out more clearly than the problem of area the great range of applications of the integral calculus. The essential difficulty in the former problem is that the force of attraction varies with the distance of the mass in the rod from the unit mass. Hence the total effect of the rod cannot be obtained by a simple addition of the forces exerted by parts of the rod. Exactly the same difficulty occurs in a variety of physical situations. If an object is pushed or pulled up a hill of varying slope, the work done, which by definition is the product of force and distance through which the force acts, cannot be obtained as a simple product, because the force required to overcome the gravitational force down the plane varies with the slope (chapter 13). The pressure of the air on, say, a square foot of the earth's surface cannot be obtained by multiplying the weight of a cubic foot of air by the height of the atmosphere above the surface of the earth, because the weight of the atmosphere varies with height. Even the calculations of such fundamental quantities as weight and mass require the calculus. Consider the weight of an ellipsoid (the shape of a football). If the solid ellipsoid is made of the same material throughout, the weight per unit volume will be constant. Hence what one needs is the volume. But almost all volumes bounded by curved surfaces can be found only by the use of the calculus.

Our experience with the gravitational attraction exerted by a rod may enable us to appreciate also why Newton hesitated to assume that the gravitational force exerted by a sphere on other masses acts as though the mass of the sphere were concentrated at its center. In this particular case the raw intuitive guess happens to be correct, but the experienced mathematician, knowing that crude intuition may be woefully in error, will not trust it. The particular problem of the force exerted by a sphere is much more difficult mathematically than that of the rod, because the sphere is a three-dimensional body. Yet Newton was able to solve it by the method of the calculus.

In the presentation of this work in his *Mathematical Principles of Natural Philosophy,* Newton proved several interesting theorems that have both amusing and practical implications. He considered first the case of a hollow earth; that is, physically, the earth was to consist of a thin spherical shell with nothing inside. Mathematically, the earth would in this case be idealized as a spherical surface with its mass distributed uniformly over the surface. He then asked, What gravitational force does this hollow earth exert on an object inside? The surprising theorem is that no force is exerted.

It is not hard to follow Newton's own argument. Consider a spherical surface (fig. 135) and a unit mass located at the point P in the interior. We may decompose the sphere into pairs of cones, one of which is shown in the figure. The two cones of any pair have equal vertex angles at P. Consider the attraction exerted on P by the mass distributed along the base A and the base a of the two cones. If the areas A and a are small, then it is approximately cor-

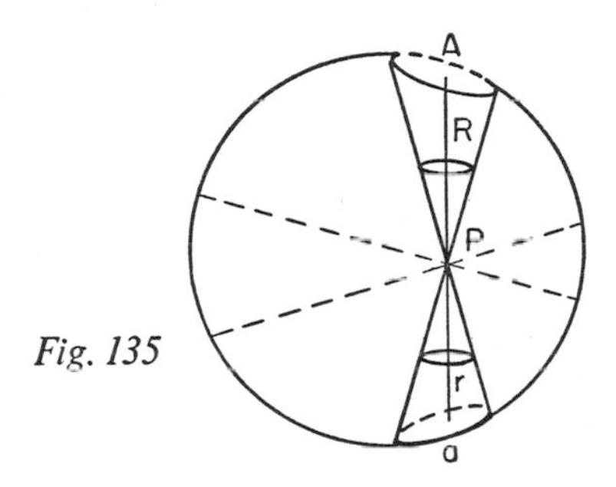

Fig. 135

rect to regard them as spherical caps or zones on spheres of radii R and r respectively. Since these caps correspond to the equal vertex angles at P, the areas A and a are to each other as R^2 is to r^2. But the entire mass is uniformly distributed over the sphere; hence the masses M and m of A and a are proportional to the areas A and a. That is,

$$\frac{M}{m} = \frac{A}{a} = \frac{R^2}{r^2}. \tag{12}$$

But the force F with which M attracts the unit mass at P is

$$F = G\frac{1 \cdot M}{R^2}, \tag{13}$$

and thc forcc f with which m attracts the unit mass at P is

$$f = G\frac{1 \cdot m}{r^2}.$$

If we take the value of M given by (12), namely $m\, R^2/r^2$, and substitute it in (13) we find that $F = f$. Hence these two caps A and a exert equal and opposing forces on the unit mass at P. Likewise, all the other cones into which the sphere is divided exert forces that cancel each other in pairs, and the total force at P is zero. Of course the foregoing argument is a loose one and can be expressed more rigorously by using the integral calculus to express the sum of the forces on P and to evaluate this sum.

It follows from our conclusion that if the earth were a hollow shell, and a person were just to step through a hole in it, that is, enter with no initial acceleration or velocity, he would be able to stand just where he stepped. If, however, he jumped through the hole, that is, if he gave himself some initial velocity, he would continue right through at that fixed velocity to the other side of the earth.

The interest in hollow shells may seem academic, but it is not at all so. Let us suppose that positive electric charges are uniformly distributed on the surface of a hollow metal sphere. What force of attraction (or repulsion) would they exert on a charge inside? Since each charge on the surface attracts a charge inside in accordance with the very same law as the law of attraction between masses, we need do no further mathematics to answer the question. The electrical force inside is zero.

The attraction of a hollow spherical shell on a mass *outside* the shell obeys the same law as if the sphere were solid; that is, the force acts as though whatever mass the shell has were concentrated at the center of the sphere, and the force then varies inversely with the square of the distance from that center to the mass outside the sphere.

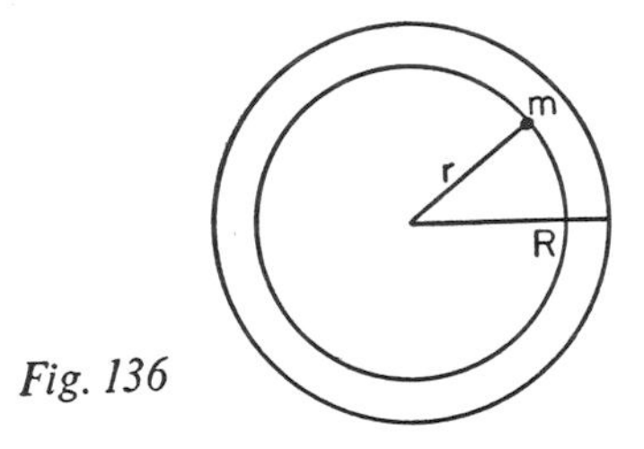

Fig. 136

Newton next considered the attraction that a solid earth would exert on a bit of mass inside the earth. Suppose the mass m is at the distance r from the center of the earth (fig. 136). Then insofar as that portion of the sphere is concerned, which is at a distance greater than r from the center, the mass m is inside, and this outer portion of the sphere acts like a set of hollow shells on a mass inside. The gravitational force this portion of the sphere exerts on m is therefore zero. However, insofar as that portion of the sphere which is less than r from the center is concerned, the mass m is external, and this inner portion of the sphere attracts m as though the mass were concentrated at the center. But the mass of this inner portion is the fraction

$$\frac{\frac{4\pi r^3}{3}}{\frac{4\pi R^3}{3}} \text{ or } \frac{r^3}{R^3}$$

of the mass M of the entire sphere. Hence the force of gravitation acting on m is

$$F = \frac{Gm\left(\frac{r^3}{R^3}M\right)}{r^2} = GmM\frac{r^3}{R^3} \cdot \frac{1}{r^2}$$

or

$$F = \frac{GmM}{R^3} r. \tag{14}$$

Since G, m, M, and R are constant, the force of attraction of the entire sphere on the mass m inside varies directly with r, the distance of m from the center.

This mathematical fact leads to an interesting conclusion. Suppose a hole were dug through the solid earth to a point on the opposite side, and a person stepped into this hole. What would his subsequent motion be? Formula (14) leads to the answer. Since the force acting on m is given by (14), the acceleration given to m, in view of Newton's second law of motion, is GMr/R^3. Moreover, we know that the acceleration acts so as to decrease r because the mass m is being pulled toward the center of the earth. Hence the acceleration strictly is

$$a = -\frac{GM}{R^3} r. \tag{15}$$

The important point about formula (15) is that the acceleration is a negative constant times r, r being the distance or displacement of the mass m from the center of the earth. This formula is exactly the same, except for the value of the constant, as the formula for the acceleration of a bob on a spring or on a pendulum.

Hence we have the answer to the question of what will happen to the man who steps into the hole through the earth. When the bob on the spring is pulled down and then released it moves up and down indefinitely. During each oscillation it reaches a point as high above the rest position as the starting point is below. So the man who steps through the hole will oscillate indefinitely from one end of the hole to the other. Since, given the acceleration, we know how to compute the period—compare formulas (7) and (11) of chapter 18—we could readily show that the man would perform one complete journey to the other side of the earth and back in one hour and twenty-five minutes. Let us not be surprised if, in this age of passion for speed, the method just described for getting to the other side of the earth in 42½ minutes presently becomes the practical use of Newton's theory.

The problem of what happens to a man who falls through a shaft in the earth is so intriguing that Leonhard Euler, one of the supreme mathematicians, reconsidered it. Euler concluded that the man, on reaching the earth's center, would then be forced back to his starting point. Euler had made a mistake in his reasoning, but, because he habitually placed full confidence in what his mathematical arguments told him, even when they violated common sense, he insisted on the conclusion. When Euler came into contact with Voltaire at the court of Frederick the Great of Prussia, Voltaire upbraided him for his obstinacy in relying upon his mathematics even when the conclusions were patently false. He made Euler promise "that he does and always will blush at the violations of common sense he had been guilty of in drawing conclusions from his formulas."

24: DIFFERENTIAL EQUATIONS—THE HEART OF ANALYSIS

Nature is pleased with simplicity, and affects not the pomp of superfluous causes.

—ISAAC NEWTON

MAN'S expectation that he can understand the multifarious workings of nature has rested upon the belief that nature is designed not only rationally but simply. Whereas rational design implies only that the diverse phenomena presented to the senses should be deducible from a few fundamental principles, the belief in simplicity has meant that these principles would themselves be readily comprehended and that their logical consequences would reveal simple patterns in nature's apparent disorder. The conviction that nature is simply designed has been asserted and even emphasized by leading thinkers from Greek times onward. One of the most famous rules for the conduct of science, known as Ockham's razor and due to William of Ockham (died about 1349), says that hypotheses are not to be multiplied beyond necessity, for nature loves simplicity. It was because Copernicus and Kepler were convinced that God had designed nature simply that they dared to uphold the heliocentric theory of planetary motion. The legend at the head of this chapter is but one of many in which Newton asserted his belief in simplicity, and Leibniz was impelled to assert that our world is the best of all possible ones partly because the simplest of physical premises accounted for a great wealth of phenomena. In our own time we find Albert Einstein saying that our actual experience confirms belief in the mathematical simplicity of nature. Quotations from the writings of great men that exhibit their faith in the simplicity of nature could be multiplied beyond necessity.

The assertions of scientists are not mere avowals of hopes and desires; they rest upon evidence. Since Greek times this evidence has been patent. From the simple axioms of Euclidean geometry thousands of properties of geometrical figures were deduced. Clearly, simple principles and their rational consequences account for the organization of forms and space. By adding to the axioms of algebra and geometry a few simple statements about motion and gravitation, Galileo and Newton drew thousands of additional phenomena into one comprehensible mathematical scheme and thereby offered most impressive evidence of simple design.

While mathematicians and scientists were demonstrating that more and more natural happenings were in accord with the logical consequences of simple systems of axioms, a change was gradually taking place in the doctrine of the simplicity of nature. Simplicity continued to mean the derivation of more and more phenomena from a few principles, but the basic principles themselves and the mathematical procedures for deriving their implications were becoming increasingly complex. Whereas the Greeks and such Renaissance scholars as Copernicus and Kepler managed to get along with Euclidean geometry and a more or less crude system of arithmetic, by 1700 the mathematical pursuit of nature's design had utilized algebra, coordinate geometry, and the calculus. Though mathematicians, scientists, and philosophers gradually became aware of the complexity of nature's simplicity they persisted in the goal of deriving as many facts and phenomena from as few principles as possible. And they were willing to pay the mathematical price. In the eighteenth century the price became the mastery of differential equations.

What is a differential equation? We have already observed in our examination of the calculus that as Kepler, Galileo, Huygens, Newton, Leibniz, and others investigated new phenomena of motion, or as they improved on older theories, they found themselves obliged to treat velocities and accelerations that varied with time and place. These instantaneous rates of change were expressed mathematically as derivatives, velocity being the derivative of the distance with respect to time, and acceleration being the derivative of velocity. Given the rate of change of a function, it is often necessary to find the function itself, and this was possible in problems we have examined by reversing the process of differentiation. Thus, to recall a trivial case, given that the velocity of a falling body is

$32t$ and knowing that the velocity is the rate of change of distance compared to time, one can write that

$$\dot{d} = 32t. \tag{1}$$

By reversing the process of finding the instantaneous rate of change one finds that

$$d = 16t^2. \tag{2}$$

Equation (1) is a differential equation, which means that it states an equality involving the rate of change of a function. It so happens that (1) is a very simple differential equation, and it is therefore possible to obtain formula (2) from it merely by reversing the process of differentiation. However most differential equations are not so simple as (1), and the problems of solving a differential equation, that is, of finding the function itself for which the asserted differential equation holds, is usually much more complicated.

It is of course not clear as yet why scientists are obliged to employ complicated differential equations in order to embrace and investigate larger classes of natural phenomena. The point will be clearer in a few moments. Let us note, however, that in the study of motion or, more broadly, mechanics, the fundamental physical axioms are Newton's second law of motion and the law of gravitation. The law of gravitation states that the gravitational force varies with distance, and the second law of motion states that the force equals the mass times acceleration. Thus the basic knowledge these laws grant is information about acceleration, which is an instantaneous rate, as a function of distance. The knowledge one generally seeks is the formula for velocity or distance. In mathematical terms, we start with differential equations and we wish to solve these differential equations.

To be more concrete about the need for differential equations, let us reconsider one of the simplest problems of motion, the motion of an object thrown straight up into the air. In our earlier discussion of this problem we confined ourselves to motion near the surface of the earth so that the acceleration caused by gravity could be taken to be constant and in fact equal to 32 ft/sec^2. However, we know that the acceleration due to gravity, if the object should travel far up into space, will decrease as the distance from the earth increases. The mathematical problem of finding the distance traveled by the object is altered considerably when this fact is taken into consideration.

We shall let M denote the mass of the earth, m the mass of the object that is thrown up, and r the variable distance between the center of the earth and the mass m. We know from many past discussions—see, for example, formula (4) of chapter 15—that the variable acceleration gravity imparts to the mass m is given by the expression

$$a = -G\frac{M}{r^2},$$

the minus sign being inserted here to express the fact that the acceleration is negative because it causes a loss in velocity of the object thrown up with some positive velocity. Now the acceleration a is the instantaneous rate of the velocity v; that is, $a = \dot{v}$. Hence

$$\dot{v} = -G\frac{M}{r^2}. \tag{3}$$

Equation (3) is a differential equation and indeed no longer of the simple type $\dot{v} = -32$, which holds for objects traveling up or down in the immediate vicinity of the earth. Equation (3) says that the instantaneous rate of change of velocity compared to time is a function of the variable distance r of the mass m from the center of the earth.

Before proceeding to solve for the velocity, let us note that we know what the quantity GM is numerically. When the mass m is at the surface of the earth, $r = 4000$ and the acceleration $\dot{v}$ is -32 ft/sec^2 or -0.006 miles/sec^2. Hence

$$-.006 = -\frac{GM}{(4000)^2}$$

or

$$GM = (.006)\ (4000)^2 = 96{,}000.$$

Hence in place of (3) we may write

$$\dot{v} = -\frac{96{,}000}{r^2}. \tag{4}$$

To determine the velocity v we must solve the differential equation (4). We shall not present the purely technical process by which the solution is achieved, but we shall note one fact that will make the solution more understandable. Equation (4) tells us something about the velocity that gravity will impart to the mass m. If the mass is thrown up into the air, the velocity imparted by the hand must also enter into the expression for the variable velocity v. Let us denote by V the (constant) velocity that the hand imparts to the mass. Then solution of the differential equation yields the formula

$$v^2 = V^2 + 2 \cdot 96{,}000 \left(\frac{1}{r} - \frac{1}{4000}\right). \tag{5}$$

Let us now see if we can utilize the solution (5) to learn something about the motion of the mass. We know, of course, without calling upon formula (5), that the mass m will start upward with the initial velocity V given to it by the hand at the surface of the earth but will lose velocity continually because gravity causes an acceleration downward. Of course the numerical value of the acceleration will decrease because, as the right side of equation (4) says, the distance r of the mass m from the center of the earth continually increases. Hence the mass will lose less and less velocity as time elapses, though it will always lose and never gain. If, however, we make the initial velocity V large enough, the mass may still have enough velocity to reach any point in space we please. Let us send it to the moon and let us ask the question, How large must V be for the mass to reach the moon?

We shall, for the present at least, ignore the fact that the moon will attract the mass m more and more as the mass approaches it. Let us suppose that the moon is just a point in space 240,000 miles away, and that the mass has lost all its velocity when it just reaches the moon. Hence v in equation (5) is 0 when the object reaches the moon and r will be 240,000 miles. If, therefore, we substitute 0 for v and 240,000 for r in equation (5) we have

$$0 = V^2 + 2 \cdot 96{,}000 \left(\frac{1}{240{,}000} - \frac{1}{4000}\right).$$

A bit of arithmetic and algebra tells us that

$$V = \sqrt{2 \cdot 96{,}000 \left(\frac{1}{4000} - \frac{1}{240{,}000}\right)}$$

or that

$$V = 6.87.$$

Hence an initial velocity of 6.87 miles per second is required to send a mass from the surface of the earth to a point precisely 240,000 miles up. Incidentally, the time required for this journey, assuming that the earth and moon stay fixed, would be about four days and twenty-one hours.

A journey to the moon may enable one to rid himself of the problems that beset him on earth, but it may very well be that life would be no different there, particularly if, as is now likely, many people from earth should undertake to transfer their habitat there. We can well understand, therefore, a person's desire to get away from it all and leave the solar system altogether. We can help him. Mathematically, the problem is that of giving the man enough initial velocity V so that he just reaches infinity. What this means as far as equation (5) is concerned is that v should be zero when r is infinite. To say that v should be zero when r is infinite means, in more precise (though still not precise) language, that v should approach 0 as $1/r$ approaches 0. Let us therefore set $v = 0$ and $1/r = 0$ in equation (5). We find that

$$V = \sqrt{2 \cdot 96{,}000 \left(\frac{1}{4000}\right)} = 6.93.$$

This velocity of 6.93 miles per second with which an object must leave the earth to travel to infinity is known as the escape velocity, for reasons somewhat as suggested above.

If an object were to leave the earth with a velocity greater than the escape velocity it would still have some velocity when r is infinite. Just what it would do with that velocity is a nonmathematical question. Perhaps the nonmathematical answer is that it would enter another universe. Let us confine ourselves to this one.

The problem we considered above took into account the decrease in the gravitational force that the earth exerts on a mass as the mass travels farther and farther away from the surface of the earth.

However, it did not take into account the resistance of the air. Whereas Galileo was content and no doubt wise to ignore this factor in his ground-breaking studies of motion, we shall now try to be realistic and include it. After all, an airplane traveling forward in space, a projectile shot up into the air, a man parachuting from an airplane, and a ship progressing through water do move in resisting mediums. However, to make our problem somewhat simple let us consider motions that take place near the surface of the earth so that the acceleration caused by gravity can be taken to be 32 ft/sec^2. Moreover, let us consider the motion of an object that falls straight down.

How shall we express mathematically the resistance of the medium? By observation and experimentation scientists have found that the resistance of mediums such as air, water, and other gases and liquids varies with the velocity of the object moving through them. This resistance may increase directly with the velocity, that is, be a constant times the velocity, or be even stronger and increase directly as the square or cube of the velocity. Let us take, however, the simplest case, namely, that the resistance increases directly with the velocity, and let us see how this factor enters into the determination of the motion.

Suppose, then, that an object is thrown downward in a resisting medium whose resistance varies directly with the velocity the falling object possesses at any instant. What can we say about the subsequent motion? The forces acting on the object are, first, the force of gravitation, which is the mass times 32 if the motion takes place near the surface of the earth. Since the medium opposes this downward motion, there is another force acting which is some constant times the velocity. It is convenient to express this constant as the product of the mass of the falling object times a constant k whose value depends upon the size and shape of the object and the nature of the medium. This second force is, then, mkv. The net force acting is the pull of gravity minus the resistance of the medium, that is,

$$32m - mkv.$$

According to Newton's second law this net force must equal the mass times the actual or net acceleration of the object. If we write the acceleration in calculus terms as $\dot{v}$, then Newton's law says that at any time t

$$m\dot{v} = 32m - mkv.$$

If we divide by m then

$$\dot{v} = 32 - kv. \tag{6}$$

Here, then, is the differential equation of an object falling in a resisting medium whose resistance is proportional to v.

The mathematician trained in the theory of differential equations would immediately proceed to solve this differential equation and then deduce some consequences. But we can learn a good deal just by studying the differential equation itself. The object is thrown downward. This means that the object is given an initial velocity. Suppose this initial velocity happens to be the value L, such that

$$32 - kL = 0. \tag{7}$$

Now equation (6) holds at all instants of time during the motion, including the instant $t = 0$ when the object is thrown downward. But, in view of (7), at the initial instant $\dot{v} = 0$. We have, then, an object that starts its downward motion with an initial acceleration of 0 and a velocity L. By Newton's first law of motion the object must continue to move at the velocity it already has, namely L. Then, according to (6) and (7), the net acceleration will continue to be 0, and the velocity will remain constant. We conclude that the object will fall downward at the constant velocity L. We now know the full story of the subsequent motion of the object without solving the differential equation.

Of course the value of the initial velocity chosen in (7) was a very special one. Suppose now that the initial velocity V is less L, so that

$$32 > kV. \tag{8}$$

The inequality (8) says that the acceleration due to gravity is greater than acceleration arising from the resistance of the medium at the outset of the motion. Equation (6), then, says that there is a net positive acceleration, and the object will gain velocity as it falls. However, as the velocity v increases it may reach the value for which kv equals 32. We are at this instant back to the first case considered, and the object will fall with a constant velocity thereafter, specifically at the velocity L for which (7) holds.

This second case is very important in practice. If an object is just dropped in a resisting medium—for example, when a man falls out of an airplane—the initial velocity V is zero. Then equation (8) certainly holds. We know, then, that this object will fall with increasing velocity until it reaches the velocity L for which (7) holds, and thereafter it will fall with this constant velocity. This quantity L is called the limiting velocity. Raindrops, which start with zero velocity, will attain a constant velocity and fall at this rate thereafter. This is fortunate; otherwise, since they often fall from high clouds, they would hit the earth with enormous speeds.

The third case we can consider is when the initial velocity V is such that

$$32 < kV. \tag{9}$$

Equation (6) now says that there will be a negative acceleration at the outset so that the object will lose velocity as it falls. But if it loses velocity, then in sufficient time the velocity v will decrease from the original value V to the value L satisfying (7). Thereafter, just as in the second case, the object will continue to fall with the limiting velocity.

It must be noted, for the sake of mathematical accuracy, that theoretically the limiting velocity is not actually attained in the second and third cases, but is approached more and more closely as the time of fall increases. Practically, however, this limiting velocity is attained in a finite amount of time, and the object continues to fall with this velocity. The limiting velocity, as is evident from (7), depends upon the value of k, and this in turn depends upon the size and shape of the object and the nature of the medium.

We could show that of two objects composed of the same material the smaller one has a smaller limiting velocity. For example, the particles of a mist are so small that the limiting velocity is small, whereas drops of rain fall with considerable limiting velocity. Likewise, of two objects the same size and shape the lighter one will fall with a smaller limiting velocity. Thus a rubber ball falling from a great height will hit the ground with far less velocity than a piece of lead of the same size and shape. We should contrast all of the foregoing results with what Galileo obtained by neglecting air resistance. His result for an object falling in a vacuum with initial velocity V was

$$v = V + 32t,$$

and of course this result says that all objects fall with constantly increasing speed, indeed a speed that becomes infinite with infinite time. Galileo was certain that bodies falling in a resisting medium reach a limiting velocity, but without the calculus he could not establish this fact mathematically. We see here, as in so many other instances, that by extending our mathematical concepts and techniques we can undertake more realistic and unfortunately more complicated problems.

We have made a series of deductions from the differential equation (6) by pursuing simple mathematical implications. Were we to solve the equation, that is, find the formula for v as a function of t, we could make further deductions. Of course v is itself the rate of change of distance compared to time, $\dot{d}$, and so one could regard the formula for v as a differential equation for d and by solving it find the formula for distance fallen in terms of time. We shall not proceed to find these functions here, because we wish to avoid too much reliance upon technical knowledge. Instead, let us look into some other types of differential equations.

Let us reconsider the motion of a bob on a spring (chapter 18). We may recall observing that when the bob is pulled downward from its rest position and released the tension in the spring pulls the bob back to the rest position, and thereafter the bob oscillates about the rest position. By applying Hooke's law and Newton's second law of motion we found that the acceleration of the bob is given by the formula

$$a = -\frac{k}{m}y, \tag{10}$$

where y is the vertical displacement, m is the mass of the bob, and k represents the stiffness of the spring. At this point we appealed to some rough analysis to find the formula expressing the displacement y as a function of the time of motion t. No modern self-respecting mathematician would behave in this way.

He would point out that acceleration is the derivative, or instantaneous rate of change, of velocity with respect to time so that (10) is really a differential equation whose solution is the velocity. Since the velocity is the derivative of distance or, in this case, displacement, the expression for velocity is in turn a differential equation whose solution is the function relating distance and time. To indicate that the acceleration is the derivative of the derivative of distance, in other words, that the acceleration is the second derivative of distance, he would write $\ddot{y}$ for a and then note that equation (10) can be written as

$$\ddot{y} = -\frac{k}{m} y. \tag{11}$$

Of course one must be worthy even of self-respect, and the mathematician who calls attention to the fact that (10) is really the explicit differential equation (11) must now show that he can do something with the differential equation. But anyone who learns the elementary technique of this subject can write down at once that

$$y = D \sin \sqrt{\frac{k}{m}}\, t, \tag{12}$$

where D is the amplitude of the motion, that is, the maximum displacement. Thus a purely mathematical process gives us the functional form that relates the displacement y and the time t. We see that the ability to solve a differential equation enables one to replace a complicated and perhaps costly physical experiment, involving all the uncertainties arising in measurement, by a straightforward and unquestionable mathematical argument.

We may recall also that an equation of the form (10) governs not only the motion of a bob on a spring but the motion of a bob on a pendulum (chapter 18) and the motion of a man who steps into a hole bored through the center of the earth to the other side (chapter 23). In these latter applications the constants k and m have different physical significance, but, as we have repeatedly emphasized, the physical meaning of the constants does not enter into the mathematics proper. As far as mathematics is concerned equation (11) is of the form

$$\ddot{y} = -Ky \tag{13}$$

where K is some positive constant. Hence the solution

$$y = D \sin \sqrt{K}\, t \tag{14}$$

applies to all three cases. Moreover, once one sees from equation (14) that the motion is sinusoidal with a frequency of $\sqrt{K}$ in 2π units of t, he knows that in all cases where (13) applies the frequency of the motion will be $\sqrt{K}$. Then the period will be—formula (11) of chapter 18—

$$2\pi \sqrt{\frac{1}{K}}. \tag{15}$$

Just to illustrate how the solution of an abstract, mathematical problem immediately conveys knowledge about dozens of situations to which the original equation applies, let us reconsider the motion of a man who steps into a hole bored through the earth.

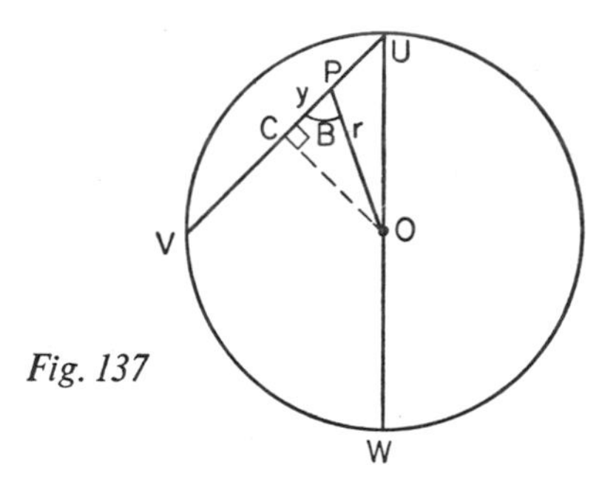

Fig. 137

This time, however, instead of boring straight through the center, we shall join any two points on the surface by a straight tunnel UV (fig. 137). We know—chapter 23, formula (14)—that if a mass m is located at a point P in the interior of the earth the gravitational force acting on m is

$$F = \frac{GmM}{R^3}\, r, \tag{16}$$

where M is the mass of the earth, G is the gravitational constant, R is the radius of the earth, and r is the distance of P from the center of the earth. This force F pulls the mass m toward the center O of the earth. However, the mass m is obliged to move along the path UV, which may be regarded as an inclined plane. Hence the force acting along UV is the component of F down the plane. We know from our study of motion along inclined planes that the component of F along UV is $F\cos B$.* If we let y represent the displacement of the mass m from the center C of its path, then

$$\cos B = \frac{y}{r}.$$

Then the force acting along UV, that is, $F \cos B$, is

$$\frac{GmM}{R^3} r \cdot \frac{y}{r}$$

or

$$\frac{GmM}{R^3} y.$$

By Newton's second law the acceleration acting on m is

$$a = -\frac{GM}{R^3} y, \qquad (17)$$

the minus sign entering because the acceleration is opposite to the displacement.

Equation (17) is precisely of the form (13) where

$$K = \frac{GM}{R^3}.$$

Hence we know at once several facts about the motion of m. It will be sinusoidal, that is, the mass will oscillate about the point $y = O$, which is the center C of its path. The period of the motion, in view of (15), is

$$2\pi \sqrt{\frac{R^3}{GM}}.$$

This period, moreover, is precisely the same as if the mass had traveled the path UW, because formula (15) of chapter 23 shows that the constant K is the same there as here. Finally, if the mass starts its motion at U, the amplitude of the motion will be CU. Hence the mass will oscillate between the two points U and V on the surface of the earth.

* In chapter 13, where we discussed motion along inclined planes, we expressed the component along the plane as $F\sin A$, where A is the complementary angle to B.

The differential equations we have examined thus far represent straight-line motion. The motion of the pendulum, which is also represented by equation (13), may seem to be an exception, but we must remember that we introduced an approximation in order to arrive at this equation for the motion in question. When we attempt to treat motion along curves by starting with differential equations—for example, the motion of a planet under the gravitational attraction of the sun—we encounter greater difficulties. Nevertheless, Newton expressed this problem in terms of differential equations and deduced all three of Kepler's laws from the equations. These deductions, in particular the elliptical motion of each planet, were a considerable technical achievement and, as we stated earlier (chapter 16), most weighty in implications.

However, this piece of work, as Newton well realized, was but a start on the problems of astronomy. The planets do not move in elliptic orbits that repeat themselves precisely year after year. The elliptic orbit holds only if there is just one planet subject to the gravitational attraction of the sun. But there are many planets in the heavens, and several of these have satellites. All of these bodies attract each other in accordance with the law of gravitation. It is no wonder, then, that the astronomers of the seventeenth century observed all sorts of irregularities in the motions of the planets, and that they were concerned about the effects of these irregularities on the future motions of the planets. There was, moreover, a practical motive for pursuing further the heavenly motions. In Newton's time, when the problem of determining longitude at sea had not been solved, scientists believed that an accurate knowledge of the moon's position at all times would furnish the basis for a method of reckoning longitude which would be a considerable improvement on the use of star positions.

Newton, himself, tackled some of these deeper astronomical problems and obtained noteworthy results. From the ratio of the masses of the moon and sun and the knowledge of their distances from the earth Newton was able to account for the variation of the tides with the phases of the moon, for the inclination of the plane of the moon's orbit to the plane of the earth's motion, and for the regression of the moon's nodes, that is, the change in position of the points in which the moon cuts the plane of the earth's orbit around the sun. As the moon moves around the earth in an almost elliptical path, the apogee (that is, the point farthest from the earth in the moon's orbit) shifts about 2° in each revolution. Newton showed that this shift was due to the sun's attraction of the moon. Newton worked so hard on the various problems connected with the moon's motion that, as he complained to his friend Halley, the astronomer, his head ached. The problems kept him awake and he resolved he would not bother with them any more.

Newton, Clairaut, and other mathematicians worked on the problem of the shape that a partly fluid mass would assume under rotation and attraction by the sun, and showed consequently that the earth is an ellipsoid of revolution, a sphere flattened at the poles. Newton and others also computed the paths of comets and concluded that some of these have periodic orbits. Thereby comets were recognized to be normal heavenly bodies rather than missiles of destruction hurled by God at offending people.

Other problems continued to plague the mathematicians. The precession of the equinoxes, the slow periodic change of the direction in which the north-south axis of the earth points, had been known for several thousand years but not accounted for on theoretical grounds. Again, while the moon always presents the same face to the earth, more or less near the edges becomes periodically visible. Increased observational accuracy had revealed that the length of the mean lunar month decreases by about 1/30 of a second per century (such was the order of accuracy that observation and theory had come to handle). Most important, Jupiter's average angular velocity around the sun had been observed to be continually increasing while that of Saturn had been observed to be continually decreasing. Since the times of revolutions of the planets were changing, then, according to Kepler's third law, the mean distances of the planets from the sun must also be changing. Hence Jupiter was approaching the sun and Saturn departing, so that ultimately Jupiter would fall into the sun and Saturn would wander off into space. These movements threatened the very existence of the solar system.

Because all of these variations had not been explained on the basis of the law of gravitation, and they seemed to deny the mathematical order of the heavens, some of the ablest mathematicians of Newton's day—Huygens, Leibniz, John Bernoulli, and others—attacked the law of gravitation.

These problems are mathematically special cases of the problem of determining the motion of *n* bodies, each attracting all the others in accordance with the law of gravitation. This problem has been tackled by the greatest mathematicians from Newton's time to the present day. However, there is no expectation that it can be solved exactly even for three bodies. Hence mathematicians from 1700 on have dealt with particular cases and have used approximations. With such maneuvers two of the greatest eighteenth-century mathematicians made phenomenal progress in treating the irregularities in the motions of the planets and their moons.

The first of these, of French and Italian extraction, was Joseph Louis Lagrange (1736–1813). As a boy Lagrange was unimpressed with mathematics, but while still at school he read an essay by Halley on the merits of Newton's calculus and became excited about the subject. At the age of nineteen he became professor of mathematics at the Royal Artillery School of Turin, the Italian city of his birth. He soon contributed so much to mathematics that even at an early age he was recognized as the greatest mathematician then active. Though Lagrange worked in many branches of mathematics—the theory of numbers, the theory of algebraic equations, the calculus, and the calculus of variations—and in many branches of physics, his chief interest was the application of the law of gravitation to planetary motion.

At the age of twenty-eight Lagrange tackled the problem of the motion of the moon under the attraction of the earth and sun and derived many excellent results. In a prize essay of 1764 he accounted for the fact that the moon's period of rotation and its period of revolution around the earth are the same, and showed that the slight variation in the portion of the moon that is visible from the earth is caused by the equatorial bulges of the earth and moon. He showed later that the attraction of the earth by the sun and the moon perturbed the earth's axis of rotation, and thus the wandering of the earth's axis of rotation, with the consequent precession of the equinoxes, was shown to be a mathematical consequence of the law of gravitation. Lagrange made another step in his mathematical analysis of the motions of the moons of Jupiter. The analysis showed that observed irregularities of the motions were also an effect of the law of gravitation. All these results were incorporated in his *Mécanique Analytique,* a work that extended and crowned Newton's work on mechanics. Lagrange once remarked that Newton was a most fortunate man because there is but one universe and Newton had already discovered its mathematical laws. However, Lagrange had the privilege of revealing the perfection of the Newtonian theory.

Another set of results, equally astonishing in both the mathematical complexities involved and in their physical significance, was contributed by the greatest scientist of the late eighteenth and early nineteenth centuries, Pierre Simon Laplace (1749–1827). Laplace was born in Beaumont, France, and educated at a school run by a priory in that town. At sixteen he proceeded to the University of Caen, where he studied for five years and wrote his first mathematical paper. The mathematician D'Alembert, who had himself made important contributions to the theory of precession of the equinoxes, secured a position for Laplace at the École Militaire in Paris. Later Laplace taught with Lagrange at the École Normale. His superb contributions to mathematics and astronomy brought him recognition from the several regimes that governed France during his lifetime. Napoleon made him a count, and the Bourbons, who returned to power after the downfall of Napoleon, made Laplace a marquis.

Laplace created a number of new mathematical methods that are today branches of mathematics, but he cared for mathematics only insofar as it helped him to study nature. When he encountered a mathematical problem in his physical research he solved it almost casually and merely stated "It is easy to see that . . ." without bothering to show how he had worked out the result. He confessed, however, that when he tried to reconstruct his own work it was not easy to see.

Laplace devoted his life to the application of the law of gravitation to problems of the solar system. Among the most important of his achievements was his demonstration that the increasing speed of the motion of the moon, which causes a decrease in the mean length of the month of $1/30$ of a second per century, is caused by a very slow decrease in the eccentricity of the earth's orbit. Laplace's crowning achievement was to explain the irregularities observed in the motion of Saturn and Jupiter. Lagrange had tackled and failed to solve this problem. Laplace showed that the observed increase in the mean speed of Jupiter and decrease in the mean speed of Saturn are periodic, and are due to the effect of one on the other. This periodicity meant that no planet would wander into space or crash into the sun. The universe was stable. The eccentricities of the various planets—that is, the sharpness of the ellipticity of their orbits—had also been observed to change, and Laplace showed that these variations, too, were periodic. This result and many others Laplace published in his *Traité de Mécanique Céleste,* a work of five volumes composed and published over a period of twenty-six years. Laplace's own conclusion as to the universal validity of the law of gravitation was expressed in the words, "Such has been the fate of this brilliant discovery that each difficulty which has arisen has become for it a new subject of triumph—a circumstance which is the surest characteristic of the true system of nature." Laplace died one hundred years after Newton.

The climax of almost two centuries of most difficult work in differential equations applied to the heavenly motions was the purely theoretical prediction of the existence and location of the planet Neptune. Unexplained aberrations in the motion of the planet Uranus were conjectured to be due to the existence of an unknown planet whose gravitational pull on Uranus was causing the irregularities. Two astronomers, John Couch Adams in England and U. J. J. Leverrier in France, used the observed data on Uranus and the general astronomical theory to calculate the orbit of the supposed planet and directed observers to the location of it at a particular time. The planet, called Neptune, was located in 1846 by the German astronomer Galle. It was barely observable through the telescopes of those days and would hardly have been noticed were astronomers not looking for it. Its discovery was an amazing achievement of mathematical prediction and was widely proclaimed as a proof of the universal application of Newton's law of gravitation. It was accompanied by a minor tragedy. Bode's law (chapter 3) failed to hold and was abandoned. The simplicity of nature could no longer tolerate a major law involving only the whole numbers.

The perfection of the astronomical theory created by the mathematicians of the Newtonian era is revealed paradoxically by a discrepancy between theory and observation that has had momentous significance in our century. We have already noted that, strictly speaking, the path of each planet is not an ellipse because the other planets disturb the purely elliptical motion which the sun's attraction produces. Such a disturbance affects, of course, the motion of Mercury. As this planet moves around the sun it is closest to the sun at one point in its practically elliptical path. This point is called the perihelion. The direction of this perihelion changes from year to year and this effect is due to the attraction exerted by the other planets. In other words, the effect of the other planets is to rotate slightly the ellipse on which Mercury would move if the other planets were not present. Newtonian theory does predict the motion of the perihelion of Mercury but falls short of observation by less than one-half of one second of angle per year. Small as this discrepancy may appear to be, it is more than can be tolerated in modern astronomy. When Einstein advanced his general theory of relativity in 1915 the only experimental support he could claim at that time was that his theory predicted the proper amount of change in the direction of Mercury's perihelion.

As this brief review of some of the major results attained by eighteenth- and nineteenth-century scientists suggests, the application of differential equations to astronomy has been enormously rewarding. Astronomy is in fact the most successful and most exact branch of science. It has not only established the mathematical pattern of the heavenly motions but it is able to predict, to the fraction of a second, sunrise and sunset, high and low tides, the appearance and disappearance of planets and stars, and eclipses. In view of the fact that the heavenly bodies are remote and that we cannot experiment with them, the state of that science is a tribute to the remarkable mathematical skill of the men we have come to know in these pages. It is also remarkable that the best evidence for the mathematical design of the universe comes from the contemplation of the heavens and not from the pursuit of mundane problems. The regard for law, which predisposes men to attribute all phenomena, even unexplained ones, to regular rather than to abnormal behavior of nature, and the mental habit of substituting law for supernatural intervention were developed by looking *away* from man's immediate problems and studying the motion of the most distant stars.

The differential equations we have considered thus far express some fact about the rate of change of one variable with respect to another—for example, velocity compared to time—and the solution yields the formula for the dependent vairable in terms of the independent variable. What characterizes all these equations is that only one independent variable, which has been time in most of our examples, is involved. Such differential equations are called ordinary, though more than ordinary intelligence is required to formulate and solve them.

In the study of a phenomenon such as sound, however, one may be interested in the motion of the sound wave through space and time. Specifically, one may wish to know the displacement of the air as a function of the time t and of x, y, z, which represent positions in space. The differential equations that express the basic physical principles of such phenomena are called partial differential equations, the *partial* indicating that there are two or more independent variables. For example, Maxwell's equations, which we treated rather summarily earlier, are partial differential equations. Naturally, partial differential equations are more complicated in their very statement, to say nothing of methods of solution, than ordinary differential equations, and we cannot enter into any presentation of them here. It is desirable to know, however, that this vast branch of mathematics, partial differential equations, serves the purpose of expressing the basic physical principles of such prevalent and vital phenomena as sound, heat, the various forms of electromagnetic waves, water waves, vibrations in rods, the flow of fluids and gases, and so forth, and that from these differential equations we can deduce by mathematical methods alone a vast amount of information about these phenomena. In fact, it is fair to say that the subject of differential equations is today the heart of mathematics and it is certainly the most useful branch for the study of the physical world.

So many physical principles are most effectively formulated as differential equations that nature and God are often credited with speaking in terms of them. Unfortunately, as the domain of differential equations has been extended to include more and more phenomena, it has proved increasingly difficult to solve the varieties that arise. Over a century ago Augustin Jean Fresnel appreciated this point and remarked that the simplicity of nature is chimera. "Nature," he said, "is not embarrassed by difficulties of analysis." But neither are mathematicians deterred. The search for solutions goes on daily, as does the search for simplicity—in the more sophisticated sense in which this term must now be understood.

25: FROM CALCULUS TO COSMIC PLANNING

The spectacle of the universe becomes so much the grander, so much the more beautiful, the worthier of its Author, when one knows that a small number of laws, most wisely established, suffice for all movements.

—MAUPERTUIS

ACCORDING to one of the legends of history, Dido, of the ancient Phoenician city of Tyre, ran away from her family to settle on the Mediterranean coast of North Africa. There she bargained for some land and agreed to pay a fixed sum for as much land as could be encompassed by a bull's hide. But Dido made her own interpretation of this bargain. Endowed with as much shrewdness as beauty she cut the hide into very thin strips, tied the strips end to end, and proceeded to enclose an area having this length of hide as its perimeter. Her second bright idea was to use this length to bound an area along the sea. Because no hide would be needed along the seashore she could thereby enclose more area. She then faced the mathematical problem of what shape to choose for the bounding curve so as to have it enclose the most area. According to the legend, Dido thought about the problem and discovered that the length of hide should form a semicircle. On the land so enclosed she founded the city of Carthage and became its queen. Thus Dido solved a difficult problem in the subject now known as the calculus of variations.

It would be pleasant to relate that Dido, having learned how much can be gained by delving into a little mathematics and having discovered a new field of mathematical investigation, plunged enthusiastically into the subject. But Dido's interest in mathematics was soon diverted to welcoming and entertaining Aeneas, the courtly refugee from Troy. Though she did her best to induce Aeneas to settle in her country, and indeed in her very household, Aeneas was a man with a mission, and he soon departed to found a new civilization in Rome. Dejected and distraught, Dido could do no more for Aeneas than to throw herself on a blazing pyre so as to help light his way to Italy. Of course mathematics was thereby deprived of talent, but the moral of this story is that if a man starts out to do something with rigidly preconceived notions of what the outcome must be, he will be blind to more advantageous lines of endeavor. And he may do more harm than good: Rome made no contributions to mathematics whereas Dido might have.

As it turned out, progress in the calculus of variations had to be delayed about 2500 years until two famous Swiss brothers, James (1654–1705) and John (1667–1748) Bernouilli relaunched the subject. In 1696, John proposed a problem as a challenge to the mathematical world. Suppose, he said, that an object starting from

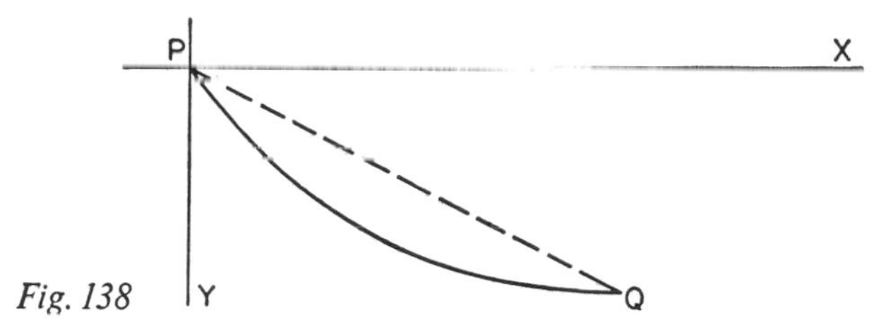

Fig. 138

rest slides along some curve (fig. 138) from the point *P* to the point *Q*; the only force causing the object to slide is gravity; along what curve must the object slide in order that the time of travel be least? To see what this problem amounts to physically, we may suppose that a chute is to be constructed from the point *P* to the point *Q* so that objects can be slid from one point to the other in the least possible time; the question then is, What should the shape of the chute be? The question might readily occur to anyone who has watched truckmen slide packages from a sidewalk to a cellar down such a chute.

The answer intuition suggests is that the curve joining *P* and *Q* should be a straight line. But we know that intuition can be faulty and should be checked by a little reasoning. In this case intuition is likely to identify the least time of travel with the shortest distance and hence accept the straight line too readily. Let us therefore examine the problem more closely.

The force that will make the object slide from *P* to *Q* is the force of gravity. But this force pulls straight down. Hence only that component of the force which is in the direction of the curve will cause the object to follow the curve. Of course this component will give the object its acceleration and, consequently, its velocity. Now let us compare the straight-line path from P to Q with the curve shown in figure 138. Along the straight-line path, the force acting is the component of gravity in the direction from P to Q. If, however, the object is allowed to slide along a path that, at the outset at least, is more nearly vertical, the component of the force of gravity in this direction, and therefore the acceleration of the object, will be larger and the object will gain velocity more rapidly. It is of course true that if the object begins to slide along a path that is steeper than the straight-line path from P to Q the path will have to curve around toward Q and be longer than the straight one. This increase in length might offset the advantage of the larger initial acceleration.

Our analysis thus far shows that we must weigh the advantage of the curvilinear path, which affords more acceleration and therefore more velocity to start with, against the advantage of shortest distance which the straight-line path possesses. Though this analysis does not lead to the answer to John Bernoulli's problem it may have made us aware that the straight-line path is not necessarily the one that will require least time. We are now at the stage where intuition can no longer be helpful and there remains only the method of mathematics.

John Bernoulli solved this problem by a trick that utilized the mathematical identity of motion under a variable force, which is the case for the component of gravity along a curve from P to Q, and the motion of light in a variable medium. He knew how to solve the latter problem and thereby solved the problem of the path of quickest descent. Though we shall be concerned primarily with a more general method of solving this and other problems of the calculus of variations, we might at this point satisfy our curiosity as to the answer John obtained. The solution, which surprised both Bernouilli brothers, is none other than an arc of the cycloid passing through P and Q. We may recall (chapter 18) that the cycloid is the path traced out by a point on a wheel that rolls along a straight line. When the cycloid lies in relation to the points P and Q as shown in figure 138 the curve serves as the path of quickest descent, or the brachistochrone, as it is called.

The properties of the cycloid had been investigated by Huygens in his study of pendulum motion, wherein he had found that the cycloid is the path of a pendulum whose period of oscillation is independent of the amplitude of its swing. The curve had also been investigated by other mathematicians who were concerned with purely mathematical properties such as the length of an arch (the portion of the cycloid generated by one revolution of the wheel), the area under an arch, the center of gravity of an arch, and so forth. Hence John Bernoulli was able to take advantage of known properties of the cycloid in obtaining his solution of the problem. Here, again, as in thousands of other instances, the mathematics created to solve one problem proved to be useful in an entirely different situation. Though John, as a mathematician, was well aware of this fact, he could not restrain himself. He wrote, "With justice we admire Huygens because he first discovered that a heavy particle falls along a cycloid in the same time always, no matter what the starting point may be. But you will be petrified with astonishment when I say that exactly this same cycloid, the tautochrone of Huygens, is the brachistochrone we are seeking."

As we have already noted, John Bernoulli solved the brachistochrone problem by a special device. But his brother James, Newton, Leibniz, and others solved the problem by a less elegant but potentially more general procedure. We shall examine this more general procedure because it will provide us with a better understanding of the present nature of the calculus of variations and of the uses to which it is put.

One considers any curve joining P and Q and, after introducing X- and Y-axes, as shown in figure 138, one represents the curve by an arbitrary relation between y and x. We may think of this relation as being in the form y equal to some expression in x, just as $y = x^2$ represents a parabola. In the coordinate system that has been set up, P will have the coordinates $(0,0)$ and the point Q, the coordinates (x_1,y_1). By reasoning that involves no more than Newton's second law of motion and some simple facts of the calculus one can show that the expression which gives the time T of descent from P to Q along the arbitrary curve is

$$(1) \qquad T = \frac{1}{\sqrt{2 \cdot 32}} \int_0^{x_1} \sqrt{\frac{1 + \dot{y}^2}{y}}\, dx.$$

In this integral, $\dot{y}$ is the derivative of the function of x describing the path from P to Q, and the 32 is the acceleration due to gravity. The details of this expression are not important; what matters, first of all, is that the time of descent is given by an integral so that the calculus is certainly involved. But the special feature of this integral, and of calculus of variations problems generally, is that we do not know the function of x, which y stands for in the integral. If we knew this function, then equation (1) would give the time of descent, which we could calculate by reversing differentiation. However, we do not care to know the time of descent; rather, we wish to find the function of x, or the expression for y in terms of x, that makes the value of the integral least.

The word *least* may recall another aspect of the calculus. We saw in our examination of the differential calculus that one may use the concept of rate of change of a function to find the value of x at which a function is least or greatest. In the present problem we wish to find that *function* or, geometrically speaking, that curve for which T is least. Hence one might suspect that a rate-of-change process should be applied here too, with the essential difference that the independent variable is not just x itself but a function of x. Instead of an equation that in the usual calculus problems yields the value of x at the minimum or maximum, we should expect an equation that yields the function describing the path of least time. But now we should ask ourselves what kinds of equations yield functions as solutions. The answer is, Differential equations! Hence, in some way not clearly specified as yet, a rate-of-change process applied to the integral (1), in which the independent variable is itself a function of x, should lead to a differential equation. The solution of this differential equation should yield the particular function of x for which T is a minimum.

The man who first systematized these thoughts and derived the form of the differential equation whose solution minimizes an integral of the type in (1) was Leonhard Euler (1707–1783), the most astonishing figure in all of mathematics, one of the greatest of the eighteenth-century mathematicians, and the creator of some of the richest mathematical ideas. Born near Basel, Switzerland, in 1707, to a preacher who wanted Leonhard to study theology, he entered the university at Basel and completed his work at the age of fifteen. While at Basel Euler learned mathematics from John Bernoulli. He decided to pursue the subject and began to publish papers at the age of eighteen. Through the younger Bernoullis, Nicholas and Daniel, sons of John, Euler secured an appointment in Russia at the St. Petersburg Academy. There he continued to do an amazing amount of research while helping the Russian government on many physical problems. At the age of thirty-four he took a position at the Berlin Academy, where he continued research and, at the request of Frederick the Great of Prussia, worked on state problems of insurance and the design of canals and waterworks. Twenty-five years after leaving Russia he returned to St. Petersburg at the invitation of Catherine the Great.

Euler's mathematical productivity is incredible. He wrote texts on mechanics, algebra, the calculus, mathematical analysis, and the calculus of variations, and these remained standard works for a century and more. Each of these books contained some original feature. His mechanics, for example, was based on analytical rather than on geometrical methods. In the case of the calculus of variations, he gave the first significant treatment of the subject. He published original research papers of high quality at the rate of about eight hundred pages a year during most of his life. The quality of these papers may be judged from the fact that he won so many prizes for them that these awards became an almost regular addition to his income. He carried on his writings despite the fact that he lost the sight of one eye while a youth and at the age of sixty lost the sight of the other. Some of his books and four hundred research papers were written after he became totally blind. His memory was phenomenal: he could carry out complicated numerical calculations in his head that competent mathematicians had difficulty in carrying out on paper. A current edition of his collected works contains some seventy volumes.

Equally amazing is the versatility of his interest. Euler contributed to practically every field of mathematics and physics. His major mathematical fields were the calculus, the theory of curves, the theory of numbers, series, and the calculus of variations. In physics his research covered mechanics, electricity and magnetism, astronomy, hydrodynamics, light, and sound. In addition, he made contributions to navigation, the design of canals and ships, systems of insurance and annuities, and telescopic design.

One might suspect that such a volume of activities could be carried on only at the expense of all other interests. But Euler married and fathered thirteen children, and he was always attentive to his family and its welfare. He also loved to express himself on matters of philosophy, but here he exhibited his only weakness. He advanced his ideas without sufficient thought; consequently, he made many blunders. He was often chided by Voltaire, and one day was forced to confess that he had never studied any philosophy and that he regretted having believed that one could understand the subject without studying it. But Euler's spirit for philosophic disputes remained undamped, and he contained to engage in them. He even rather enjoyed the sharp criticisms that he evoked from Voltaire.

Among the many books Euler wrote perhaps the outstanding one is his *Calculus of Variations* (1744). It brought him immediate recognition, and from the date of its publication he succeeded John Bernoulli as the greatest living mathematician. At the very outset of his work on the calculus of variations Euler recognized clearly that mathematics had taken up a new class of problems, namely, that of maximizing or minimizing some variable whose value depended not on another ordinary variable but on a class of functions, or, if we wish to speak geometrically, on a class of curves. Thus, in the case of the brachistochrone, the time of descent depends upon the class of all possible curves that join the two points P and Q. In Dido's problem the quantity to be maximized or minimized is the area which depends upon the class of curves having a given length, the length of the hide she had formed into a thong. In all these problems the variable to be maximized or minimized is expressed as an integral of the type exhibited in (1) above, and involving a typical member of the class of functions; the problem is to determine that function for which the integral has a maximum or minimum value.

Inspired by the work of the Bernoullis, Euler showed in 1736 that the desired function or curve is obtained by solving a differential equation which is still called Euler's equation and is fundamental in the calculus of variations. If the function that maximizes or minimizes the integral is expressed as a relation between the variables y and x, then Euler's differential equation is a second order ordinary differential equation, involving therefore y, $\dot{y}$, and $\ddot{y}$, whose solution is y or the function of x. Of course the precise terms that appear in this differential equation depend upon the problem being solved.

The theory of the calculus of variations was further improved by Lagrange. He supplied a better analytical formulation and method of approach, which, however, led to the same Eulerian differential equation as the key to the solution. Both Euler and Lagrange formulated many new types of problems, some of which are still being explored.

To be clearer about the nature of the calculus of variations, and to gain some inkling of the vast importance of the subject, let us

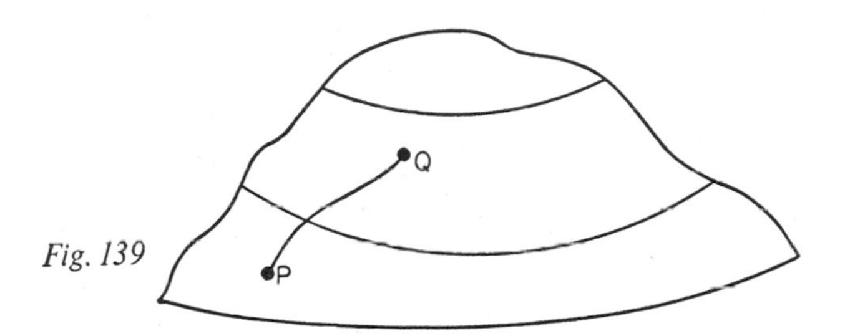

Fig. 139

examine some standard problems. Let us suppose that a person climbing a hill wishes to proceed from point P (fig. 139) to point Q. What path is the shortest? Mathematically, the hill is a surface, and the mathematical problem at hand is to find that curve, among all curves on the surface which join P to Q, whose length is least. The variable, then, that is to be minimized in this problem is length, and this length depends upon the curves which join P to Q. As one might infer from our survey of the integral calculus, the analytical expression for the length of a curve is an integral, and, of course, this integral involves the function representing the curve. To determine that function or curve for which the integral has the least value, we would solve Euler's differential equation, in which the terms would depend upon the form of the integral expressing the length of any curve joining P and Q.

The problem of determining the curve of least length joining two points on a surface, first considered by John Bernoulli, is an important one both mathematically and practically. These paths of least length are investigated so often that they have been given a special name, *geodesics.* Navigation obviously is concerned with geodesics since the choice of the shortest path saves time and money. If the surface of the earth is treated as a perfect sphere, and this may be assumed for navigation on the oceans, then the geodesic between two points on the surface is the shorter arc of the great circle passing through these points. A great circle is by definition one whose center is at the center of the earth. For example, a circle of longitude is a great circle, but a circle of latitude, other than the equator, is not. Hence a ship sailing from New York to a city in Europe on the same circle of latitude, for example, Bari, Italy, will not travel due east along the circle of latitude but will travel, insofar as it is physically possible, along the great circle joining these two cities. Air navigation uses the same mathematics because practically all planes fly at a constant altitude and hence travel along a sphere concentric with but larger than the surface of the earth. If, however, one wishes to take into account that the earth is a flattened sphere, technically, an oblate spheroid, then the geodesic between two points is no longer an arc of a great circle but depends upon the location of the points. The shapes of geodesics in mountainous regions of the earth's surface depend, of course, on the shapes of the mountains themselves.

Among surfaces the plane is the simplest and most important. What is the geodesic on a plane? We all know the answer and in a sense know it too well. Most people regard the fact that the straight line is the shortest distance between two points as obvious. Indeed, many textbooks on geometry include this statement among the axioms just because it is supposed to be obvious. But the statement implies that the straight-line path between two points is shorter than *any* curve, whether or not it is readily visualizable, between these two points. Since our intuition can and does often deceive us, the assumption that the straight line is the geodesic in the plane is not nearly as obvious as people generally believe. The statement should not be and is not an axiom of geometry in any mature presentation of the subject. But the statement is a correct theorem of mathematics, proved by the method of the calculus of variations.

Let us consider another illustration of the type of problem considered in the calculus of variations. Let C be any curve that joins the points P and Q (fig. 140). Suppose that the curve C is now revolved around a line, the X-axis of figure 140. The path this curve describes is a surface, called a surface of revolution for ob-

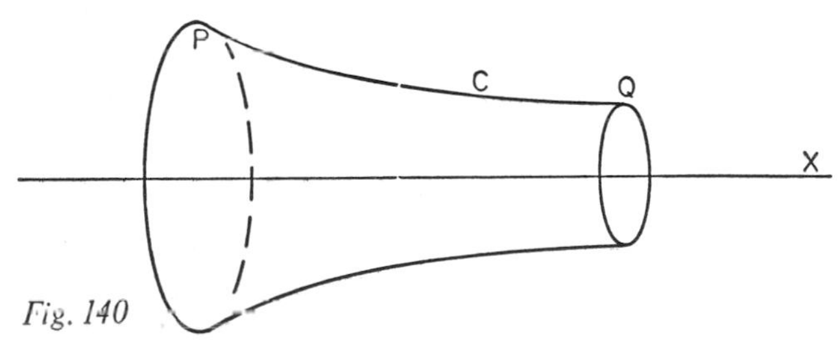

Fig. 140

vious reasons. A question one may raise concerning such surfaces is, What is the shape of the curve C for which the area of the surface is least? In this problem, the area of the surface of revolution depends upon the curve joining P to Q, and we seek that curve for which the area is a minimum.

The curve C that generates the surface of minimum area has been found by the method of the calculus of variations, and it is called a catenary; the surface generated is called a catenoid. Fortunately, we can visualize what this curve looks like. If a rope of arbitrary length is hung between the two points P and Q, the shape it will assume is a catenary. Among the family of such catenaries one will be the minimizing curve. It is certainly worth noting that intuition could hardly have suggested this fact; indeed, the result seems contrary to intuition, because one would expect that the surface of revolution generated by the straight-line segment joining P and Q, the surface of a truncated cone, would have the minimum area. Certainly the distance from P to Q along this surface is less than along the catenoid. But one soon sees that the latter surface is smaller around the middle section than the truncated cone is and therefore may well be the minimum surface of revolution.

Typical of a large class, called isoperimetrical problems, is the problem Dido solved. She found that, among all curves of the same perimeter which bounded area along the seashore, the semicircle bounded maximum area. A slightly simpler isoperimetrical problem is that of finding among all curves of the same perimeter the one which by itself bounds maximum area. This question was raised as far back as Greek times, when Zenodorus, who lived about the time of Christ, tried to prove that the circle bounds more area than any other curve of the same perimeter. Greek geometry was too limited in power to prove this theorem. One can show by geometry, as we did in an earlier chapter, that the circle contains more area than any *rectangle* of the same perimeter, but the general result one seeks is that the circle bounds more area than *any other* polygonal or curvilinear shape. This general problem is quite difficult, and though mathematicians were long ago convinced that the circle is the answer, a rigorous proof was unknown until the late 1800's.

The study of the calculus of variations shows us that man has the knowledge to maximize areas with given perimeter, to find the path of descent requiring least time, and to determine the shortest paths on surfaces. Patently, this knowledge is power. Problems of this sort arise continually in engineering ventures and their solutions enable men to design products so as to utilize minimum materials or to maximize output and profits. For example, a rather simple modification of the least-time problem arises when an airplane, leaving the ground, seeks to reach a given altitude in the shortest possible time. The usual reason for seeking altitude quickly is that once a plane leaves the ground it is safer at high altitudes, for should it have to land again it has more room in which to maneuver. In the case of fighter planes sent to attack an enemy plane there are tactical reasons for wishing to reach the altitude of that plane as soon as possible. This problem is not, as in the case of the brachistochrone problem, that of going from one point to another in the shortest possible time but rather that of going in the least possible time from a point on the ground to any point on a specified horizontal line. One might suppose that this problem is solved simply by having the plane climb straight up. But this solution ignores essential difficulties. Except for some experimental models, planes do not fly straight up; they cannot, in fact, climb at very steep angles. Moreover, the steeper the angle of climb, the less acceleration and therefore the less speed the plane can gather. Hence there is some question as to which path will require least time. The answer, incidentally, differs for different types of planes.

But man's efforts and activities are puny compared to those of the largest entrepreneur in the universe. Nature moves around millions of planets and stars, among which the earth itself is a feather by comparison with others. Nature creates more light and heat in one minute than man will probably create in his entire existence on earth. The gravitational force between two minor planets is far larger than any man-made machine exerts. Certainly nature would be wise to perform these varied activities in the most efficient way and hence learn something about maximizing or minimizing variables. Apparantly nature knows all about the calculus of variations. In fact the real significance of the subject lies not in showing man how to maximize his profits but in revealing a basic principle in the design of the universe.

Let us consider some trivial examples first. Suppose one blows air into a soap bubble, as children do with pipes. The soap bubble assumes a spherical shape. Now the sphere is the smallest surface that can enclose a given volume. Thus the soap bubble assumes the shape that requires the least area to contain the volume of air blown into it. Why does the bubble do this? As air is blown into the bubble the soap film is being stretched. The less surface required to contain the air, the less tension will there be in the soap film. Acting on instructions from nature, the soap film seeks to extend itself least and thus assumes the spherical shape.

The tendency of soap films to extend themselves as little as possible, or to contract as much as possible, can be exploited to obtain visual indications of the solutions of many mundane problems of the calculus of variations. Suppose that *ABC* is a circular fence (fig. 141) forming part of the boundary of a region, and that one wishes

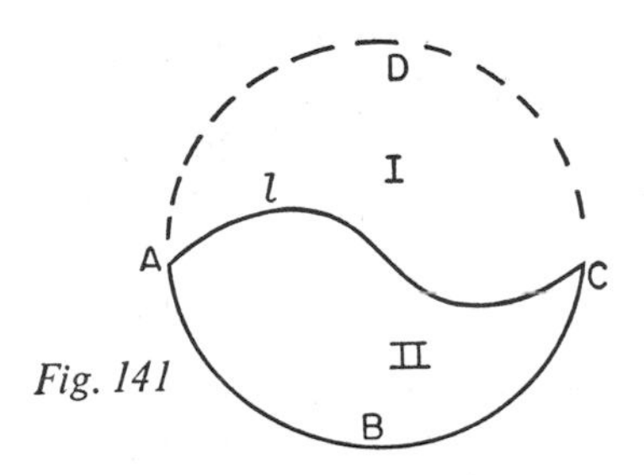

Fig. 141

to complete the boundary by building a fence of given length l from A to C. The question one may raise here is, What should the shape of the fence of length l be so that together with ABC, it bounds maximum area? Of course the length l must be larger than the straight-line distance from A to C, else there would be no latitude in the choice of the shape. Also, the fence would not be placed below the line AC because it can just as well be placed above and enclose more area.

In the light of past experience with isoperimetrical problems one might guess that the desired fence should be a circular arc of length l passing through A and C. Before trying to prove such a conjecture one might use soap films to test it. The proper experiment is carried out by first of all completing the circle of which ABC is an arc and then constructing a circular wire frame of this size, or to scale. One then takes a chord of length l (or its equivalent in the scale adopted) and ties one end to A and the other to C on the wire frame. The entire frame and chord are dipped into soapy water and then withdrawn. A film will form over the entire circle $ABCD$. If one now punctures the film in region II the remaining film will contract as much as possible and so will pull the chord toward D. Of course the less area there is in region I the more there will be in region II. The experiment would show that the chord forms an arc of a circle.

Soap films can be used to show that the catenoid is the minimum surface of revolution generated by a curve. If one dips a circular frame of wire into a soap solution and then withdraws it gently a film will cover the area bounded by the wire. If a second smaller circle of wire is now placed on the soap film and then pulled away while being kept parallel to the first circle, a soap film filling out the catenoidal surface will extend between the two circles. It is interesting to note that as the smaller circle is pulled farther and farther away, the film will ultimately break up and just fill out the two circular areas of the wire frames. This phenomenon illustrates beautifully what the mathematics of this problem predicts, namely, that when Q is far enough from P (fig. 140) there is no minimizing catenary joining P and Q and the minimal surface degenerates into the areas of the two circular disks.

But soap films are nature's toys. The major phenomena of nature behave so as to make some important physical quantity a maximum or minimum. Let us go back for a moment. We may recall Heron's discovery (chapter 6) that light traveling from one point to another by means of reflection from a mirror takes the shortest path. However, when light travels from a point in one uniform medium to a point in another it does not take the shortest path; the actual path, as we have seen, is a broken line (chapter 7). It appears as though reflection and refraction of light are governed by different principles. But this is not so. We pointed out in chapter 22 that in both cases light selects the path requiring least time. Apparently it is the latter principle that determines the path. This conclusion is borne out by the behavior of light from the sun as it travels to a point on the earth's surface through the earth's atmosphere. The atmosphere is very thin far up above the earth, and the speed of light is greater there than near the surface of the earth. The light therefore takes a roundabout path that lies in the upper atmosphere far more than the straight-line path does and then veers toward the earth; yet it takes less time for its journey (fig. 43). Apparently a light ray starting from one point and headed for another considers all possible paths and then wisely chooses the one that economizes most on time.

On the basis of these and other observations on the behavior of light Pierre de Fermat affirmed his Principle of Least Time, which says that of all possible paths that light might take in traveling from a point P to a point Q, it takes the one requiring least time. Because the principle united a number of seemingly unrelated behavioral properties of light it was widely acclaimed. Of course this achievement was in line with the general goal of the seventeenth-century mathematicians and scientists to deduce as many happenings of nature as possible from a small group of physical and mathematical principles.

Fermat's work suggested to others that perhaps his principle could be broadened to include other natural events. The time for the next move became ripe when Newton's work showed how the laws of mechanics apply to the universe. To a limited extent even Newton's own first law of motion states a minimum principle. That falling bodies take the shortest path was observed in the late Middle Ages, and the then current adage was that nature acts by the shortest path. But Newton said more generally that a body in motion and undisturbed by forces will continue to travel in a straight line at a constant speed. The natural preference for straight-line paths is clearly a preference for shortest paths. However, Newton's first law of motion was not nearly as significant as the second law and the law of gravitation. Hence, a minimum principle that might embrace mechanical phenomena would somehow have to include these properties or their consequences.

The mathematician and versatile scientist Pierre L. M. de Maupertuis (1698–1759), a keen student of Newton's work on gravitation, made the next decisive step. Like Euler, Maupertuis studied under John Bernoulli. He then became a soldier in the army of Louis XV of France. At the age of thirty-eight he led an expedition to Lapland and measured the length of a degree of longitude there, confirming the flattening of the earth at the poles. This expedition made his scientific reputation.

Maupertuis prided himself most on his achievement that bears on our story. After having worked in the theory of light and gravitation, he announced, in 1744, a new minimum principle, the Principle of Least Action, from which he claimed he could deduce the behavior of light and masses in motion. The principle asserts that nature always behaves so as to minimize an integral known technically as action, and amounting to the integral of the product of the mass, velocity, and distance traversed by a moving object. From this principle he deduced the Newtonian laws of motion. With sometimes suitable and sometimes questionable interpretation of the quantities involved, Maupertuis managed to show that optical phenomena, too, could be deduced from this principle. Hence, to an extent at least, he succeeded in uniting the optics of the eighteenth century and mechanical phenomena. Despite some vagueness in the formulation of his principle, Maupertuis opened up a new direction for research that supplied many problems to mathematics, and he advanced a theme that was to stir the imaginations of the greatest mathematicians and physicists.

Maupertuis advocated his principle for theological reasons. It seemed to him that the laws of behavior of light and matter must indicate the perfection of God's creation. The least action principle satisfied this criterion because it showed that nature was being economical. He therefore proclaimed his principle to be not only a universal law of nature but also the first scientific proof of the existence of God, for it was "so wise a principle as to be worthy only of the Supreme Being."

Euler, like Maupertuis, believed that the existence of a minimum principle such as least action was no accident, and therefore defended all of Maupertuis's claims for it. Euler himself sharpened the mathematical formulation of the principle, and, as noted earlier, derived the form of the differential equation that the paths of particles and light rays making the action least must satisfy. He also made some applications of the principle. Lagrange then generalized the principle so as to make it applicable to more dynamical problems and showed that the least action principle and the law of conservation of energy may replace Newton's laws of motion as the complete basis for dynamics.

The outstanding English mathematician of the nineteenth century, Sir William Rowan Hamilton (1805–1865), improved upon and generalized the minimum principles of Maupertuis, Euler, and Lagrange. After showing precocity in languages as a youth, Hamilton turned to mathematics. He studied at Trinity College, Dublin, and at the age of twenty-one began to contribute original research. At that same age, while still an undergraduate, he was appointed professor of astronomy at the college and became automatically Royal Astronomer of Ireland, a title he held for the rest of his life.

Hamilton's first big step was to show that all problems of optics can be solved by one method. He arrived at the conclusion by starting from the least time principle of Fermat and the least action principle of Maupertuis. He then proceeded to apply this method to problems of mechanics, and as a consequence was able to formulate a new minimum principle that included Maupertuis's as a special case.

To understand Hamilton's principle, we must utilize the concepts of potential and kinetic energy. Let us consider a simple example. Suppose a ball is held in the air. We know that if released it will fall. Hence the ball has the capacity to do work, and since work technically is the product of the force and the distance through which it acts, the work the ball of mass m is capable of doing is the force it can exert—that is $32m$—times the distance y through which it can fall. This capacity to do work is the potential energy of the ball.

If the ball is released it acquires velocity. The energy the ball possesses by reason of the fact that it is in motion is called kinetic energy; the mathematical expression for it is $mv^2/2$. Of course since y decreases as the ball falls it loses potential energy, but since it falls with increasing velocity it acquires more kinetic energy.

Hamilton's principle considers any motion and presumes that we can write down the kinetic energy T and the potential energy V which an object or a mass possesses at any time during that motion. It then states that the integral of the difference between the kinetic and potential energies over the interval of time during which the motion takes place is a minimum for the path actually taken by the object. In symbols, if I represents the integral, then

$$(2) \qquad I = \int_{t_0}^{t_1} (T - V)\,dt$$

must be a minimum for the path traversed by an object during its motion from time t_0 to time t_1.

In the case of a ball falling under the action of gravity the integral (2) becomes

$$(3) \qquad I = \int_{t_0}^{t_1} \left(\frac{mv^2}{2} - 32my\right) dt$$

and the path taken by the ball must be the one that makes this integral a minimum. Were we to write Euler's differential equation for the minimizing path and solve this equation, we would obtain the well-known law for bodies falling straight down under the action of gravity.

Hamilton's principle yields the paths of falling bodies, the paths of projectiles, the elliptical paths of bodies moving under the law of gravitation, the laws of reflection and refraction of light, and the more elementary phenomena of electricity and magnetism, which were the only ones known about the last two subjects in his time. The principle can also be used to solve new problems of mechanics, optics, electricity and magnetism. However, the chief significance of the principle lies in showing that the phenomena of all these branches of physics satisfy a minimum principle. Since it relates these phenomena by a common mathematical law, it permits conclusions reached in one branch to be reinterpreted for another. Hamilton's principle is the final form of the least-action principle introduced by Maupertuis, and because it embraces so many actions of nature it is the most powerful single principle in all of mathematical physics.

It is interesting to note that whereas Maupertuis and Euler believed that God designed the universe in accordance with a minimum principle, Hamilton, and even Lagrange before him, no longer attached metaphysical or theological significance to their principles. With exceptions, such as the famous twentieth-century physicist Max Planck, the creator of quantum theory, scientists have kept apart from their professional work the religious convictions that formerly motivated some of the finest research. From the modern viewpoint the inspiration the seventeenth- and eighteenth-century mathematicians and scientists drew from religious beliefs seems strange.

To the scientists of 1850, Hamilton's principle was the realization of a dream. We have already had numerous occasions to mention that from the time of Galileo scientists had been striving to deduce as many phenomena of nature as possible from a few fundamental physical principles. As we know, they made striking progress in this endeavor. But even before these successes were achieved Descartes had already expressed the hope and expectation that all the laws of science would be derivable from a single basic law of the universe. This hope became a driving force in the late eighteenth century after Maupertuis's and Euler's work showed that optics and mechanics could very likely be unified under one principle. Hamilton's achievement in encompassing the most developed and largest branches of physical science, mechanics, optics, electricity, and magnetism under one principle was therefore regarded as the pinnacle of mathematical physics.

Max Planck has expressed the regard of scientists for Hamilton's principle. Speaking of physical science, he says, "It has as its highest and most coveted aim the solution of the problem to condense all natural phenomena which have been observed and are still to be observed into one simple principle, that allows the computation of past and more especially of future processes from present ones.

. . . Amid the more or less general laws which mark the achievements of physical science during the course of the last centuries, the principle of least action is perhaps that which, as regards form and content, may claim to come nearest to that ideal final aim of theoretical research."

Hamilton's principle still is a basic one for many branches of physics. But we must now recall that after Hamilton's time, electromagnetic theory entered science as a new branch and altered completely the mathematical theory of light. The theory of relativity generalized Newtonian mechanics and set it upon a totally different foundation. And quantum theory has altered completely our approach to microscopic mechanical and electromagnetic phenomena. Hence, the minimum principle that unified the knowledge of light, gravitation, and electricity of Hamilton's time no longer suffices to relate these fundamental branches of physics. Within fifty years of its creation, the belief that Hamilton's principle would outlive all other laws of physics was shattered. Minimum principles have since been created for separate branches of physics such as electromagnetism and hydrodynamics, that is, the flow of fluids, but these are not only restricted to the separate fields but seem to be contrived rather to express some natural physical principle such as least time or least action. The hope of revising the principle so that it will achieve the unification of modern physics still drives mathematicians. This is the problem to which the greatest mathematical physicists of our day have devoted themselves and the problem to which Einstein devoted the last years of his life. Stripped of the theological associations, the belief in a minimum principle still activates physical science.

Whether or not the search for such a principle will be successful, the very effort points up how seriously mathematicians and physicists still pursue the program of Galileo and Newton. A single minimum principle, a universal law governing all processes in nature, is still the direction in which the search for simplicity is headed, with the price of simplicity now raised from a mastery of differential equations to a mastery of the calculus of variations.

26: NON-EUCLIDEAN GEOMETRIES

One must regard nature reasonably and naturally as one would the truth, and be contented only with a representation of it which errs to the smallest possible extent.

—JOHN BOLYAI

TOWARD the ends of their lives, Euler, D'Alembert, and Lagrange agreed that the realm of mathematical ideas had been practically exhausted and that no new great minds were appearing on the mathematical horizons. Of course, these men had grown old and their vision was already dimmed, for Laplace, Legendre, and Fourier were in young manhood. In one respect, however, these elder statesmen of the mathematical kingdom were correct in their evaluation of the state of affairs; their immediate successors continued to explore and polish the very same ideas which the mid-eighteenth century had pursued.

But history shows that the human mind is fertile, ingenious, and creative beyond all possible anticipations. It is true that even the richest vein of thought is ultimately exhausted, and then, indeed, a period of stagnation may ensue. Inevitably, however, there arise brilliant new conceptions and new periods of feverish and rewarding research. Euler and his contemporaries failed to reckon with history.

The man who was to change the course of mathematics was but six years old when Euler and D'Alembert died in 1783, and though the youngster was precocious—he used to say jokingly that he could reckon before he could talk—he could hardly have attracted attention at that age. Karl Friedrich Gauss's brilliance was noticed by his elementary school teachers, and they helped him to secure a good education. He was so bright that even his university teachers were not fully able to keep up with his work. There is a story that at Göttingen he approached one of his university teachers, A. G.

Kästner, with a proof that the 17-sided regular polygon is constructible with straightedge and compass—at that time one of the outstanding construction problems. Kästner was incredulous and, seeking to dismiss Gauss as university teachers today dismiss angle-trisectors, said the construction was unimportant because practical constructions were known. Of course, he knew that the problem was of theoretical interest, but he did not believe Gauss could solve it. Gauss then tried to interest Kästner by pointing out that he had solved a seventeenth degree algebraic equation to perform the construction; the teacher replied that the solution was impossible. Then Gauss explained that he had reduced the solution of the seventeenth degree equation to one of lower degree and then solved the latter. "Oh well," said Kästner, "I have already done this." For this rebuff, Gauss repaid Kästner, who prided himself on being something of a poet, by lauding him as the best mathematician among poets and the best poet among mathematicians.

It was the solution of the construction problem mentioned above at the age of nineteen that convinced Gauss, who had displayed remarkable ability in the study of languages also, that he should make mathematics his life's work. He did, and became the leading creative mathematician of the star-studded nineteenth century. The regard in which he was held by his contemporaries is amply shown by another story. Laplace was asked one day who was the greatest mathematician in Germany, and he replied, "Pfaff." The interrogator, surprised, remarked that he had been led to believe that it was Gauss. "Oh," replied Laplace, "Pfaff is the greatest mathematician in Germany, but Gauss is the greatest mathematician in Europe." Despite the fact that he was the unquestioned master of his day, Gauss was extremely modest. His observation, "If others would but reflect on mathematical truths as deeply and as continuously as I have, they would make my discoveries," not only exhibits this modesty but offers much encouragement to aspiring mathematicians.

Gauss is always described as a mathematician, but it would be more appropriate to describe him as a student of nature. He adopted this motto:

Thou, nature, art my goddess; to thy laws
My services are bound . . .

Of course, Gauss appreciated that mathematics offers the supreme method for the investigation of nature, and he coined the now well-known phrase, Mathematics, Queen of the Sciences, to emphasize this evaluation. In view of this devotion to nature, one should not be surprised to find that while he contributed mightily to such branches of mathematics as the theory of numbers, topology, differential geometry, the theory of functions of a complex variable, and non-Euclidean geometry, he also made major contributions to the physical sciences of astronomy, electromagnetism, terrestrial magnetism, and geodesy. The concreteness of his interests extended to inventions—he was one of the inventors of the electric telegraph—the conduct of land surveys, map-making, systems of insurance, and the statistical study of errors of measurement. Gauss is commonly ranked with Archimedes and Newton. It is more than pertinent to the understanding of the role and meaning of mathematics that all three of these men were as much devoted to physical research as to mathematics.

Of all of Gauss's creations, the most revolutionary in concept and the most weighty in affecting the course of mathematics and science is non-Euclidean geometry. To understand just what this creation is and why it proved to be so significant requires going back a bit. As of 1800, mathematics rested upon two foundations, the number system and Euclidean geometry. In Gauss's time, mathematicians would have emphasized the latter because many facts about the number system, and about irrational numbers especially, were not logically established nor clearly understood. Indeed, those properties of the number system that were universally accepted were still proved by resorting to geometric arguments, much as the Greeks had done 2500 years earlier. Hence, one could say that Euclidean geometry was the most solidly constructed branch of mathematics, the foundation on which many other branches were erected, the surest body of knowledge man possessed. Nevertheless, there was an undercurrent of dissatisfaction with at least one of the axioms on which Euclidean geometry and therefore all the subsequent developments of mathematics rested.

As stated by Euclid, the axiom runs as follows: If two lines in one plane intersect a third in such a way as to make the sum of the interior angles on one side of the third line less than two right angles, then these two lines will meet on that side of the third line on which these angles lie. In terms of figure 142, if the sum of angles 1 and 2 is less than two right angles, then the lines l and m will meet somewhere to the right of the third line n. The modern

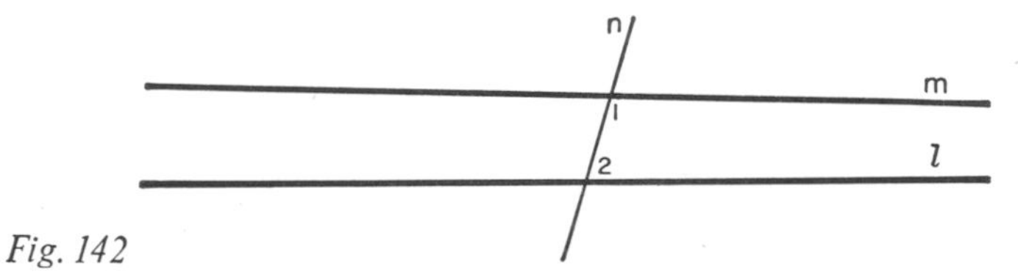

Fig. 142

student of mathematics may not recognize this axiom because it is customary today to replace it by the following more familiar one, namely, given a line l (fig. 143) and a point P not on that line, then there exists one and only one line m in the plane determined by P and

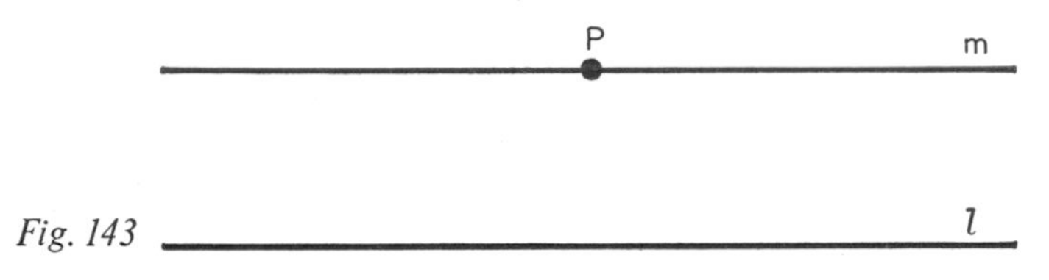

Fig. 143

l which passes through P and does not meet l. The axiom asserts, in other words, that there is one and only one line through P, which is parallel to l. This alternative statement is equivalent to Euclid's in that it may be used in conjunction with the other Euclidean axioms to deduce all the theorems of Euclidean geometry.

Let us look for a moment at the second version. No student of mathematics hesitates to accept this assertion as an unquestionable truth, partly because he meets it for the first time when he is still a little too young to be critical, and partly because it does seem to correspond with experience. Yet, more careful examination shows us that it should be regarded critically. The axioms of Euclidean geometry were accepted as self-evident truths because they asserted facts about physical space that observation and experience immediately confirmed. But this axiom asserts that the lines l and m will never meet despite the fact that these lines extend indefinitely far on either side or, as we say, extend to infinity. Since experience and observation are confined to a limited region about the earth's surface, there is some question as to what happens indefinitely far out in space. It may not be correct to assume that the idealized space which is supposed to describe our physical world contains parallel lines.

Euclid tried to be much more cautious. His version does not assert that two lines will never meet, but rather states a condition under which they will meet at some finite point. In this respect his version is superior. On the other hand, it is a more complicated statement than one would like to accept as a supposedly self-evident truth. Hence, we can see why some mathematicians from Euclid's time on were not wholly satisfied with his statement and tried to remedy its deficiency by either of two procedures. Euclid had stated ten axioms for his geometry. Some mathematicians tried to deduce Euclid's parallel axiom from the other nine axioms. Had they succeeded, then, of course, the Euclidean axiom would have become a theorem deduced from the other nine unquestionable axioms; and since deduction gives unquestionable conclusions, there could have been no doubt about the correctness of the Euclidean statement. This procedure failed, though for a while many mathematicians thought they had succeeded. But each made some assumption that was later recognized to be unjustified.

The other procedure attempted by mathematicians was to replace the Euclidean parallel axiom by a simpler, more intuitively appealing statement that would nevertheless yield the theorems already proven. In this case, one would still have ten axioms, but each would be unquestionably applicable to the physical world. A number of alternative axioms were indeed proposed, including the one above (fig. 143), which is due to John Playfair. At first blush, each of these proposed axioms seemed preferable to Euclid's, but when examined more closely, was found to be questionable for the same reason that Playfair's axiom is, namely, these alternative axioms made assertions about what happens indefinitely far out in space. As we have seen, such assertions are not supported by our experience with or our perceptions of physical space.

The history of these attempts to overcome the objection to Euclid's parallel axiom is a lengthy one extending over 2200 years. The greatest mathematicians participated in it. So lengthy and so prodigious were these efforts that in 1767 D'Alembert called the problem of the parallel axiom the scandal of geometry. This history is so readily accessible in other sources that we shall not dwell on it here, except to call attention to the persistence of mathematicians and to the extraordinary care with which they analyze every concept and assertion falling within their province.

The persistence finally brought results. It is true that it took the genius of one of the best mathematicians, but the fruitless efforts of countless predecessors no doubt helped Gauss to see the correct solution.

Gauss himself had tried without success to replace Euclid's parallel postulate by other more acceptable assumptions. But even before, and while making these efforts, in fact from the time that he was about fifteen years old, he suspected that the task was impossible and that it was equally impossible to deduce the Euclidean axiom from the other nine. But this statement puts the matter negatively. What Gauss appreciated was that to obtain the theorems of Euclidean geometry one had to include an axiom which made an assertion, either Euclid's or an equivalent one, about how straight lines behaved at infinity. If such an *independent* statement is needed to construct Euclidean geometry, then it should be *logically* possible to make a different assertion about the behavior of straight lines at infinity. In other words, there could be a consistent, logical development of a geometry in which a parallel axiom contradictory to Euclid's holds. This was the bright new concept. Gauss actually saw farther than this, but let us explore this thought first.

It should be possible to adopt a parallel axiom that contradicts Euclid's and yet develop a consistent geometry. Gauss actually constructed such a geometry, but unfortunately did not publish his results. The whole question of publication was a sore point in Gauss's work. Hosts of new ideas continually flooded his mind, and as he explored them he jotted down on loose sheets and in a scientific diary brief notes of what he proved. To write up these ideas carefully for publication in scientific journals would have required additional weeks and even months of labor. Gauss felt so pressed to explore new ideas that he would not take time for the preparation of publications. What he did publish, in itself extensive, he rewrote many times, polishing and repolishing. He wanted his publications to be logically complete and precise. On his seal he expressed this devotion to perfected work by the motto *Pauca sed matura* (few but ripened). These polished papers were, however, not easy to read. The simpler steps of the proofs were omitted, as were the intuitive meaning, motivation, and larger goals of the work. Even good mathematicians encountered considerable difficulties in understanding Gauss's papers.

Since Gauss did not make known many of his brilliant creations, other mathematicians worked toward and arrived at the same results independently. This duplication of effort was not only a loss to mathematics but it often caused considerable frustration to those who found, after having labored for years on a new creation, that Gauss had anticipated them. One of these men, Carl G. J. Jacobi, was especially distressed because on several occasions when he came to Gauss to relate some new discoveries the latter pulled out from his desk drawer some papers that contained the very same discoveries. Jacobi resolved to get even. He visited Gauss one day and showed him some new work. Just as Jacobi had feared, Gauss opened his desk drawer and pulled out some papers covered with the very same results. Jacobi then remarked, "It is a pity that you did not publish this work since you have published so many poorer papers."

In the case of non-Euclidean geometry, Gauss had an additional reason for not publishing his work. In the early nineteenth century, geometry meant the geometry of physical space, and Euclid's version was accepted as an absolute, unchallengeable body of truths. Hence, anyone professing to present another geometry conflicting with Euclid's would undoubtedly have been judged insane not only by the general public, which was ignorant of all geometry, but by mathematicians, scientists, and philosophers. In a letter to the mathematician Friedrich Wilhelm Bessel, written in 1829, Gauss says explicitly that he feared the clamor of the Boeotians, a metaphorical reference to one of the dull-witted Greek tribes. There are limits to allowable heterodoxy even in mathematics, and these limits are soon reached in every era. Gauss's work on the new geometry, which he first called anti-Euclidean, then astral geometry, and finally non-Euclidean geometry, was found among his papers after his death.

The first men to publish a non-Euclidean geometry were Nicholas I. Lobatchevsky (1793–1856), a Russian who was at one time professor of mathematics at the University of Kazan, and John Bolyai (1802–1867), a Hungarian who became an officer in the Austrian army. Both men seem to have appreciated by 1823 that there can be geometries different from Euclid's. Lobatchevsky published his first paper on the subject in 1829–30, and Bolyai's work appeared in an appendix to his father's text on mathematics, published in 1833. That Gauss, Lobatchevsky, and Bolyai should arrive independently at one of the greatest advances in the history of thought appears at first to be one of the remarkable coincidences of history. Actually, the whole subject of what to do about Euclid's parallel axiom was being actively pursued in various European centers, and there were many other mathematicians who, before and after 1830, proved a number of the theorems that are the technical essence of the work of these three men but without quite drawing the correct implications. Moreover, Gauss had written about his ideas to a number of friends, among them Wolfgang Bolyai, John's father, and Lobatchevsky had studied under Bartels, Gauss's friend and countryman. Hence, the suggestion to work on this subject, and perhaps the most important point of all, the realization that there can be non-Euclidean geometries, may have come from Gauss. In fact, knowing of Gauss's interest in this subject, Wolfgang sent his son's work to Gauss for approval. Gauss replied:

"If I commenced by saying that I am unable to praise this work, you would certainly be surprised for a moment. But I cannot say otherwise. To praise it would be to praise myself. Indeed, the whole contents of the work, the path taken by your son, the results to which he is led, coincide almost entirely with my meditations, which have occupied my mind partly for the last thirty or thirty-five years."

Wolfgang was pleased with this reply because his son was being praised. But John, like many others who found that Gauss had anticipated them, never overcame his dislike for Gauss.

The technical content of the non-Euclidean geometry created by Gauss, Lobatchevsky, and Bolyai is readily described. Their chief idea was the one we have already mentioned, namely, that Euclid's parallel axiom is an independent assumption about parallel lines and hence that it is at least logically possible, whether or not it serves any scientific or practical purpose, to replace it by a con-

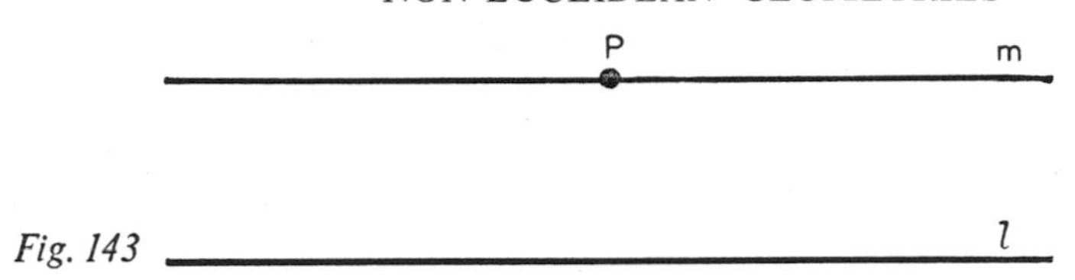

Fig. 143

tradictory axiom. What alternative axiom did these men adopt? Given the point P and line l (fig. 143), Euclid had assumed that there was one and only one line through P (in the plane of P and l) which did not meet l. One alternative would be to assume that every line through P met l; that is, there are no lines through P which are parallel to l. This possibility had been tried by several mathematicians whose objective at the time was to deduce Euclid's parallel axiom from the other nine, and they had found that theorems deduced from this axiom and the other nine axioms of Euclid contradicted each other. This outcome meant that such an alternative to Euclid's parallel could not be entertained, since a body of inconsistent results certainly made no sense. Hence, the only alternative open to the creators of non-Euclidean geometry was to suppose that

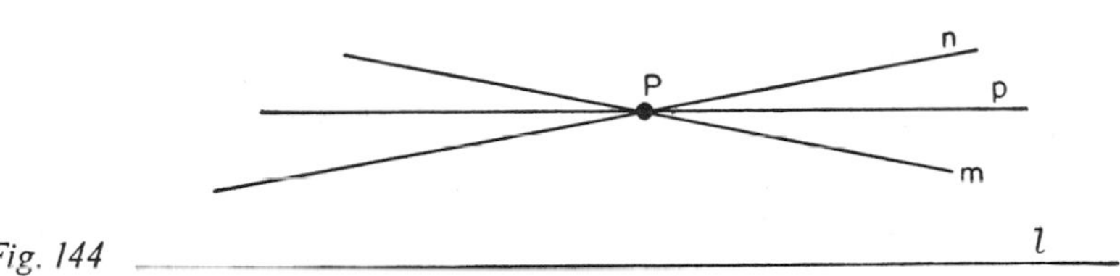

Fig. 144

through P there pass at least two lines which do not meet l. Of course, if m and n are two lines (fig. 144) through P that do not meet l, then any other line through P, such as p, lying between m and n, could likewise not meet l; to do so, p would have to intersect m and n, at some other point in order to reach l, and this would be impossible since two lines can intersect in at most one point. Hence, the assumption that there are at least two parallels to l through P really amounts to assuming that there are an infinite number of such lines. This is the assumption made by Gauss, Lobatchevsky, and Bolyai.

This assumption seems silly and physically false. Most people would protest at once that while there may be one parallel to l through P, any other line through P would have to meet l. That is, most people are so accustomed to the Euclidean parallel axiom, and this axiom accords so well with their physical intuition, which is, incidentally, partially the product of experience and training, that any other assumption seems ridiculous. But at this point one must face squarely the most vital feature of mathematics. The subject is concerned with the logical development of the implications of sets of axioms. In the cases of the development of number and Euclidean geometry, the axioms were intuitively acceptable, indeed seemingly self-evident, and it therefore seemed reasonable to explore their implications. On the other hand, the axiom adopted by Gauss, Lobatchevsky, and Bolyai does not seem to be intuitively acceptable; hence, one believes the axiom to be absurd and reasoning with it impossible. But there actually is no barrier to the course set by these men, any more than there is to investigating an ideal society composed of thoroughly honest leaders, despite the fact that such a society does not fit our experience.

Let us try to put past experiences and intuition aside for the present and see what follows from the new set of axioms, that is, the nine axioms of Euclid and the new parallel axiom. Since the methods of proof used to obtain theorems in the new geometry are precisely those used in Euclidean geometry, we shall not look into them but rather concentrate on the theorems themselves. Some of these are identical with the Euclidean theorems since their proofs do not depend upon the parallel axiom. Thus, the theorem that the base angles of an isosceles triangle are equal holds in both geometries. However, the use of the new parallel axiom gives rise to hundreds of theorems differing radically from those in Euclid. We shall note a few of these.

Perhaps the most striking of the non-Euclidean theorems states that the sum of the angles of a triangle is always less than 180°. Moreover, this sum is not always the same quantity but depends upon the area of the triangle. The smaller the area, the closer the angle sum is to 180°. Another surprising theorem states that if the angles of one triangle are equal respectively to the angles of another, then the triangles are *congruent*. In Euclidean geometry, of course, the equality of corresponding angles would imply only that the triangles are similar; that is, they have the same shape but may differ enormously in size. Another interesting non-Euclidean theorem states that if m and l are parallel lines (fig. 145) and PQ is the perpendicular from P to l, then angle 1 at P is an acute angle; in Euclidean geometry it is a right angle. Moreover, the size of this angle increases as the perpendicular distance from m to l becomes shorter. Thus angle 2 is larger than angle 1 because $P'Q'$ is shorter than PQ. In this new geometry the Pythagorean theorem is replaced by a more complicated theorem. Clearly this non-Euclidean geometry contains many strange theorems, as might be expected from the strangeness of the parallel axiom on which they rest.

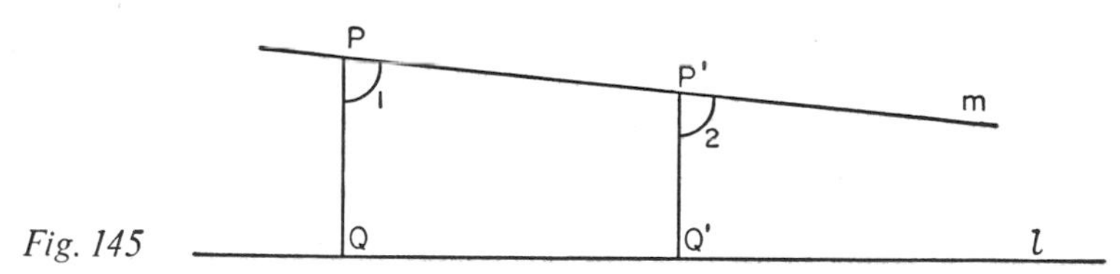

Fig. 145

And now that we have this mathematical development, what shall we do with it? Let us assume for a moment that it is no more than an aberration of human thought. And with the perversity so often found in human beings, let us defy common sense still more. Gauss, Bolyai, and Lobatchevsky had shown that it is possible to construct a new geometry by adopting a new parallel axiom. Why confine the change in axioms to just the parallel axiom? What would happen if we changed other axioms in addition? Of course, we should be judicious. If we should adopt a set of axioms that led to contradictory theorems, our work would be totally meaningless. Also, if we are going to change any more of Euclid's axioms, we should have some reason for making this change. There might then be the possibility of learning something significant about Euclidean geometry even if the new geometry proved to be useless in and for itself. Thus, the non-Euclidean geometry just examined shows us that any change in Euclid's parallel axiom produces an enormous change in the theorems. In any event, one should not look for profit or gain in undertaking a scientific venture. The pursuit of an idea may or may not bring rewards, but the chase itself may be exciting.

Some twenty years after the work of Bolyai and Lobatchevsky, another profound mathematician, Georg Friedrich Bernhard Riemann, advanced good reasons for changing another Euclidean axiom. Riemann, like many other mathematicians, began his studies with the intention of entering the ministry, and in this instance, too, the strife between science and religion for the possession of brilliant minds ended with a victory for mathematics. At the University of Göttingen, the shy and gifted Riemann studied mathematics under Gauss and physics under W. Weber, Gauss's collaborator in the invention of the telegraph. Like Gauss, Riemann was deeply interested in physical science, and from that source drew the inspiration for his mathematical investigations.

In the field of non-Euclidean geometry, Riemann wrote one of the classic mathematical papers, entitled "On the Hypotheses Which Underlie Geometry," and he began by calling attention to a distinction that seems obvious once it is pointed out: the distinction between an unbounded straight line and an infinite straight line. The distinction between unboundedness and infiniteness is readily illustrated. A circle is an unbounded figure in that it never comes to an end, and yet it is of finite length. On the other hand, the usual Euclidean concept of a straight line is also unbounded in that it never reaches an end but is of infinite length.

Riemann emphasized that experience suggests the unboundedness of the straight line rather than its infiniteness. Hence, he proposed to replace the infiniteness of the Euclidean straight line by the condition that it is merely unbounded. He also proposed to adopt a new parallel axiom, namely, given a line and a point not on that line, then every line through the point and in the plane of the given point and line meets the given line. In brief, there are no parallel lines. This last mentioned axiom had been tried, as mentioned earlier, in conjunction with the infiniteness of the straight line and had led to contradictions. However, by assuming the unboundedness of the straight line, the new parallel axiom, and other axioms of Euclid, Riemann found that he could construct another consistent non-Euclidean geometry.

Before proceeding to illustrate the theorems Riemann established, we might concede that to the novitiate Riemann's axioms are as disturbing as the parallel axiom of Gauss, Bolyai, and Lobatchevsky. But the argument here is again that which we insisted on in discussing the preceding non-Euclidean geometry. We are logically free to adopt and pursue the implications of Riemann's axioms even though our intuition tells us that these axioms do not make physical sense. Let us therefore favor our minds rather than our senses and note some of the theorems that follow from Riemann's axioms.

He proves readily that the straight line must have the structure of the circle and like the circle have finite length. The sum of the angles of any triangle turns out to be *greater than* 180° and *increases* with the area of the triangle. As in the earlier non-Euclidean geometry, two triangles with equal corresponding angles must be congruent. In this non-Euclidean geometry, too, the Pythagorean theorem is replaced by a more complicated theorem. And there are many more such strange theorems.

We must come to grips now with the question we rather cavalierly shouldered aside before. Of what use are these non-Euclidean geometries? Of course, the very fact that the concept of geometry was broadened by the creation of these geometries is in itself something of a gain. Mathematicians had very naturally thought that there was but one geometry, the Euclidean idealization of physical space, and they now realized that they could create other consistent geometries. But this gain might be regarded as a minor one, a paper profit.

As a matter of history, the attitude of almost all mathematicians for many years was that the new geometries were just logical toys, and that Euclidean geometry was indubitably the representation of physical space. But this view did not prevail. Here, too, Gauss led the field. In 1817, Gauss wrote to a friend, "I am becoming more and more convinced that the necessity of Euclidean geometry cannot be proved." And from this time onward, Gauss took seriously the possibility that non-Euclidean geometry—he knew just one—might be as applicable to the physical world as Euclidean geometry was. In fact, he acted on this thought. The test of the applicability of a geometry is whether the theorems fit physical facts. Now Gauss's geometry asserts that the sum of the angles of a triangle is less than 180° and Riemann's, that the sum is greater than 180°. Of course Euclidean geometry asserts that the sum is exactly 180°. Gauss thereupon undertook to measure the sum of the angles of a triangle. He stationed an observer on each of three mountain peaks and directed each one to measure the angle between his lines of sight to the other two observers. The sum of the measures of the angles of the triangle formed by the three peaks was within 2″ of 180°.

This test would at first glance appear to decide the issue in favor of Euclidean geometry. But we are overlooking something. In both Gauss's and Riemann's geometries there is a theorem which states that the smaller the area of a triangle, the closer the sum of the angles is to 180°. Small is of course a vague word, and certainly, compared to the triangle formed by the sun, moon, and the earth, Gauss's triangle was a small triangle. Hence the sums that these new geometries call for could be expected to be very close to 180°. A departure of 2″ from 180° might be just enough to encompass either of these geometries. To make matters worse, all measurements are subject to error and in Gauss's method the error was appreciably greater than 2″, so that the correct sum for his triangle could very well have been significantly more or less than 180°. Thus the test was not decisive. Had Gauss been able to measure the angles of a really large triangle, the angle sum might have proved to be considerably less than or more than 180° and thus favor one of the non-Euclidean geometries.

There are, however, other arguments for and against the applicability of non-Euclidean geometry to physical space, and these should be considered. For example, both of the non-Euclidean geometries assert that two similar triangles must be congruent. It seems as though one could construct physically a very small triangle and a very large one, with equal corresponding angles. These triangles would be similar but not congruent, contrary to what the non-Euclidean geometries assert. Hence, the conclusion would appear to be that these geometries cannot describe physical space. But such a construction would actually prove nothing. How can one be sure that he has constructed two triangles with identical angles? All physical constructions are approximate. One might indeed make the effort to construct two exactly similar triangles, but their corresponding angles might actually differ by an amount too small to measure. Thus, the argument based on this construction fails.

The staunch Euclidean advocate might make one more rebuttal. "Perhaps one can't reject the possible applicability of non-Euclidean geometry by using the theorems as a test. But the axioms certainly do not apply, and hence there is no reason for the theorems to apply. I accepted the axioms previously only for the purpose of seeing what might follow logically from them, but there never was any doubt in my mind that these axioms had no connection with the physical world." To answer this argument, we need only recall the reason mathematicians found for questioning the Euclidean parallel axiom. It makes an assertion about what happens infinitely far away. And experience does not tell us what happens in such regions. By the same token, experience does not preclude either of the new parallel axioms, since their assertions differ from the Euclidean statement only in what happens over large distances.

The upshot of the entire attempt to decide which geometry fits physical space is that all do equally well insofar as ordinary experiences in limited regions are concerned. There remains only the possibility that in really large regions one might find that one geometry fits better than the others. Gauss and Lobatchevsky saw this point almost from the outset. Both considered the possibility of their form of non-Euclidean geometry applying to space among the stars, and Gauss, as we noted, once called his new geometry astral geometry.

We have insisted thus far that the non-Euclidean geometries can be applied to physical space as readily as Euclidean geometry can be. Implicit in the entire discussion was the understanding that the mathematical straight line was to be interpreted physically as the ruler's edge or the stretched string. This interpretation is, of course, the natural one, for, in fact, the mathematical straight line was suggested by these physical counterparts. For practical work conducted on or near the earth's surface and in limited regions, the stretched string, stretched tape measure, or ruler's edge is indeed the most convenient and useful interpretation of the straight line. Let us consider, however, the very simple problem of triangulating a region. Surveyors begin by adopting a base line whose length is measured off by laying down a yardstick or tape measure. This might be *AB* of figure 146 (page 458). But how are *AC* and *CB* determined? By sighting points *C* and *B* through a telescope located at *A*, the angle *CAB* is determined. Likewise, angle *ABC* is determined. Then *AC* and *BC* are calculated by trigonometric formulas. In this process, the surveyor assumes that the triangle *CAB* is the one that could be formed by strings stretched between

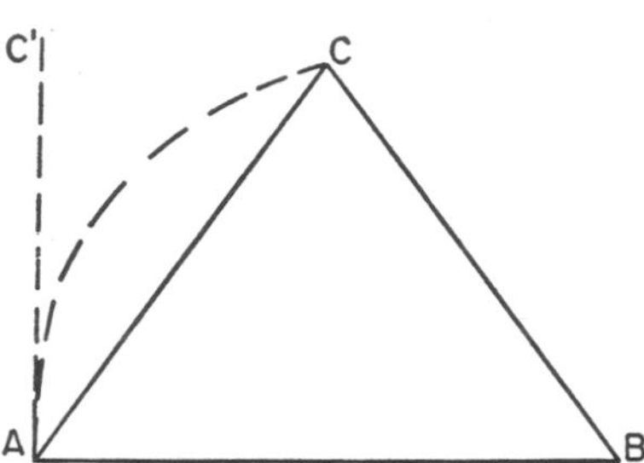

Fig. 146

C and A and between B and A, and hence that he is justified in applying Euclidean trigonometry.

Actually, the point C is observed at the point A because a light ray travels from C to A. Hence, the surveyor is really assuming that the light ray travels along the stretched-string path between C and A. It may not. The path of the light ray may actually be the broken line shown in the figure. At point A, the light ray appears to come from the direction $C'A$ and so the angle the surveyor really measures is angle $C'AB$. Whether or not the light ray does follow the stretched-string path CA—we shall have more to say about the paths of light rays shortly—the fact is that in obtaining the distance CA, the surveyor did not use the stretched string for that distance but the path of a light ray; and he assumed that the light ray followed the stretched-string path.

And now let us consider the most precise of all sciences—astronomy. In determining the distance, size, and path of any heavenly body, the astronomer necessarily uses light rays to obtain the direction of the body. Hence, the astronomer, as well as the surveyor, actually uses the path of a light ray as his physical straight line. He may or may not find it reasonable to assume that this path is also the stretched-string path, but he cannot escape the use of the light ray. Hence, before we consider the question of what paths light rays actually follow, we should be clear that the natural choice of the physical straight line for many scientific purposes, and indeed fundamental scientific purposes, is not the stretched string but the path of the light ray.

What are the paths of light rays? Generally they are not straight. If we restrict ourselves to very small regions in which the atmosphere is uniform, then light rays do follow straight paths. But rays that travel through the atmosphere for any considerable distance—and certainly light rays that reach us from the sun, stars, and planets—are not straight (chapter 25).

Let us accept, then, the facts that scientists use light rays to determine directions and distances, and that light rays are not straight in the sense of stretched strings. What have these facts to do with non-Euclidean geometry and its applicability to the physical world? Why can't we recognize the curvature of light rays and still use the usual stretched-string interpretation of the straight line and, of course, the only geometry that makes sense; namely, Euclidean geometry? But the very point of the entire discussion of light rays is that they are the only tool we have for bridging large distances. They are the only practical straight lines we have except in the case of the very short distances where stretched strings or yardsticks can be used. If, then, the actual straight lines are light rays, we must now ask, What axioms do these light rays obey? If l is a light ray (fig. 147) lying in one plane, say the plane of the paper, and P is any point in this plane, then could there be many light rays through P that do not meet l, no matter how far out these rays extend? Or

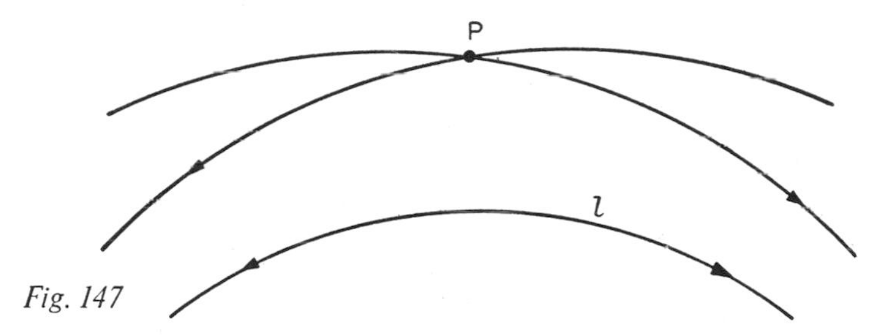

Fig. 147

is it possible that every light ray through P meets l somewhere? Perhaps to save embarrassment on all sides, we shall not answer these questions. All that matters now is that we recognize that there is a question about what axioms light rays satisfy and hence about which geometry applies to the physical world.

The subject of non-Euclidean geometry abounds in fascinating ideas, some of interest primarily to the mathematician, others bearing on the applicability of the subject to the physical world. Were we to pursue these ideas we should find that the entire question of the applicability of non-Euclidean geometry is no longer academic. There are many other varieties of non-Euclidean geometry, more complicated than the ones we have examined, and known as Riemannian geometries. One of these Riemannian geometries is used in the theory of relativity and thereby furnishes the best scientific theory for large-scale phenomena and for masses moving with high velocities. Gauss and Lobatchevsky would have relished this theory.

Those mathematicians who, toward the end of the eighteenth century, were fearful that the subject had exhausted itself, failed to reckon with non-Euclidean geometry, for this creation revolutionized the status and prospects of mathematics and science and shattered many long established philosophical doctrines. Indeed, it is no exaggeration to say that non-Euclidean geometry was the most weighty intellectual creation of the nineteenth century, or, at worst, might have to share honors with the theory of evolution. We must, however, confine ourselves here to the effects the new geometries exerted on the future of mathematics.

Mathematics had always been regarded as a sort of special science. Like science, it sought knowledge about the real world, but it confined its search to quantitative relationships and geometrical properties. But unlike those of science, the conclusions of mathematics had been regarded as deduced from basic truths. Therefore mathematical knowledge was distinguished from scientific knowledge by the certainty of its results. In fact, the very reason that mathematicians persisted for so many centuries in attempting to find simple equivalents for Euclid's parallel axiom, instead of entertaining contradictory possibilities, is that they could not conceive of geometry being anything else than the true geometry of physical space. On the other hand, though science seeks to penetrate the superficial impressions of the physical world that our senses receive, and to obtain a deeper and more coherent understanding of the underlying mechanisms, on the whole, scientists had recognized that their researches produced only theories which do not provide a veridical description of what takes place, but must be constantly altered to fit new facts or to supply simpler explanations than earlier ones. The current scientific theories in all eras were understood to be merely the best connected accounts of the particular phenomena embraced by the theory. Science, in other words, is a grand detective story that takes the clues obtained by observation and experimentation, builds up an account of how all these clues are related, and predicts the murderer. If later evidence proves to be inconsistent with the theory, it is revised even if the supposed murderer has already been hanged.

The creation of non-Euclidean geometry showed first of all that mathematics could no longer be regarded as a body of unquestionable truths. Certainly not more than one of the three geometries we have examined, to say nothing of the Riemannian geometries created later in the nineteenth century, could be the truth. But then the question arose as to whether any one of them was. Mathematics retained its deductive method of establishing its conclusions, but it was soon appreciated that mathematics offers only certainty of proof on the basis of uncertain axioms.

Insofar as the study of the physical world is concerned, mathematics acquired the same status as any branch of science. It offers nothing but theories. As long as any geometry agrees with measurements and predicts correctly, it is the accepted theory of space. But if new facts show that the theory is not accurate, a better one must be constructed to replace the existing one. It so happens that the non-Euclidean geometries were constructed before difficulties with Euclidean geometry were encountered and hence for a long time afterward mathematicians did not question seriously the infallibility of Euclidean geometry. But the greatest nineteenth-century minds, Gauss and Riemann, saw the handwriting on the wall. The present view was neatly expressed by Einstein: "So far as the theories of mathematics are about reality, they are not certain; so far as they are certain they are not about reality."

What was the effect of non-Euclidean geometry on the future progress of mathematics? One might expect that the loss of the quality of truth might have caused a decline of interest in the subject. But the actual effect was quite the opposite. Mathematics passed from serfdom to freedom. Up to the time of non-Euclidean geometry mathematicians were fettered to the physical world. Mathematics was devoted to the study of nature and was intended to yield new truths about nature, either by itself or in conjunction with science. But it became clear, or at least clearer, in the years after 1830, that mathematicians could devote themselves to any ideas which attracted them, however wild and unrelated to the physical world they might appear to be on first sight. Had not the history of non-Euclidean geometry shown that seemingly absurd ideas may prove to be not only illuminating but of actual use to science? The mathematicians were finally liberated to create what their hearts and minds desired. Mathematicians found their house burned to the ground only to discover gold under the floor boards. They suffered a great tragedy only to rise from the experience enriched in character. The enormous expansion in mathematical activity in the last century is largely the consequence of this new birth of freedom.

If truth and evident physical applicability were no longer restrictions on mathematical creation, what guided and controlled mathematical creation after 1830? What criteria determined significant mathematics as opposed to cheap, idle speculations? What value was there in any mathematical work that could no longer be trusted to give truths? These questions are the largest one can ask about the nature of current mathematical activity. We shall answer them shortly.

The abolition of the truth of mathematics raised far more pressing questions for the scientist and engineer. The conviction that nature was mathematically designed and that man was uncovering that design had given assurance that the laws of mathematics were infallible guides to the operation of the universe. But this conviction was shattered, and the successes achieved with the aid of mathematics now appeared to be accidental. Hence all of the human undertakings and predictions based on mathematics became questionable. Since some of the most complicated and daring constructions had been based on Eucildean geometry, was there not some chance that the bridges and buildings would collapse? Would not machinery fly apart? Might not the planets, whose paths Laplace not too long before had proved to be stable, suddenly wander off into space? These questions are genuine ones. One might think that the scientists, at least, and perhaps the mathematicians also, would have become frantic or despairing.

But fortunately there is a law of inertia in the world of ideas as in the physical world. New ideas are not readily put into circulation. Human minds tend to be closed, and they dismiss the new ideas as nonsense or as the work of eccentrics until the realization that these ideas are important finally dawns upon them. Even the mathematicians of the late nineteenth century did not take non-Euclidean geometry seriously for physical applications, though they derived a great deal of pleasure from exploring the new concepts and relating them to other domains of mathematics. The scientific world did not awaken to the reality of non-Euclidean geometry until the creation of the special theory of relativity in 1905.

But now the basic question must be faced. Mathematics is a creation of the human mind and yet possesses extraordinary power to represent the processes of nature and to predict the happenings of nature. Why does mathematics work? What is the secret of its power? This question also remains to be discussed.

We might draw one more lesson from the history of non-Euclidean geometry. The assertions contained in Euclid's axiom appealed to hundreds of generations, throughout the time that Western culture had existed, as self-evident and beyond all doubt. Nevertheless, these axioms were finally seen to be questionable as affirmations about the real world. We see, therefore, that we must constantly re-examine our firmest convictions. Mathematics derived its seemingly surest principles from nature, but it is now abundantly clear that we must be always critical of what we accept, even from nature.

27: MATHEMATICS AND NATURE

How can it be that mathematics, a product of human thought independent of experience, is so admirably adapted to the objects of reality?

—ALBERT EINSTEIN

THERE is a well-known story, largely apocryphal, concerning the visit of the French Encyclopedist, Denis Diderot, to the court of Catherine the Great. During his stay there he shocked many members by his atheistic views. To silence him the Empress decided to play a royal trick. She informed him that a great mathematician, Leonhard Euler, would present to the court a mathematical demonstration of the existence of God and invited Diderot to be present. Diderot accepted. Euler appeared, rattled off some meaningless mathematical formulas, and concluded, "So God exists." He then contemptuously challenged Diderot to refute him. According to the story, which is incorrect in this detail, Diderot knew no mathematics and was so disconcerted that he withdrew shamefaced from the court. Presumably the joke was on Diderot; actually the joke was on Euler and the entire mathematical world.

Every mathematician of the eighteenth century was sure that his subject proffered the most certain body of truths man possessed. Euler was perhaps the cockiest of this group. At the age of twenty, with no practical experience to guide him, he wrote a theoretical paper on the design of ship masts, which, incidentally, won a prize offered by the Paris Academy. He stated in his conclusion that he did not think it necessary to check his results by experiments because they had been deduced from the surest foundations in mathematics and mechanics, hence their correctness could not be questioned. The conviction was so widespread that mathematics contained within itself the profound inner design of the universe that many mathematicians, including Euler, believed that the existence of God could be proved as a necessary consequence. Indeed,

Euler spent many months in trying to establish this conclusion by divers mathematical arguments.

But, as Francis Bacon had already observed, "Nature is more subtle many times over than the senses and understanding of man." The development of non-Euclidean geometry showed that man's mathematics did not speak for nature, much less lead to a proof of the existence of God. It became clearer, too, that man institutes the order in nature, the apparent simplicity, and the mathematical pattern. Nature itself might have no inherent design, and perhaps the best one could say of man's mathematics was that it offered no more than a limited, workable, rational plan.

But if mathematics lost its place in the citadel of truth, it still was at home in the physical world. The major, inescapable fact, and the one that still has inestimable importance, is that mathematics is the method par excellence by which to investigate, discover, and represent physical phenomena. In some branches of physics it is, as we have seen, the essence of our knowledge of the physical world. If mathematical structures are not in themselves the reality of the physical world, they are the only key we possess to that reality. The discovery of non-Euclidean geometry not only did not destroy this value of mathematics or confidence in its results but rather, paradoxically, increased its usefulness, because mathematicians felt freer to investigate radically new ideas, and yet found these to be applicable. In fact the role of mathematics in the organization and mastery of nature has expanded at an almost incredible rate since 1830. And the accuracy with which mathematics can represent and predict natural occurrences has increased remarkably since Newton's day. We seem, therefore, to be confronted with a paradox. A subject that no longer lays claim to truth has furnished the marvelously adaptable Euclidean geometry, the pattern of the extraordinarily accurate heliocentric theory of Copernicus and Kepler, the grand and embracing mechanics of Galileo, Newton, Lagrange, and Laplace, the physically inexplicable but marvelously applicable electromagnetic theory of Maxwell, and the sophisticated theory of relativity of Einstein. All of these highly successful developments rest on mathematical ideas and mathematical reasoning. The question becomes inescapable. Why does mathematics work?

From the time of the Greeks, when the power of mathematics to reveal new knowledge of the physical world was first glimpsed, this question was considered and answered according to the light of the age. Let us examine some of these answers. They are not only of interest but are still the explanations offered by some people. Insofar as the axioms of mathematics are concerned the dominant view until recent times was that they were self-evident truths about the physical world. That there were truths in general was unquestionable, and the truths of mathematics were so apparent that one needed no experience to recognize them. Plato went further than most philosophers by asserting that the true reality consisted of ideas and relationships among them; the physical world was merely an imperfect representation of these ideas and relationships. Among philosophers who accepted truths, Descartes was most critical. But he was confident of the mathematical truths and accounted for his certainty with the argument that God planted within us the ideas of numbers and figures, and from the knowledge of these ideas their basic properties were apparent. Though Galileo certainly tried to rid himself of preconvinced notions he, too, was sure that there are truths and that we can discover them by paying attention to what nature says. Of such truths the axioms of mathematics were the ones most readily grasped.

The greatest eighteenth-century philosopher, Immanuel Kant, was as certain as these other men of the truths of mathematics and of the axioms in particular; however, Kant rested his case on the thesis that the human mind supplies the concepts and axioms with which to organize experience. Hence there was a necessary precise correspondence between what the mind accepts and experience demonstrates. Kant did make a distinction between the knowledge that the mind builds up with the sensations it receives and the physical world that, so to speak, lies beyond the sensations and is unknowable, but he was confident that the mind's organization of knowledge gave us reliable and indeed infallible knowledge.

Though the axioms of mathematics were accepted self-evident truths, there remained the larger question of why the deductive arguments performed by human reason yielded truths. In other words, why should there be agreement between the theorems and physical experience? Might not the reasoning lead away from truths? Prior to the advent of non-Euclidean geometry the almost universal answer to this question was that the universe was rationally and indeed mathematically designed. Of course the designer was God. Had Plato, for example, written the Bible it would undoubtedly have opened with the statement: "In the beginning God created mathematics and then created heaven and earth according to the laws of mathematics." Copernicus and Kepler saw the hand of God in the marvelous correspondence between their mathematical scheme and the actions of nature. Galileo expressed his belief in the mathematical design of nature in a classic statement: "Philosophy is written in that great book which ever lies before our eyes—I mean the universe—but we cannot understand it if we do not first learn the language and grasp the symbols, in which it is written. This book is written in mathematical language, and the letters are triangles, circles, and other geometrical figures, without which it is humanly impossible to comprehend a single word." Newton saw everywhere in the universe evidence of God's majestic design and of His constant and continuous concern to keep the universe running according to plan. As James Thomson, in his memorial poem on Newton, put it, Newton,

> *from motion's simple laws*
> *Could trace the secret hand of Providence,*
> *Wide-working through this universal frame.*

Leibniz accounted for the agreement between mathematical reasoning and facts by "the pre-established harmony between thought and reality." Nature was the art of God.

Of course the mathematical design of nature had to be uncovered by man's continual search. God's ways appeared to be mysterious but it was certain that they were mathematical, and that man's reasoning would in time discern more and more of the rational pattern which God had utilized in creating the universe. The fact that man reasoned exactly in the way in which God had planned seemed readily understandable on the ground that there could be but one brand of correct reasoning. The possibility that man's reasoning might not conform to God's ratiocinations occurred to Descartes, but he dismissed it with the argument that God would not deceive us by allowing our mental faculty to function falsely.

Despite the persistence of the belief in a rational universe designed by God, some contrary views cropped up even before the discovery of non-Euclidean geometry. The intensely religious and yet occasionally wavering Blaise Pascal remarked that nature proves God only to those who already believe in Him. When Napoleon, on being handed a copy of Laplace's *Celestial Mechanics,* remarked to the author that there was no mention of God in this great and extensive work, Laplace replied, "I have no need of this hypothesis." However, Laplace had not considered the question of how it was that his superb mathematical reasoning should fit the universe so remarkably well.

The creation of non-Euclidean geometry thrust to the fore the question of how it is that mathematical reasoning gives us knowledge of the physical world. Since reasoning on the basis of axioms contradicting Euclid's yielded theorems that applied to the physical world, it was no longer possible to assert that mathematics yielded truths, for truth is unique. Nevertheless, because the applicability of mathematics to the physical world remained as effective as before, and new developments added to the extent and depth of these applications, some great mathematicians and scientists, notably Sir James Jeans and Sir Arthur Stanley Eddington, continued to affirm God's mathematical design of nature. About the only difference in this later view from those of the seventeenth and eighteenth century is that one can no longer affirm that the present mathematical account of the heavenly motions, for example, or of electromagnetic waves, is the final one but rather an excellent approximation which, by continued improvement, may become the correct formulation.

It seems fair to say, however, that within the last century the belief in the mathematical design of nature has receded. Yet the problem of accounting for the remarkable insight into the behavior of nature that mathematics offers still remains. Peculiarly enough, the effect of non-Euclidean geometry on our aggravated problem makes it easier to approach an answer, even if the resulting explanation may not be wholly satisfying. What we have to explain now is not how mathematics produces truths but rather the correspondence between physical reality and the mathematical representation.

We shall attempt to answer this question in the light of our twentieth-century understanding of the nature and role of mathematics. Mathematics begins with the selection of certain concepts that appear to be instrumental in studying the physical world, for example, the concepts of number and geometry. Having chosen his concepts the mathematician seeks next some sound fundamental facts on which to base his reasoning. There is not much doubt that these premises or axioms are inferred from observations and experience. Actual experience with collections of physical objects shows that $3(4 + 5) = 3 \cdot 4 + 3 \cdot 5$. Consequently, the mathematician adopts as an axiom about number the general statement, called the distributive axiom, which says that for any numbers a, b, and c, $a(b + c) = ab + ac$. To men whose experience with physical space was limited to small regions it seemed clear that a straight line could be extended indefinitely in either direction. This experience was so common that it did not occur to man to question, before the development of non-Euclidean geometry, whether such an axiom applies to the entire three-dimensional universe.

In the most fruitful applications of mathematics to the physical world, some nonmathematical axioms also enter. The Newtonian system of mathematical mechanics depends as much on the Newtonian laws of motion and gravitation as it does on the axioms of mathematics. The experiential basis for the physical axioms is more evident. Yet one cannot really distinguish the physical axioms from the mathematical ones on logical grounds. Both types are suggested by observation and experience and are abstractions from experience. It is true that the axioms of number and geometry were obtained first and have existed, therefore, for many more centuries. They may even be in some vague sense more fundamental, that is, the quantitative and geometrical properties may be the basic ones, and the others of secondary importance, just as food is a primary need, whereas shelter, at least in some climates, is secondary. If anything, we might say that the mathematical axioms are more obvious, whereas the correct physical axioms require a deeper and more penetrating analysis of the physical world to be uncovered. But this distinction is unimportant because both types are needed to make any real progress in the study of the physical world. Nevertheless, we shall stress the role of the mathematical axioms because our problem at the moment is to account for the effectiveness of mathematics.

The third element in the mathematical approach to nature is the obtainment of new conclusions. In this phase of his work the mathematician may indeed be distinguished from the physical scientist. The mathematician, as we have already noted many times, insists on establishing his conclusions by deductive arguments, whereas the scientist will feel free to resort to observations and experimentation and subsequent inductive generalizations to arrive at his conclusions. With concepts and axioms in mind, whether solely mathematical or a combination of physical and mathematical ones, the mathematician retires to a corner and deduces new conclusions about the physical world. These conclusions generally depend upon hundreds of steps of pure reasoning and yet yield such knowledge as the distance to the sun and sometimes such totally unexpected phenomena as the existence of radio waves. It may indeed be true that the deduced facts are necessary consequences of the axioms, and the latter derive from the physical world, but that reasoning of a highly intricate type should produce physically serviceable knowledge is the mystery which demands resolution. Why should the physical world conform to the pattern of man's reasoning?

One widely accepted view is that man learns to reason by studying nature. Let us see just what this means. One of the basic laws of reasoning says that an assertion cannot be both true and false. One cannot assert that the sun is both red and non-red. In the mathematical realm one says, for example, that two lines cannot be both parallel and nonparallel. This law is known in logic as the law of contradiction. What prompts us to accept this principle of reasoning? The answer is that this is what we observe in nature. Objects do not possess contradictory qualities at the same time. Of course the reader may protest, "How could it be otherwise? Is it not nonsense to believe that an object can be both red and non-red?" It is true that we cannot conceive of such an object, but this inability may be due to the superficiality of our thinking or to our training. The mathematicians and scientists who identified Euclidean geometry and space could not conceive of any other geometry. Their argument was that there is but one space and one body of laws about that space and the two agree perfectly. Many educated people say today that 2 + 2 must be 4. But there are algebras, physically useful algebras, in which this statement does not hold.

If at the present time at least the law of contradiction appears to be inescapable, rather than an inference from experience, there are other logical principles that appear more clearly to have an empirical origin. It is correct to argue that if all angels are white and no salamander is white, then no salamander is an angel. This argument is deductive and valid. Nevertheless, until one acquires experience with the logical principle used in this argument, and in fact even tests the conclusions of arguments made with the principle, he is not usually inclined to accept it. The doubt is understandable. Let us note that the seemingly similar argument, if all angels are white and no salamander is an angel, then no salamander is white, is not valid. Hence it seems very likely that man has learned to reason by studying what happens in nature, and it is therefore not surprising that his reasoning yields results that accord with nature.

History affords some confirmation of this thesis. The laws of reasoning were formulated abstractly by Aristotle in the fourth century B.C., centuries after man had practiced precise reasoning. Moreover, the mathematical reasoning of the preceding centuries concerned readily visualizable geometrical facts so that observation and measurement could readily have been employed to correct or improve poor deductive reasoning.

Whether or not man's reasoning is consistent with the behavior of nature, he has one more means at his disposal to make his mathematics fit the physical world. If his theorems do not fit, he is free to change his axioms. This recourse seems farfetched, but scientists have adopted this procedure in our time. The increased range of astronomical observations and the study of motions with speeds approaching that of light caused Einstein to adopt a non-Euclidean geometry for his theory of relativity. The mathematician may have to change his axioms still further as he attempts to apply them to more diverse or to more accurately measured phenomena.

There are other philosophical explanations of the remarkable correspondence between mathematical reasoning and the behavior of the physical world, but it must be admitted that no one of these is final. It may be that the effectiveness of the mathematical representation and analysis of the physical world is as unexplainable as the very existence of the world itself and of man. It is nonetheless a phenomenon that we can accept and utilize to great advantage. Having found that this gift from an unknown donor can be so profitably employed, our civilization now seems set to exploit it to the hilt. If the attempt to understand why it works leaves us with an enigma, as does the smile of the Mona Lisa, this merely means that we have an intriguing subject for further study and contemplation.

The dominant view today as to the nature of mathematical activity is, then, that the concepts and axioms are derived from experience, the principles of reasoning used to deduce new conclusions were most likely derived from experience, and, insofar as the applicability of mathematics to the physical world is concerned, the conclusions must be checked against experience. Except for the fact that mathematics insists on deductive proof of its conclusions, it would appear that this subject hardly differs from any branch of physical science. Shall we conclude, then, that mathematics is just a special kind of science?

The answer is no. The description we have just given of the subject and the aspects we have emphasized throughout this book afford too narrow a view. Even the Greek mathematicians, who certainly believed that they were pursuing truths of nature, felt free to investigate ideas that had no immediate application but nevertheless fitted their conception of legitimate mathematical activity. A classic example is afforded by the conic sections. Fifteen hundred years before Kepler used the ellipse in astronomy and Galileo used parabolas to describe trajectories, the Greeks had explored these curves fully.

The history of non-Euclidean geometry provides another example of the exploration of an idea that was not intended for use in science or engineering. The hundreds of investigators who sought to replace Euclid's parallel axiom by an intuitively more acceptable one were not trying to correct any errors in Euclidean geometry and certainly did not even conceive of a new geometry. Their concern was largely to perfect the structure of geometry. And even though Gauss, Bolyai, and Lobatchevsky had suspected that Euclidean geometry was not necessarily the geometry of physical space, and even though they mentioned the possibility of their new geometry being applicable to astronomy, they explored this radical idea more for the sake of what light it might shed on Euclidean geometry. The history of such creations shows that while nature is the womb from which mathematical ideas are born, these ideas can be studied for themselves while nature is left behind. Hence one must recognize that mathematicians have always felt free to pursue ideas that appeared to them to be relevant to their subject, regardless of whether the resulting work would be of immediate value to science.

It is true of the work we have just discussed that the mathematicians undoubtedly thought that it at the very least rounded out knowledge of subjects important for the study of nature. Nevertheless, this work shows that one must include in the range of mathematical investigations ideas that are only indirectly related to ones which prove useful. The freedom to explore such ideas has proved to be a boon in several respects. The very ideas we mentioned above did ultimately yield a new and richer insight into nature. Mathematicians have thereby not only anticipated the needs of science but have suggested the directions science should take. It seems unlikely that Kepler would have invented the ellipse to describe planetary motion, because that task, together with the one he actually performed of fitting the ellipse to data, would have been superhuman. Most likely he would have modified the theory of epicycles still further. Likewise, Einstein took full advantage of the already existing non-Euclidean geometry to create the theory of relativity. Undoubtedly the tasks of conceiving of such a geometry, which in itself demanded the genius of Gauss, and applying it to a radically new physical theory would have been beyond the powers of one man.

But it would be wrong to suppose that mathematicians have felt obliged to limit themselves to ideas possessing potential or indirect bearing on science. The individual can be motivated by forces that have little to do with the broader social and historical movements directing the larger currents of thought. Many mathematicians pursue the subject simply because they like it. To them the subject offers intellectual challenge and values that draw them to it far more strongly than money or power attract people generally. They enjoy the excitement of the quest for new results, the thrill of discovery, the satisfaction of mastering difficulties, and the pride in achievement. There are, moreover, delights and aesthetic values to be derived from surveying orderly chains of reasoning such as occur in most proofs, from the contemplation of the results themselves, and from grasping the ideas that make the proofs work. Those portions of mathematics which prove valuable in the study of nature offer the additional satisfaction of unifying a multitude of seemingly disorganized facts and of comprehending nature's ways.

One finds evidence of the aesthetic drive and the appreciation of mathematical beauty in the writings of many mathematicians. Archimedes, for example, though famous for his scientific work and inventions as well as for his original and powerful mathematical work, valued the latter more. He did not deign to leave any written account of his inventions. Isaac Newton spoke of God as interested in the preservation of cosmic harmony and beauty and regarded the mathematical design of the universe as an expression of that beauty.

The mathematician is free to create beautiful theorems and theories. But one might well ask whether there is any test of this beauty. Unfortunately there are no truly objective criteria, any more than there are in art or literature. It is fair to require that a valuable mathematical creation use effectively the axioms on which it rests; that is, one must not use a blunderbuss to kill a flea. The reasoning should be sufficiently powerful or complex; a trivial or immediately obvious argument hardly warrants being written down. The most desirable proofs contain some underlying scheme or display some principle or method that is applicable to other problems than the one in question. And there should be in addition some idea or device that produces surprise or gratification that a genuine difficulty has been overcome. But all of these criteria are basically subjective, and a piece of work that the creator or some readers may regard as superb can be dismissed contemptuously by others. Since mathematicians are human beings there is no doubt that jealousies, rivalries, and even closed-mindedness have caused some to deprecate fine creations. A standard joke, "Trivial is what the other fellow does," jibes at the all too common failure of mathematicians to be sufficiently appreciative of the work of others.

To recapitulate the substance of our last few remarks, we may say that the mathematician creates theory that is designed to solve current problems of science, theory that sheds light on the physically applicable results though not directly of use, theory that has potential use in science, and theory that offers aesthetic values. We see, therefore, that the scope of the mathematician's work is considerably broader than what has been presented in this book.

Mathematicians are free to pursue their own inclinations and to cater to their own tastes, but there is, of course, the danger that ideas pursued solely for their intrinsic appeal may deviate so far from normal human interests that they will attract only an esoteric group. There is also the danger that ideas concocted solely by the human mind and bearing no relation to the meaningful and weighty phenomena of the physical world may become thin, impoverished, and insignificant however facile the reasoning which develops the ideas. In view of the vast extent of mathematical knowledge and the proliferating demands of science, there is much already done and yet to be done that offers all the satisfactions of arbitrarily chosen directions of research and yet has more substance and more moment for our civilization. Certainly the lesson history teaches, and the reason for the great emphasis placed on mathematics today, is that mathematics provides the supreme plan for the understanding and mastery of nature.

Mathematics may be the queen of the sciences and therefore entitled to royal prerogatives, but the queen who loses touch with her subjects may lose support and even be deprived of her realm. Mathematicians may like to rise into the clouds of abstract thought, but they should, and indeed they must, return to earth for nourishing food or else die of mental starvation. They are on safer and saner ground when they stay close to nature. As Wordsworth put it, "Wisdom oft is nearer when we stoop than when we soar."

INDEX

ABOUT THE AUTHOR

DR. MORRIS KLINE is a prominent teacher, research mathematician, and writer. He has taught mathematics at the undergraduate and graduate levels since 1930 and has been a guest lecturer at a number of universities in the United States and Europe. He is currently Professor of Mathematics at New York University. His research career was effectively launched during two years as a research assistant at the Institute for Advanced Study of Princeton.

A good deal of Dr. Kline's time is still devoted to the direction of the Division of Electromagnetic Research of New York University. During the year 1958–59 he was a John Simon Guggenheim Memorial Fellow and a Fulbright lecturer at the Technical University in Aachen, Germany.

Dr. Kline is married and has three children; the Klines live in Brooklyn, New York.